PLASTIC WASTES

PLASTIC WASTES

Management, Control, Recycling, and Disposal

by

U.S. Environmental Protection Agency

Office of Solid Waste
Office of Water
Washington, DC

T. Randall Curlee
Sujit Das

Oak Ridge National Laboratory
Oak Ridge, Tennessee

NOYES DATA CORPORATION
Park Ridge, New Jersey, U.S.A.

363.728
P715

Published in the United States of America by
Noyes Data Corporation
Mill Road, Park Ridge, New Jersey 07656

10 9 8 7 6 5 4 3 2 1

Library of Congress Cataloging-in-Publication Data

Plastic wastes : management, control, recycling, and disposal / by
    U.S. Environmental Protection Agency, T. Randall Curlee, Sujit Das
    (Oak Ridge National Laboratory).
        p.   cm. -- (Pollution technology review, ISSN 0090-516X ; no.
    201)
      Includes index.
      ISBN 0-8155-1265-1 :
      1. Plastic scrap. I. Curlee, T. Randall. II. Das, Sujit.
    III. United States. Environmental Protection Agency. IV. Series.
    TD798.P56   1991
    363.72'88--dc20   90-23207
                                               CIP

Foreword

Plastic wastes in the municipal solid waste (MSW) stream and in the industrial sector are discussed in the studies in this book. Quantities of plastic wastes generated, characterization of the wastes, their environmental impact, and management of the waste stream are described. Also covered is the effect of improper disposal of plastic wastes on the marine environment.

As plastics use, and particularly plastic packaging, have grown more prevalent, so have concerns about plastic wastes disposal. Discarded plastics, besides being highly visible, are a rapidly increasing percentage of solid waste in landfills. These problems have made plastic wastes a major focus in the management of solid waste.

The book has two parts. Part I, emphasizing plastic wastes in the MSW stream, provides a technological review of the plastics industry, production and consumption statistics, definitions of major end use markets, and disposal paths for plastics. It examines management issues and environmental concerns on landfilling and incineration, the two primary MSW disposal methods currently in use. Plastic litter and its impact are covered. Source reduction and recycling are considered, and recommended actions for industry and other concerned groups are identified.

Part II addresses issues related to plastics recycling in the industrial sector: manufacturing and post-consumer plastic waste projections, the estimated energy content of plastic wastes, the costs of available recycling processes, institutional changes that promote additional recycling, legislative and regulatory trends, the potential quantities of plastics that could be diverted from the municipal waste stream and recycled in the industrial sector, and the perspectives of current firms in the plastics recycling business. Several scenarios are presented which assume specific technical, economic, institutional, and regulatory conditions.

v

The information in the book is from the following documents:

> *Methods to Manage and Control Plastic Wastes—Report to Congress,* prepared by the U.S. Environmental Protection Agency, Office of Solid Waste and Office of Water, February 1990.
>
> *Plastics Recycling in the Industrial Sector: An Assessment of the Opportunities and Constraints,* prepared by T. Randall Curlee and Sujit Das of Oak Ridge National Laboratory for the U.S. Department of Energy, November 1989.

The table of contents is organized in such a way as to serve as a subject index and provides easy access to the information contained in the book.

> Advanced composition and production methods developed by Noyes Data Corporation are employed to bring this durably bound book to you in a minimum of time. Special techniques are used to close the gap between "manuscript" and "completed book." In order to keep the price of the book to a reasonable level, it has been partially reproduced by photo-offset directly from the original reports and the cost saving passed on to the reader. Due to this method of publishing, certain portions of the book may be less legible than desired.

NOTICE

The materials in this book were prepared as accounts of work sponsored by the U.S. Environmental Protection Agency and the U.S. Department of Energy. On this basis the Publisher assumes no responsibility nor liability for errors or any consequences arising from the use of the information contained herein.

Mention of trade names or commercial products does not constitute endorsement or recommendation for use by the Agency or the Publisher. Final determination of the suitability of any information or product for use contemplated by any user, and the manner of that use, is the sole responsibility of the user. The book is intended for information purposes only. The reader is warned that caution must always be exercised with potentially hazardous materials such as plastic wastes, and expert advice should be obtained before implementation of processes involving these wastes.

All information pertaining to law and regulations is provided for background only. The reader must contact the appropriate legal sources and regulatory authorities for up-to-date regulatory requirements, and their interpretation and implementation.

Contents and Subject Index

PART II
RECYCLING IN THE INDUSTRIAL SECTOR

Part I

Wastes in the Municipal Solid Waste Stream

The information in Part I is from *Methods to Manage and Control Plastic Wastes— Report to Congress,* prepared by the U.S. Environmental Protection Agency, Office of Solid Waste and Office of Water, February 1990.

Executive Summary—Findings and Conclusions

This report was developed in response to Section 2202 of the 1987 Plastic Pollution Research and Control Act, which directs EPA to develop a report to Congress on various issues concerning plastic waste in the environment. Specifically, EPA is required to:

- Identify plastic articles of concern in the marine environment,

- Describe impacts of plastic waste on solid waste management, and

- Evaluate methods for reducing impacts of plastic wastes, including recycling, substitution away from plastics, and the use of degradable plastics.

In this report, EPA has examined two other methods for reducing impacts associated with plastic wastes in addition to those specified in the statute. These are: (1) source reduction of plastic waste (this is broader than substitution away from plastics) and (2) methods for controlling the sources of plastic marine debris.

SCOPE OF THE REPORT

The report focuses primarily on plastic waste in the municipal solid waste (MSW) stream, that is, post-consumer plastic waste. The only exception to this focus is the consideration of plastic pellets, which are the raw materials used in the processing and manufacture of plastic products. Pellets are included in the report because they have been found in high concentrations in the marine environment and they pose ingestion risks to some forms of marine life.

SUMMARY OF MAJOR FINDINGS AND ACTION ITEMS

PRODUCTION AND USE OF PLASTICS

Plastics are resins, or polymers, that have been synthesized from petroleum or natural gas derivatives. The term "plastics" encompasses a wide variety of resins each offering unique properties and functions. In addition, the properties of each resin can be modified by additives. Different combinations of resins and additives have allowed the creation of a wide array of products meeting a wide variety of specifications.

U.S. production of plastics has grown significantly in the last 30 years, averaging an annual growth rate of 10%. Continued growth is expected. The largest single market sector is plastics packaging, capturing one-third of all U.S. plastics sales. Building and construction (25% of U.S. sales) and consumer products (11%) follow.

2

Plastics production and use has grown because of the many advantages plastics offer over other more traditional materials. A few of the desirable intrinsic properties of plastics include: (1) design flexibility -- plastics can be modified for a wide variety of end uses, (2) high resistance to corrosion, (3) low weight, and (4) shatter resistance. Table ES-1 provides information on some of the major classes of plastic resins, their characteristics, and examples of product applications.

PLASTICS IN THE MARINE ENVIRONMENT

EPA has identified several articles of concern in the marine environment due to the risks they pose to marine life or human health or to the aesthetic (and related economic) damage they cause. These articles of concern are: plastic pellets, polystyrene spheres, syringes, beverage ring carrier devices, uncut plastic strapping, plastic bags and sheeting, plastic tampon applicators, condoms, fishing nets and traps, and monofilament lines and rope.

Many other items of marine debris (made from plastic as well as other materials) have been identified during the development of this report. Taken as a whole, all components of marine debris are unsightly and offensive to many people.

Specific sources for each debris item are not well known; however, the major land-based sources appear to be:

- Combined sewer overflows (CSOs) and sewage treatment plants

- Stormwater runoff and other non-specific sources

- Plastic manufacturing and fabrication and related transportation activities (for pellets)

The major marine-based sources appear to be:

- Commercial fishing vessels

- Offshore oil and gas platforms

Recreational littering (on land and from vessels) also contributes to marine debris.

The following are EPA's **major action items** for reducing and controlling the sources of marine debris:

COMBINED SEWER OVERFLOWS --

- EPA will ensure that all permits for CSO discharges include technology-based limitations for the control of floatable discharges.

Table ES-1
PLASTIC RESIN CHARACTERISTICS, MARKETS, AND PRODUCTS

Resin Name	Characteristics	Primary Product Markets	Product Examples
Low–Density Polyethylene (LDPE)	Moisture–proof; inert	Packaging	Garbage bags; coated papers
Polyvinyl Chloride (PVC)	Clear; brittle unless modified with plasticizers	Building and construction; packaging	Construction pipe; meat wrap; cooking oil bottles
High–Density Polyethylene (HDPE)	Flexible; translucent	Packaging	Milk and detergent bottles; boil-in-bag pouches
Polypropylene (PP)	Stiff; heat- and chemical– resistant	Furniture; packaging	Syrup bottles; yogurt tubs; office furniture
Polystyrene (PS)	Brittle; clear; good thermal properties	Packaging; consumer products	Disposable foam dishes and cups; cassette tape cases
Polyethylene Terephthalate (PET)	Tough; shatterproof	Packaging; consumer products	Soft drink bottles; food and medicine containers

- EPA is developing guidance for States and local communities on effective operation and control of a combined sewer system. Information on low-cost control mechanisms, which may be helpful in reducing releases of floatable debris, will be included.

- EPA will sample a limited number of CSO discharges to pinpoint which articles are frequently released from CSOs.

STORMWATER DISCHARGES --

- EPA is developing a Report to Congress on stormwater discharges. Floatable discharges will be considered in this report. The report is expected to be completed by mid-1990.

- A subsequent report will be prepared on control mechanisms necessary to mitigate the water quality impacts of discharges examined in the initial Report to Congress. A final report is targeted for the end of 1991.

- EPA will sample and study a limited number of stormwater discharges to better pinpoint which articles are released from these sources.

VESSELS --

- EPA recommends that Federal and State agencies should enter into agreements with the U.S. Coast Guard to enforce Annex V of MARPOL, which prohibits the discharge of plastic waste at sea.

- EPA recommends that port facilities, local communities, industry, and interested Federal agencies should coordinate efforts to develop recycling programs for plastic waste that is brought to shore in compliance with Annex V of MARPOL.

- EPA will support the National Oceanic and Atmospheric Administration's (NOAA) investigation of methods to reduce the loss and impacts of fishing nets and gear by providing related information, such as information on degradable plastics.

LITTER PREVENTION AND RETRIEVAL --

- EPA will continue to support and conduct a limited number of harbor and beach surveys and cleanup operations.

- EPA will continue to work with NOAA and other Federal Agencies to distribute educational materials to consumers on marine debris.

■ EPA is developing an educational program for consumers that describes the proper method for disposing of household medical waste.

MANAGEMENT OF PLASTIC WASTES

Most post-consumer plastic waste is landfilled along with municipal solid waste. A small percentage (approximately 10%) of municipal solid waste is incinerated, and 10% is recycled. Only 1% of post-consumer plastic waste is recycled.

Plastic waste accounts for a large and growing portion of the municipal solid waste stream. Plastics are about 7% (by weight) of municipal solid waste and a larger percentage by volume. Current waste volume estimates range from 14 to 21 percent of the waste stream. The amount of plastic waste is predicted to increase by 50% (by weight) by the year 2000.

Half of the plastic waste stream is packaging waste. The rest of the plastic waste stream includes non-durable consumer goods such as pens and disposable razors and durable goods such as furniture and appliance casings.

Management of plastics in a landfill

Plastic wastes have not been shown to create difficulties for landfill operations. The structural integrity of a landfill is not affected by plastic wastes.

Plastics wastes affect landfill capacity because of the large and growing amount of plastic waste produced, not because the wastes are not degradable. Some have claimed that plastic waste affect landfill capacity even more than other larger volume wastes (e.g., paper) because plastics do not degrade in a landfill. While it is true that plastic wastes are very slow to degrade in landfills, recent data indicate that other wastes, such as paper and food waste, are also slow to degrade. Degradation of waste, therefore, has little effect on landfill capacity.

Data are too limited to determine whether plastic additives contribute significantly to leachate produced in municipal solid waste landfills. Only certain additives have the potential for causing a problem; however, their contribution to leachate volume or toxicity is unknown.

Management of plastics in an incinerator

Plastics contribute significantly to the heating value of municipal solid waste, with a heating value of three times that of typical municipal waste.

Controversy exists regarding whether halogenated plastics (e.g., polyvinyl chloride) contribute to emissions from municipal waste incinerators. Emissions of particular concern are acid gas emissions and dioxin/furan emissions. EPA and the Food and Drug Administration are continuing to analyze these issues. Final conclusions await completion of these additional analyses.

Plastic additives containing heavy metals (e.g., lead and cadmium) contribute to the metal content and possibly the toxicity of incinerator ash. Additional investigation is needed to determine with greater accuracy the impact of plastic additives on incinerator ash toxicity (i.e., whether lead- and cadmium-based plastic additives contribute to leachable lead and cadmium in ash). Potential substitutes for these additives are examined in Appendix C of this report.

METHODS FOR REDUCING IMPACTS OF PLASTIC WASTES

Source Reduction

Source reduction is defined to include activities that reduce the amount or toxicity of the waste generated. EPA is considering <u>all</u> components of the waste stream as possible candidates for source reduction.

There are a number of ways of achieving source reduction. Examples include:

- Modify design of product or package to decrease the amount of material used.

- Utilize economies of scale with larger size packages.

- Utilize economies of scale with product concentrates.

- Make material more durable so that it may be reused.

- Substitute away from toxic constituents in products or packaging.

It is difficult to consider source reduction of plastic waste or any single component of the waste stream in isolation because the goal of source reduction is to reduce the amount or toxicity of the <u>entire</u> waste stream, not just of one component. Attempts to reduce the amount of one component may actually cause an increase in another component and possibly in the entire waste stream.

For this reason, *source reduction actions need to be carefully examined. In many cases, particularly those involving material substitution, a lifecycle evaluation should be completed.* Such an analysis includes an evaluation of the impacts of the material from production to disposal. For example, the changes in natural resource use, energy use, consumer safety and utility, and product disposal that may result from the proposed action should be considered. This type of evaluation will ensure that source reduction efforts do not merely shift environmental problems from one media or waste stream to another.

The following are EPA's **major action items** regarding source reduction:

- EPA has issued a grant to the Conservation Foundation to evaluate strategies for MSW source reduction. Under this grant the Conservation Foundation will convene a national steering committee of municipal solid waste source reduction experts to discuss source reduction opportunities and incentives to promote source reduction, including potential selection criteria for a corporate source reduction awards program. The steering committee is expected to provide recommendations by the Fall of 1990.

- Building on work done with the Conservation Foundation, EPA will develop a model for conducting lifecycle analyses. A preliminary model should be available by the end of 1991.

- EPA has partially funded an effort to analyze the environmental impact of six different packaging materials and the effects of various public policy options that are aimed at altering the mix of packaging materials. Project completion is expected by early 1991.

- EPA is continuing to evaluate the potential substitutes for lead and cadmium-based additives that are identified in Appendix C of this report.

- EPA supported the Coalition of Northeastern Governors (CONEG) in developing preferred packaging guidelines and a regional framework for encouraging source reduction actions. CONEG's Source Reduction Task Force issued a report in September 1989, which outlined packaging guidelines and recommended that a Northeast Source Reduction Council be formed with representatives from the northeastern states, industry, and the environmental community. The council will develop long-range policy to reduce packaging at the source, implement the packaging guidelines, and educate the public. EPA is working with the Council on these activities.

- EPA is examining potential incentives and disincentives to source reduction of municipal solid waste. A report is expected by early 1990.

Recycling

Most plastic recycling efforts to date have focused on polyethylene terephthalate (PET) soft drink bottles and to a lesser extent on high density polyethylene milk jugs. In total, only about 1% of the post-consumer plastic waste stream is recycled.

Plastics recycling is in its infancy. Efforts underway right now by the plastics industry and State and local governments are numerous and varied. Thus, the information presented here represents the current state of plastics recycling and will, most likely, very quickly be out of date. It is very difficult to predict the future of plastics recycling because so much depends on the research and other efforts that are now underway.

Technologies exist for recycling either single homogeneous resins or a mixture of plastic resins:

- Recycling of relatively homogeneous resins (e.g., PET from soft drink bottles) may yield products that compete with virgin resins. Such recycling offers the greatest potential to reduce long-term requirements for plastics disposal. However, a system to capture and recycle the products of such recycling must be established.

- Recycling of a mixture of plastic resins often yields products that compete with low-cost commodities such as wood or concrete. This approach may capture a large percentage of the plastic waste stream because separation of resins is not a barrier to this approach. However, because the products of this type of recycling may eventually require disposal, mixed plastics recycling may delay, but may not ultimately reduce, the long-term requirements for plastic waste disposal.

The major factors currently limiting plastics recycling are:

- *Collection and supply.* This appears to be the greatest limitation facing recycling of both single resins and a mixture of resins; however, the recycling of single resins is more severely limited by the lack of ability to separate a complex mixture of plastic wastes (such as would be collected through a curbside program). There are several methods of collection including curbside collection, drop-off centers, buy-back centers, and container deposit legislation (i.e., "bottle bills"). Curbside collection and bottle bills have received the most attention:

  -- Curbside collection of plastics (and other recyclables) can capture a great variety and amount of plastic waste. However, this strategy imposes relatively high costs for collection and is not universally applicable (e.g., not all areas offer curbside collection of municipal solid waste).

  -- Container deposit legislation, which was enacted primarily to control litter, not increase recycling, has proven effective at diverting plastic soft drink containers from disposal; however, soft drink bottles represent only a small percentage (approximately 3%) of plastic wastes. Thus, this method, as currently implemented, will not divert significant amounts of plastic wastes. In addition, recycling officials have raised concerns that container deposit systems remove the most valuable, revenue-generating material from the recycling stream. This may impair local efforts to recycle other materials (e.g., newspaper, cans, etc.) in curbside collection programs.

These two collection strategies are interrelated. Waste management officials need to carefully weigh the costs and benefits related to each strategy (described in Section 5 of this report) and, very importantly, the relationship between the two choices before selecting a collection mechanism.

- *Markets.* The markets for the products of mixed plastics recycling still face serious questions, particularly regarding cost-competitiveness. Markets for the products of single resins such as PET and HDPE appear to be large. Recycling of other single resins (e.g., PS) is only just beginning; therefore, market evaluations are difficult to make.

The following are EPA's **major action items** regarding plastics recycling:

- EPA is providing technical assistance and general information to the public on plastics recycling through a municipal solid waste clearinghouse and a peer match program. Both of these efforts offer information and assistance on recycling of all municipal solid waste components, not just plastics.

- EPA is examining potential incentives and disincentives to recycling of municipal solid waste components.

- EPA calls on the plastics industry to continue to research and provide technical and financial assistance to communities on plastics collection, separation, processing, and marketing.

Degradable Plastics

There are various mechanisms that are technically viable for enhancing the degradability of plastic. *Biodegradation and photodegradation are the principal mechanisms currently being explored and commercially developed.* The most common method for enhancing the biodegradability of plastics has involved the incorporation of starch additives. Production of photodegradable plastics involves the incorporation of photo-sensitive carbonyl groups or the addition of other photo-sensitive additives.

Before the application of these technologies can be promoted, the uncertainties surrounding degradable plastics must be addressed. First, the effect of different environmental settings on the performance (e.g., degradation rate) of degradables is not well understood. Second, the environmental products or residues of degrading plastics and the environmental impacts of those residues have not been fully identified or evaluated. Finally, the impact of degradables on plastic recycling is unclear.

EPA does not believe that degradable plastics will help solve the landfill capacity problems facing many communities in the U.S. However, there may be potentially useful applications of this technology, including agricultural mulch film, bags for holding materials destined for composting, and certain articles of concern in the marine environment (.e.g, beverage container rings).

The following are EPA's **major action items** regarding degradable plastics:

- EPA has initiated two major research efforts on degradable plastics. The first project will evaluate degradable plastics in different environmental settings and examine the byproducts of degradation. The second project will evaluate the effects of degradable plastics on post-consumer plastics recycling. Interim results are expected by mid-1990.

- EPA calls on the manufacturers of degradable plastics to generate and make available basic information on the performance and potential environmental impacts of their products in different environmental settings.

- Title I of the 1988 Plastic Pollution Control Act directs EPA to require that beverage container ring carrier devices be made of degradable material unless such production is not technically feasible or EPA determines that degradable rings are more harmful to marine life than non-degradable rings. The uncertainties regarding degradable plastics (discussed above) pose some difficulties for EPA's implementation of this Act; however, some specific information is known regarding ring carrier devices:

    -- EPA has not identified any plastic recycling programs that currently accept or are considering accepting ring carriers. Therefore, degradable rings should not impair recycling efforts.

    -- Ring carriers are usually not colored and therefore do not include metal-based pigments. Thus, concerns regarding leaching of pigments appear to be minimal for these devices.

The research on degradable plastics (see above) now underway at EPA will help resolve remaining issues. EPA will initiate a rulemaking to implement the above legislation in 1990. A final rule is expected by late 1991.

1. Introduction

In 1987 Congress passed the United States-Japan Fishery Agreement Approval Act (Public Law 100-220), which includes the Marine Plastic Pollution Research and Control Act (MPPRCA). Among other objectives, MPPRCA is intended to control disposal of plastics from ships and improve efforts to monitor uses of drift nets. Title II of the Act directs EPA to investigate and report on various issues concerning plastics waste in the environment, including:

- Articles of concern in the marine environment

- Impacts of plastics waste on solid waste management

- Methods for reducing impacts of plastics on the environment and solid waste management (e.g., recycling, substitution away from plastics, and use of degradable plastics)

This report is the compilation of data gathered by EPA in response to that directive.

The focus of the report is plastic wastes in the municipal solid waste (MSW) stream: the amount of such waste; its impact on human health, the environment, and management of the MSW stream (i.e., post-consumer plastic waste); and options for reducing these impacts (e.g., recycling). In addition, the report considers:

- Improper disposal of plastics in the marine environment (e.g., disposal from vessels) and on land (i.e., littering)

- Impacts on the marine environment of plastic pellets used in the manufacture of plastic articles; this industrial waste is considered here because the high concentrations of the pellets found in the world's oceans are a major concern

Principal findings are listed at the beginning of each section. The report is organized as follows:

Section 2 provides context for the rest of the report. It provides a technological overview of the plastics industry, statistics concerning production and consumption levels of various types of plastics in the United States, and definitions of the major end use markets and disposal paths for plastics.

Section 3 categorizes plastic articles of concern in the marine environment and the impact of these and other plastic wastes on ocean ecology.

Section 4 examines management issues and environmental concerns associated with disposing of plastics (as a part of MSW) by the two primary means used in the United States: landfilling and incineration. In addition, the section covers the problems associated with plastic litter and analyzes the relative impacts of plastic and other litter.

Section 5 examines strategies for reducing plastic waste management problems (e.g.. source reduction and recycling). The strategies chosen are geared to resolving the plastic waste management issues identified in Sections 3 and 4.

Section 6 outlines the actions to be taken by EPA as well as recommended actions for industry and other groups to address the concerns identified in the earlier sections. The objectives presented here are divided into two categories: those for improving the management of the MSW stream and those for addressing problems outside the MSW management system.

Appendix A provides an overview of the legal authorities available to EPA and other Federal agencies for improving plastics waste management.

Appendix B supports the discussion of plastic recycling efforts in Section 5 by presenting information on state recycling programs, state bottle bills, and the characteristics of various community curbside recycling programs.

Appendix C investigates the potential for substitution of less toxic additives for the lead- and cadmium-based additives used in some plastics. A summary of the status of EPA's research on this topic is presented.

2. Production, Use, and Disposal of Plastics and Plastic Products

This chapter provides a technological overview of the plastics industry (Section 2.1) and presents key statistics concerning production, import/export, and consumption levels of major types of plastics used in the United States (Section 2.2). In addition, this chapter defines the characteristics of the major plastics types (Section 2.2.6), the major end use markets (Section 2.3), and the disposal paths for plastic wastes generated in the United States (Section 2.4). This introductory material is the context for understanding and assessing the fate and impact of various plastics once they are discarded by consumers. The quantitative information about production levels as well as the descriptions of the types and uses of the diverse plastic products define the role of plastics in the economy. Additionally, information presented about the growth trends among different types of plastic help to determine the plastic waste management requirements of the future.

2.1 SUMMARY OF KEY FINDINGS

Following are the key findings of this section:

- The term "plastic" encompasses many different types of materials offering a wide variety of properties.

- Production of plastic goods involves three primary steps: 1) manufacturing resins, 2) incorporating additives, and 3) processing or converting resins (usually by a different firm, or processor) into end products for various markets, including packaging, building and construction, and consumer products.

- Additives are used in some plastics to 1) modify physical characteristics, 2) influence aesthetic properties, or 3) permit processing of the resins.

- U.S. production of plastics has grown from about 3 billion pounds in 1958 to about 57 billion pounds in 1988 for an average annual growth rate of 10.3%. During the same period, annual GNP growth (measured in constant 1982 dollars) has averaged 3.2%.

- The largest-volume market sectors for plastics are packaging and building and construction.

- The five plastics in order of greatest use in 1988 were low- and high-density polyethylene (LDPE and HDPE, respectively), polyvinyl chloride (PVC), polystyrene, and polypropylene (PP).

14

- Polymers are categorized as thermoplastic or thermoset plastics. Of these two, the former can be melted and reformed -- a characteristic that is important in recycling plastic products.

- Industry actions (i.e., announced capacity increases amounting to 25% of current capacity) and market forecasts suggest that rapid market growth should be expected for a number of years in the plastics industry.

- Plastics represented 7.3% of MSW by weight (including all waste sources) in 1986 (Franklin Associates, 1988) and are expected to increase to 9% by the year 2000. Information regarding the composition of the plastic waste stream is limited.

2.2 TECHNOLOGICAL OVERVIEW

Production of plastic goods involves three primary steps: 1) manufacturing resins, 2) incorporating additives, and 3) processing or converting resins (usually by a different firm, or processor) for various markets for disposable and durable end products, including packaging, building and construction, and consumer products.

2.2.1 Manufacturing Resins

Plastics are resins, or polymers, that have been synthesized from petroleum or natural gas derivatives (see Table 2-1). Chemicals composed of small molecules called monomers are typically produced from the crude oil or natural gas liquids and then allowed to react to form the solid polymer molecules. All polymers are composed of long chains of monomers; the chains may or may not be attached to each other. Plastics that can be softened and reformed are termed thermoplastics, and plastics that cannot be melted and reformed are termed thermosets.

- **Thermoplastics.** Because the monomer chains in these polymers are not cross-linked (that is, they comprise two-dimensional rather than three-dimensional molecular networks), these plastics can be melted and reprocessed without serious damage to the properties of the resins (Curlee, 1986). As the temperature or pressure of a thermoplastic resin increases, the molecules can flow as needed for molding purposes. The molecular structure becomes rigid, however, when the resin is cooled. This malleability is one reason thermoplastic resins comprise such a large percentage of the plastics market (see Table 2-2). Plastics manufacturers produce thermoplastics in a number of easily transportable forms, including pellets, granules, flakes, and powders. Examples of thermoplastic resins include polyethylene, polyvinyl chloride (PVC), polystyrene, and thermoplastic polyesters such as polyethylene terephthalate (PET).

Table 2-1
FEEDSTOCK CHEMICALS FOR HIGH-VOLUME PLASTICS

Feedstock Chemical	Possible Products
Acetylene	Polyvinyl chloride Polyurethane
Benzene	Polystyrene Polyurethane Acrylonitrile–butadiene–styrene (ABS)
Butadiene	Polyurethane ABS
Ethylene	High– and low–density polyethylene Polyvinyl chloride Polystyrene ABS Polyethylene terephthalate (PET) Polyurethane Polyesters
Methane	PET Polyurethane
Naphthalene	Polyurethane
Propylene	Polypropylene Polyurethane Polyester
Toluene (a)	Polyurethane foams, elastomers, and resins Polyesters
Xylenes (a)	Polystyrene PET ABS Unsaturated polyseters Polyurethanes

(a) This feedstock chemical can be used to derive benzene; see above
for the resins that can be derived from benzene.
Source: The Society of the Plastics Industry, 1988.

Table 2-2

U.S. SALES OF PLASTICS,
BY RESIN
(MILLION POUNDS, 1988)

Resin	Sales	As % of Total Sales
THERMOPLASTIC RESINS		
Low-density polyethylene (LDPE)	9,865	17.3
Polyvinyl chloride and copolymers (PVC)	8,323	14.6
High-density polyethylene (HDPE)	8,244	14.5
Polypropylene (PP)	7,304	12.8
Polystyrene (PS)	5,131	9.0
Thermoplastic polyester (incl. polyethylene terephthalate (PET) and polybutylene terephthalate (PBT))	2,007	3.5
Acrylonitrile/butadiene/ styrene (ABS)	1,238	2.2
Other styrenics (a)	1,220	2.1
Other vinyls (b)	958	1.7
Nylon	558	1.0
Acrylics	697	1.2
Thermoplastic elastomers	495	0.9
Polycarbonate	430	0.8
Polyphenylene-based alloys (c)	180	0.3
Styrene/acrylonitrile (SAN)	137	0.2
Polyacetal	128	0.2
Cellulosics	90	0.2
Total- Thermoplastic	47,005	82.6

(Cont.)

Table 2-2 (Cont.)

U.S. SALES OF PLASTICS,
BY RESIN
(MILLION POUNDS, 1988)

Resin	Sales	As % of Total Sales
THERMOSETTING RESINS		
Phenolic	3,032	5.3
Polyurethane	2,905	5.1
Urea and melamine	1,515	2.7
Polyester, unsaturated	1,373	2.4
Epoxy	470	0.8
Alkyd	320	0.6
Total- Thermosetting	9,615	16.9
Others (d)	288	0.5
TOTAL	56,908	100.0

(a) Excludes ABS and SAN. Examples include styrene–butadiene and styrene–based latexes, styrene–based polymers such as styrene–maleic anhydride (SMA), and styrene–butadiene (SB) polymers.
(b) Includes polyvinyl acetate, polyvinyl butyrol, polyvinylidine chloride and related resins.
(c) Includes modified phenylene oxide and modified phenylene.
(d) Includes small-volume resins of both the thermoplastic and thermoset type.

Note: Sales includes all sales by domestic manufacturers for domestic consumption or for export.

Source: Modern Plastics, 1989.

■ **Thermosets.**   Thermosetting plastics (thermosets) tend to be rigid, infusible, and insoluble.   Because the molecular chains are cross-linked (i.e., the resin is composed of a three-dimensional network of molecules), thermosets are stronger under high temperatures than thermoplastics.   For the same reason, however, these plastics cannot be reshaped once the molecular structures are formed.   Most thermoset resins are used in industries such as building and construction and transportation (e.g., marine craft).   Examples include phenolics, polyurethanes, and epoxy resins.

The wide variety of markets for plastics has created a demand for an equally wide array of resins (polymers).   Plastics scientists have developed a number of innovative methods to tailor polymer characteristics to specific end uses.   This research involves developing new blends of existing polymers, creating new polymers, and/or incorporating new additives for new applications in plastics engineering.

These hundreds of resins on the market, whether thermoplastics or thermosets, can be further categorized according to the level of production and market demand for each resin.   The resin categories are commodity, transitional, and engineering/performance.   See Table 2-3 for the characteristics associated with each of these categories as well as for examples of thermoplastic resins of each type.

Commodity -- Commodity resins are defined as those polymers produced in large volumes and used as the material inputs for numerous plastic products.   These polymers resemble commodities in that their basic characteristics are well-established and are not subject to refinements or differentiation among manufacturers.   These polymers are produced at the lowest cost of all plastics.   Because of the large volumes produced, these polymers are the most readily identified among plastic wastes.   Examples include various plastics used for packaging.

Transitional -- Transitional resins are polymers that are produced at a significantly lower rate than commodity resins but at a significantly higher rate than specialty resins.   Likewise, the price per pound of transitional resins is 75 cents to $1.25 -- more than the commodity resins and less than the specialty resins.

Engineering/performance -- Specific types of specialty resins are manufactured by only a few companies and have a limited range of uses.   Because both engineering and performance resins are produced in relatively small quantities for narrowly defined applications, the price per pound of these resins is high -- as much as $20 per pound.   Development of new technologies and automated production lines may eventually catalyze a larger market for these plastics (e.g., as replacements for metal), and thus a larger production volume (Chem Systems, 1987).   Typical applications for engineering/performance polymers are listed in Table 2-4.

Table 2-3

CLASSIFICATION OF
THERMOPLASTIC POLYMERS

Characteristics	Commodity	Transitional	Engineering	Performance
VOLUME (Pounds per polymer)	1.5–10 billion	0.5–1.5 billion	20–500 million	Less than 20 million
PROCESSABILITY	High	Good	Good	Least
THERMAL STABILITY	Low	Medium	High	Excellent
PRICE ($/lb, 1986)	0.25–0.75	0.75–1.25	1.25–3.00	3.00–20.00
POLYMERS INCLUDED:	LD polyethylene Polyvinyl chloride HD polyethylene Polypropylene (PP) Polystyrene Polyethylene tereph-thalate (PET, bottle grade)	ABS/SAN Acrylics PP (Glass-filled)(a) PE (Glass-filled)(a) Other styrenics-selected polymers(a)	PET (Glass-filled)(a) PBT Polyacetal Polycarbonate Modified PPO/PPE Nylon (6 and 66) SMA terpolymer(a) Other alloys and blends(a)	Fluoropolymers Liquid crystal polymers(a) Nylon (11 and 12) Polyamideimide Polyarylate Polyetheretherketone (PEEK) Polyetherimide Polyethersulfone Polyimide Polyphenylenesulfide Polysulfone

(a) Polymers under development.

Source: Adapted from Chem Systems, 1987. Production volume ranges have been updated in some cases with more recent data.

TABLE 2-4

TYPICAL APPLICATIONS FOR ENGINEERING AND PERFORMANCE POLYMERS

Polyacetal	Automotive (steering column, window and windshield wiper) components, hardware, faucets, valves, gears, disposable cigarette lighters, medical apparatus
Nylon	Wire and cable, barrier packaging film, electrical connectors, windshield wiper parts, radiator and tanks, brake fluid reservoirs, gears, impellers, housewares
Polycarbonate	Electrical/electronic components, housings, switches, aerodynamically styled headlights, glazing, appliances, medical apparatus, compact (audio) discs, baby bottles
Modified PPO/PPE	Business machine and appliance housings, TV cabinets and components, electrical/electronic components, automotive interior trim and instrument panels
PBT	Electrical/electronic components, automotive electrical components (distributor caps and rotors), automotive exterior body components, pump and sprinkler components
PET (glass-filled)	Electrical/electronic components, automotive electrical components, consumer products, office furniture components
Polyarylate	Glazing, electrical/electronic components, automotive fog lamps, microwave oven components
PEEK	Wire and cable, aerospace composites, electrical connectors and coils, bearings
Polyetherimide	Printed circuit boards, microwave oven components, frozen food trays, electrical/electronic components
Polyethersulfone	Printed circuit boards, electrical/electronic components, composites
Polyphenylene sulfide	Electrical/electronic components, halogen headlight components, carburetor components, pump components, industrial parts, composites

TABLE 2-4   (continued)

Polysulfone	Electrical/electronic components, pumps, valves, pipe, microwave oven cookware, medical apparatus
Fluoropolymers	Nonstick cookware coating, wire and cable, solid lubricating additives for other plastics
Liquid crystal polymers	Freezer-to-oven cookware, fiber optic construction, electrical/electronic components
Polyamide-imide	Valves, mechanical components, automotive engine block, industrial components

Source: Chem Systems, 1987.

2.2.2 Incorporating Additives

Some resins are used essentially as they are formed (as pure polymers), but market demands usually dictate further tailoring of the polymers' properties. Additives are used in plastics 1) to alter physical characteristics (e.g., to increase flame retardance), 2) to influence aesthetic properties (e.g., to add color), or 3) to permit processing of the resins (e.g., to increase plastic melt flow) (Radian, 1987). The type of additive determines when it is added, whether at the resin manufacturing plant or at the processing plant. (In thermoplastics, additives may be added after the resin is formed or during processing into end products.) The type of additive also determines the amount used, ranging from less than 1% to 60% by volume of the end product.

Additives are incorporated into plastics by one of two ways:

■ As solids or liquids physically mixed with the plastic polymer; because these additives do not react with the polymer but are mixed with it, there exists a hypothetical potential for these additives to leach out of the end product. Any leaching that could possibly occur depends on the characteristics of the polymers and additives, including their degree of miscibility (i.e., the degree to which they are mixed), and on the environmental exposure of the plastic (e.g., temperature conditions).

■ As substances that react with the plastic polymer; these additives generally cannot migrate out of the plastic end products into outside media without chemical breakdown of the plastic (Dynamac, 1983). (Any unreacted additives may, however, be available for leaching.)

The primary types and roles of additives are discussed in Section 2.3.7.

2.2.3 Processing Resins into End Products

Processors can select from among a variety of technologies in creating products from thermoplastic and thermosetting plastics. For most thermoplastic commodity resins, the processor usually purchases the resins in the form of small pellets. The pellets may then be coated with the additives (e.g., colorants) needed in the final products or fed into machines that melt and mix the resins. Resins can also be colored using a hot compounding technique at this juncture. The melted resins can be extruded through a die (e.g., to form pipe or fiber), pressed into a mold, or foamed by introducing a gas. After cooling, the products are trimmed to remove any excess material and sent to the next processing step. Certain thermoplastic resins, however -- most notably, PVC -- are purchased as a powder, which is pressed into pill-like shapes and then processed.

The working range between the temperature at which a thermoplastic melts and that at which it decomposes may be fairly narrow, and some decomposition may take place at even the lowest melt temperature. Thus, thermoplastics are kept in a melted state for as short a time as possible, and temperatures are generally kept low.

Thermosets, on the other hand, are often shipped to processors in liquid form. The processor then simultaneously molds or foams plastic products and then "cures" the resin. The curing process (e.g., setting into a mold) usually causes the catalytic cross-linking or other chemical reaction that permanently hardens the thermoset into a desired shape (e.g., that of the mold). Pressure and heat can both be used in processing thermosets.

2.3 PRODUCTION AND CONSUMPTION STATISTICS

2.3.1 Historical Overview

The first commercial plastics were developed over one hundred years ago, but the growth of the petrochemical industry (beginning in the 1920s) was the catalyst behind plastics becoming major consumer materials. Now plastics have not only replaced many wood, leather, paper, metal, glass, and natural fiber products in many applications, but also have facilitated the development of entirely new types of products. As plastics have found more markets, the amount of plastics produced in the United States has grown from about 3 billion pounds in 1958 to about 57 billion pounds in 1988 (Modern Plastics, 1989; see Table 2-2). This growth represents an average annual rate of 10.3%. During the same time period, GNP (measured in constant 1982 dollars) has grown at an annual rate of 3.2%.

Between 1935 and 1958, many new plastics were developed based on technology that resulted from the needs of war and the markets for new products following the war. These efforts led to production of plastic films for packaging of foods and other items, plastic sheets for windows and decoration, and upholstery materials for automobile seats and furniture.

Early plastics, however, were often inferior to traditional materials; thus, they were still not accepted as the material of choice for many durable and nondurable goods. Only after new and better plastics were developed in the 1960s and 1970s did plastics become the first choice for many items of commerce. Acceptance by manufacturers and consumers of these new plastics led to further developments in processing and synthesizing (e.g., some modern thermosets can be set without heat or pressure). Now plastics materials have become industrial commodities in the same category as steel, paper, or aluminum, and the various end use markets support a major plastics industry that ships more than $82 billion worth of products per year (Chem Systems, 1987).

2.3.2 Domestic Production of Plastics

The most important plastics in terms of 1988 U.S. production volume are listed in Table 2-2. On a weight basis, thermosets currently account for only 16.9% of the total domestic production of plastic resins. Of the thermoplastics, the most important resins on the basis of volume produced (i.e., the commodity resins) are low-density polyethylene (LPDE), polyvinyl chloride (PVC), high-density polyethylene (HDPE), and polypropylene (PP); these account for 60% of total thermoplastic production. For information concerning minor plastics not listed on this table, see the Modern Plastics Encyclopedia (1988).

2.3.3 Import/Export and Domestic Consumption

Import/export and domestic consumption data for major thermoplastics are listed in Table 2-5. There is a modest positive net trade balance (exports minus imports) for all of the resins. The only resins for which export volume exceeds 10% of U.S. production are polypropylene (PP) and acrylonitrile/ butadiene/styrene (ABS).

Plastics are also exported or imported as finished products. Such shipments are not reflected in the statistics on resins. In later sections on plastics in the solid waste stream, data are employed that capture the effect of imported plastics in final products (Franklin Associates, 1988).

2.3.4 Economic Profile of the Plastics Industry

2.3.4.1   Sector Characteristics

Two sectors of the economy are involved in the manufacture of plastic products: manufacturers of plastic resins and processors of the plastic resins into plastic products. According to available government statistics detailed below, the latter sector, plastics processing, is much larger both in terms of sales generated and workforce employed.

Resin manufacturing is dominated by large petrochemical plants. Table 2-6 presents estimates of the total nameplate capacity of major firms for the production of the most important commodity thermoplastics (on the basis of volume produced). (Nameplate capacity is the design capacity of the plant.) As the table indicates, the major petrochemical firms have sufficient nameplate capacity to satisfy the entire U.S. demand for these resins.

As indicated in Section 2.2.1, engineering/performance resins used in commerce are manufactured by a smaller group of firms. These high-performance or unusual resins may be generated in batches for specific customers rather than in standardized, large-volume production processes. Most of the resins, however, are made in the same large petrochemical plants as the commodity resins because these manufacturers are capable of investing in the research necessary for developing the polymers.

Department of Commerce statistics indicate a total of 477 resin manufacturers (as classified in SIC 2821, U.S. Bureau of the Census, 1988). The industry employed 55,500 workers and generated annual shipments valued at $23.9 billion for 1987 (SPI, 1988). Resin manufacturing plants are concentrated in the Gulf of Mexico and in the Atlantic Coast states. The states with the largest employment in resin manufacturing are Texas, New Jersey, West Virginia, Pennsylvania, Louisiana, Ohio, Michigan, and California (U.S. Bureau of the Census, 1985).

Table 2-5

**IMPORTS AND EXPORTS
OF MAJOR THERMOPLASTIC RESINS
(MILLION POUNDS, 1988)**

Resin	Domestic Production			Exports as % of U.S. Production	Imports	Total Domestic Consumption(a)	Imports as % of U.S. Consumption	Net Trade Balance (Exp-Imp)
	For U.S. Consumption	For Export	Total U.S. Production					
Low-density polyethylene	8,911	954	9,865	9.7	816	9,727	8.4	138
Polyvinyl chloride	7,854	469	8,323	5.6	133	7,987	1.7	336
High-density polyethylene	7,540	704	8,244	8.5	86	7,626	1.1	618
Polypropylene and co-polymers	6,102	1,202	7,304	16.5	33	6,135	0.5	1169
Polystyrene	4,979	152	5,131	3.0	53	5,032	1.1	99
ABS	1,016	222	1,238	17.9	50	1,066	4.7	172

(a) Domestic consumption was defined as the sum of U.S. consumption of domestic production and imports.

Source: Modern Plastics, January 1989.

Table 2-6

NAMEPLATE CAPACITY OF MAJOR
MANUFACTURERS FOR SELECTED COMMODITY RESINS
(Million pounds, 1988)

Resin	Number of Manufacturers Included	Total Nameplate Capacity 1/1/89	Total U.S. 1988 Sales	Capacity As % of Total U.S. 1988 Sales
Low- and high-density polyethylene	16	20,125	18,109	111
Polyvinyl chloride	13	9,622	8,323	116
Polypropylene	13	8,355	7,304	114
Polystyrene	16	6,140	5,131	120

Note: The number of manufacturers included was based on listings in the source.

Source: Modern Plastics, 1989.

In contrast to the relatively select group of resin manufacturers, the plastic processing and converting industry encompasses over 12,000 establishments (U.S. Bureau of the Census, 1988). These firms purchase resins and process them for all manner of packaging, consumer, or industrial uses. The plastic processing industry generated shipments of $60.5 billion in 1987, or approximately 1.5 times the sales of the resin manufacturers. The plastic processing sector also employs a workforce more than ten times the size of that for resin manufacturing (580,000 workers; SPI, 1988).

2.3.4.2   Market Conditions and Prices for Commodity Resins

Recycling programs that potentially will be developed for plastics waste (see Section 5) may compete with virgin resin production; the price movements for virgin resins may thus affect the economic viability of recycling efforts. This section examines the behavior of markets for commodity resins. As noted, manufacturing of each of the commodity thermoplastics (see Table 2-6) is dominated by ten to twenty petrochemical firms. Due to economies of scale in production, each plant must be quite large to achieve competitively low production costs. Thus, any change in capacity due to the construction of a new plant can be a significant market development.

Furthermore, because the construction of new capacity requires several years from inception to production, industry planners cannot predict with certainty the market conditions that will prevail when new capacity is brought online. At that time, therefore, the industry may face excess capacity. Because, by definition, the commodity plastic resins made by different plants are interchangeable, manufacturers with excess capacity may respond by trying to undercut the prices offered by other plants. As a result, the plastics market can experience erratic swings in product prices when substantial, discrete shifts occur in available production capacity.

Price levels are also affected by the availability of the natural gas derivatives that are the principal raw materials for plastics manufacture. This availability may be influenced by a range of factors in energy markets or by production problems in the major plants that produce the derivatives. In the latter category, for example, accidental fires at two ethylene plants in 1988 created a raw material shortage for the manufacturers of polyethylene and several other resins (Chemical Week, 1988).

It should be noted that these generalizations represent extreme simplifications of chemical industry pricing behavior. Actual industry decisions incorporate long-term contract pricing agreements, contracting and planning issues for raw material supplies, and a myriad of other factors.

In recent years, little new capacity has become available despite a period of rapid demand growth. The apparent strains on capacity have probably contributed to the price increases seen for major commodity resins (Table 2-7). As the table shows, prices for these resins have risen sharply in the past two years.

Table 2-7

PRICE MOVEMENTS FOR
SELECTED COMMODITY RESINS
(1986–1988)

Resin	Price ($/lb)	
	1986	1988
Low-density polyethylene	0.29	0.51
Polyvinyl chloride, pipe grade	0.29	0.43
High-density polyethylene	0.32	0.51
Polyethylene terephthalate (PET, bottle grade)	0.55	0.70

Note: Prices shown reflect contract or prevailing selling prices that incorporate discounts, allowances, and rollbacks from current list prices.

Source: Modern Plastics, 1989b.

In an apparent response to the price increases and projected increases in market demand (these are described further below), a number of firms have announced forthcoming capacity increases. Manufacturers of low- and high-density polyethylene, polyvinyl chloride, and polypropylene -- the four largest-volume resins -- have announced capacity additions equivalent to nearly 25% of current industry capacity (Modern Plastics, 1989).

Any future price volatility for plastic resins cannot be predicted. For this study, it should be noted that price swings in plastic resin markets could influence many aspects of the flow of plastic materials through the economy. Price changes influence, for example, the competitiveness of plastics with other products in any of the intermediate or end use markets -- and thus the rate at which plastics enter the solid waste stream. Price swings also will affect the economic return for programs in solid waste management or reduction (e.g., source reduction or recycling).

2.3.5   Forecasts of Market Growth

In 1987, The Society of the Plastics Industry (SPI) commissioned a market forecast study (Chem Systems, 1987). This research examined the historical growth rates in the major plastics markets and developed forecasts of future growth to the year 2000. Table 2-8 presents the principal forecast findings for the total plastics market and for each of eight end use markets. Historically, the average annual growth for plastic sales (measured by weight) has ranged from 3.2% per annum in electrical/electronic sales to 7.8% in industrial markets and 9.6% in "other" markets. The high growth rates in the latter categories partly reflect the new uses for engineering thermoplastic resins in a variety of technical or specialized applications. Packaging is the number one market in terms of absolute size; this market grew rapidly from 1970 to 1985, with an average annual growth rate of 7.1%. The second largest market, building and construction, also showed a relatively large average annual growth rate of 6.2%.

The SPI research estimated that future growth rates among the end use markets until the year 2000 would vary from 2.4% (adhesives, inks, coatings) to 4.0% (transportation) per annum. The forecasters assumed an average annual growth rate for the U.S. Gross National Product during this period of 2.9% per annum. Thus, the overall market growth for plastics, estimated at 3.2%, was forecast to exceed the rate for the economy as a whole. Even so, the rate predicted was almost half the actual annual growth rate between 1970 and 1985 (6.3%).

The SPI forecast study was performed in 1987 using 1985 data. Since that time, plastic markets have grown at a faster rate than had been projected. Aggregate U.S. production grew from 47.8 billion pounds in 1985 to 56.9 billion pounds in 1988; the annual average growth in recent years, therefore, has been virtually the same as that of the past two decades.

Table 2-8

PLASTIC INDUSTRY MARKET SECTOR GROWTH, 1970-2000
(Millions of Pounds)

	1970	1985	2000	Average Annual Growth Rate (%)	
				Actual Average 1970-1985	Forecast 1985-2000
Packaging	4,695	13,200	22,580	7.1	3.6
Building & construction	4,095	10,350	14,975	6.2	2.5
Other	1,585	6,175	10,220	9.5	3.4
Consumer	2,030	3,715	5,670	4.1	4.9
Electrical/electronic	1,825	2,930	5,020	3.2	3.7
Furniture/furnishings	1,275	2,635	4,100	4.9	3.0
Adhesives inks & coatings	1,475	2,525	3,585	3.6	2.4
Transportation	1,035	2,365	4,240	5.7	4.0
Total domestic demand	18,015	43,895	70,390	6.3	3.2
Total exports	1,180	3,945	5,160	8.4	1.8
Total	19,195	47,840	75,550	6.3	3.1

Source: Chem Systems, 1987.

2.3.6   Characteristics of Major Resin Types

Differences in resin properties and in the economics of production determine the manner in which resins are used in various end markets. This section summarizes the main characteristics of resins and introduces the most common product uses. The information is designed to allow relationships to be identified between resin types and the plastic products that eventually appear in the solid waste stream.

The major types of plastics are listed in Table 2-9, along with their salient characteristics and primary product markets. For production data for each of these resins, see Table 2-1; for import/export and total domestic consumption data for the four most important resins (by volume produced), see Table 2-4.

Table 2-10 presents a complete distribution of the commodity resins according to the product market in which they are used. The majority of low- and high-density polyethylene and polyethylene terephthalate resins are used in packaging. PVC and two thermoset resins (phenolic and urea melamine) are used primarily in building and construction. Other thermoplastics such as polystyrene and polyethylene terephthalate are used most commonly in categories defined for consumer and institutional products. More information about each of the end user markets is provided in Section 2.4.

2.3.7   Characteristics of Major Additive Types

Plastic additives play an important role in modifying the characteristics of virgin resins. The additives used encompass a variety of chemicals and can be as significant as the resin itself in determining product use. This section introduces the categories of additives and presents a summary of information about production levels (for each category and for selected chemicals or minerals within the category), purposes for additive use, and the patterns of use relative to the different plastic resins. Additives are examined again in Section 4, where the issue of additive toxicity is considered.

Table 2-11 summarizes the characteristics of the various categories of additives. The table lists the purposes for each type of additive as well as the kinds of final product that could contain the additives. Additives are used with resins 1) to increase the ease of processing of the resin. and/or 2) to improve the characteristics of the final product. Additives used for the first purpose include antistatic agents, catalysts, free radical initiators, heat stabilizers, and lubricants. Most additives improve on balance the characteristics of the final product.

Manufacturers produced 9.7 billion pounds of plastic additives in 1982, a quantity equal to 17% of the weight of the polymers themselves. Additives in the largest categories of use generated most of this production. Table 2-12 presents the production levels for 16 categories of additives. As can be seen from the table, fillers and plasticizers represent 75% of all additives produced. Three more categories -- reinforcing agents, flame retardants, and colorants -- account for another 19%. Production levels for the remaining categories are one or two orders of magnitude smaller.

Table 2-9

RESIN CHARACTERISTICS, MARKETS, AND PRODUCTS

Resin	Resin Characteristics	Primary Product Markets	Product Examples
THERMOPLASTICS			
Low-density polyethylene (LDPE)	Largest volume resin used for packaging. Moisture-proof, inert	Packaging	High-clarity extruded film, wire and cable coatings, refuse bags coated papers
Polyvinyl chloride (PVC)	Strength and clarity. Brittle unless modified with plasticizers	Building & construction, packaging	Construction pipe, meat wrap, blister packs, cooking oil bottles, phono records, wall covering, flooring
High-density polyethylene (HDPE)	Tough, flexible, translucent	Packaging	Milk and detergent bottles, heavy-duty films, e.g. boil bag pouches, liners, wire & cable insulation
Polypropylene (PP)	Stiff, heat & chemical resistant	Furniture & furnishings, packaging, other	Syrup bottles, yogurt and margarine tubs, fish nets, drinking straws, auto battery cases, carpet backing, office machines & furniture, auto fenders
Polystyrene (PS)	Brittle, clear, rigid, good thermal properties; easy to process	Packaging, consumer products	Disposable foam dishes & cups, egg cartons, take-out containers, foam insulation, cassette tape cases

(Cont.)

Table 2-9 (Cont.)

RESIN CHARACTERISTICS, MARKETS, AND PRODUCTS

Resin	Resin Characteristics	Primary Product Markets	Product Examples
Other Styrenics	Strong, stretchable	Adhesives, coatings & inks	Assembly and construction adhesives, pressure sensitive labels and tapes, footwear soles, roof coatings
Polyethylene terephthalate (PET)	Tough, shatter resistant	Packaging, consumer products	Soft drink bottles, other beverage, food, & medicine containers, synthetic textiles, x-ray and photographic film, magnetic tape
Acrylonitrile/butadiene/ styrene	Tough, abrasion resistant	Transportation, electrical and electronic products	Pipe, refrigerator door linings, telephones, sporting goods, automotive brake parts
THERMOSETS			
Phenolic	Heat resistant, strength, shatter resistant	Building & construction	Handles, knobs, electrical connectors, appliances, automotive parts
Polyurethane	Malleable for rigid or flexible foams	Furniture & furnishings, building & construction, transportation	Cushioning, auto bumpers & door panels, varnishes
Urea and melamine	Rigid, chemically resistant	Building & construction, consumer products	Plywood binding, knobs, handles, dinnerware, toilet seats
Polyester, unsaturated	Malleable for fabrication of large parts	Building & construction, transportation	Electrical components, automobile parts, coatings, cast shower/bath units

Sources: Chem Systems (1987); Wirka (1988); SPI (1988).

Table 2-10

RESINS DISTRIBUTED BY
MAJOR END MARKETS

Resins/Market Share (%)	Packaging	Building & Construction	Consumer & Institutional Products	Electrical & Electronic	Furniture & Furnishings	Transport-ation	Adhesives, Inks, & Coatings	All Other	Exports	Total
THERMOPLASTICS										
Low-density polyethylene	64.1	3.9	6.3	4.4	0.4	--	3.5	6.9	10.4	100.0
Polyvinyl chloride	7.6	62.6	3.3	6.5	3.6	--	1.4	9.8	5.3	100.0
High-density polyethylene	53.0	9.5	11.0	1.7	--	2.6	--	11.3	10.8	100.0
Polypropylene	21.6	0.5	14.4	4.3	18.7	4.1	--	17.9	18.5	100.0
Polystyrene	29.2	11.7	34.3	9.5	0.8	--	--	12.1	2.3	100.0
Other styrenics(a)	0.0	3.8	3.8	0.0	14.6	0.9	54.2	22.6	0.0	100.0
Polyethylene terephthalate(a)	59.2	0.0	18.7	5.3	0.0	0.0	0.0	9.9	7.0	100.0
ABS/SAN(a)	0.0	15.9	6.8	21.8	0.0	24.5	0.0	21.8	9.1	100.0
THERMOSETS										
Phenolic	--	83.4	--	4.2	--	2.2	2.3	7.2	0.6	100.0
Polyurethane	3.8	20.2	--	5.9	37.1	18.8	--	14.3	--	100.0
Urea and melamine	2.0	76.2	--	--	2.4	--	--	19.4	--	100.0
Polyester, unsaturated	--	39.7	7.2	3.5	1.6	34.7	--	12.6	0.7	100.0

Source: The Society of the Plastics Industry, 1988, except (a) which is Chem Systems, 1987.

Table 2-11

CHARACTERISTICS AND USES OF PLASTICS ADDITIVES

Additive	Examples or Types	Purpose	Typical Applications For Products w/Additive
Antimicrobials	Oxybisphenoxarsine, isothiazalone	Increase resistance of finished product to microorganisms	Roof membranes, pond liners, appliance gaskets, outdoor furniture, trash bags
Antioxidants	Phenolics, amines, phosphites, thioesters	Prevent deterioration of appearance and physical properties during processing and long term use	Numerous
Antistatic agents	Amine salts, phosphoric acid esters, polyethers	Control static buildup during processing or in final product	Films, bottles, electronics and computer room furnishings, medical equipment
Blowing agents	Azobisformamide, chlorofluorocarbons pentane	Add porosity to produce foamed plastics	Food trays, insulation, cushions, clothing, mattresses
Catalysts and curing agents	Numerous	Facilitate polymerization and/or curing of resin	Numerous
Colorants	Organic and inorganic pigments, dyes	Enhance appearance and consumer appeal of end product	Consumer products
Fillers	Minerals, e.g. calcium carbonate wood flours	Add hardness or other properties, lower production costs	Coatings, composites, flooring
Flame retardants	Aluminum trihydrate, halogenated hydrocarbons, organophosphates, antimony oxide	Reduce combustibility of plastic	Consumer, electrical, transportation & construction

(Cont)

Table 2-11 (Cont.)

CHARACTERISTICS AND USE OF PLASTICS ADDITIVES

Additive	Examples or Types	Purpose	Typical Applications For Products w/Additive
Free radical initiators	Peroxides, azo compounds	Assist in polymerization or curing	Numerous
Heat stabilizers	Organotin mercaptides, lead compounds, and barium, cadmium and zinc soaps	Prevent heat degradation or improve heat resistance of polyvinyl chloride	Construction pipe, bottles, wire & cable coatings, film, sheet and upholstery
Impact modifiers	Methacrylate butadiene styrene, acrylic polymers, chlorinated PE, ethylene vinyl acetate	Improve strength and impact-resistance	Rigid PVC applications, building & construction (pipe and siding)
Lubricants & mold release agents	Fatty acids, alcohols and amides, esters, metallic stearates, silicones, soaps, waxes	Improve viscosity of plastic or reduce friction between plastic and surrounding surfaces, including molds	Molded and extruded consumer products
Plasticizers	Phthalates, trimellitates, aliphatic di- and tri-esters, polyesters, phosphates	Soften and flexibilize rigid polymers	Garden hose and tubing, floormats, gaskets, coatings
Reinforcers	Glass fibers, wood flours	Improve physical properties of resin	Laminates
UV stabilizers	Hindered amines, carbon black, hydroxy-benzophenones, hydroxybenzotriazoles	Prevent or inhibit degradation by UV light	Building materials, agricultural films

Sources: Kresta (1982); Modern Plastics (1989); Radian Corp. (1987); Rauch Associates (1986); Seymour (1978); and Stepek–Daoust (1983).

Table 2-12

ADDITIVE PRODUCTION LEVELS, USE CONCENTRATIONS,
AND MAJOR POLYMER APPLICATIONS
(1987)

Additive	Total Production (million lb)	Additive Concentration in Plastic Products(a) (lb additive/100 lb resin)	Largest Polymer Markets
Fillers	5,586	High 10–50	PVC, Unsat. polyester
Plasticizers	1,694	High 20–60	PVC, cellulosics
Reinforcements	893	High 10–40	Unsat. polyester, epoxy
Flame retardants	513	High 10–20	Various (in building, auto)
Colorants	438	Low 1–2	Numerous
Impact modifiers	130	High 10–20	PVC, styrenics, polyolefins, engineering plastics
Lubricants	96	Low < 1	PVC, PS, polyolefins
Heat stabilizers	83	Moderate 1–5	PVC
Free radical initiators(b)	44	Low < 1	LDPE, PS, PVC, acrylics, PE
Antioxidants	42	Low < 1	Polyolefins, impact styrene, ABS
Chemical blowing agents	13	Moderate 1–5	Polyurethane, PVC, PP, PS, ABS
Antimicrobial agents	11	Low < 1	PVC, PE, polyurethane
Antistatic agents	8	Low < 1	PVC, polyolefins, polyurethane
UV stabilizers	7	Low < 1	Polyolefins, PE, PP, PS, PVC polycarbonate
Catalysts(c)	6	Low < 1	Polyurethane
Others	104	Low < 1	
TOTAL	9,668		

(a) Estimates refer to concentrations in those products where the additive is used.
(b) Includes organic peroxides only, as reported by source.
(c) Includes urethane catalysts only, as reported by source.

Source: Production estimates and polymer markets from Chemical and
         Engineering News, 1988.
         Concentration estimates developed by Eastern Research Group.

The much higher production and consumption estimates for some additives are explained partly by the manner of their use: Some additives are mixed into the plastic polymer in bulk, while others are combined only at the rate of 1% or less of the polymer. Table 2-12 summarizes the rate of use in products for the largest categories of additives. The large-volume additives may represent one-half as much weight as the resin in some applications. In contrast, colorants, and most other categories of additives, are added at very low rates into the polymers. The table also describes the resins that are most likely to be combined with an additive for a particular product use. For example, polyvinyl chloride (PVC) may be combined with any of several additives. In some cases, the additive is almost ubiquitous (e.g., colorants are employed in 70% of all products), and in others the use of the additive is determined less by the nature of the resin than by the product use (e.g., automotive or construction uses require flame retardants).

The additive categories are defined according to purpose, and a range of chemical compounds or minerals are used within each category. Table 2-13 presents consumption data for the most commonly used additives within each category. In most categories, one or two additives are preferred by manufacturers, with others used for specialized and much more limited circumstances. For example, fiberglass consumption represents approximately 80% of all use of reinforcing agents. Also, some additives are much more expensive, thus limiting their use.

2.4 MAJOR END USE MARKETS FOR PLASTICS

The major end use markets for plastics, in order of volume, are 1) packaging; 2) building and construction; 3) consumer products; 4) electrical and electronics; 5) furniture and furnishings; 6) transportation, including automobiles, vans, trucks, and aircraft; 7) adhesives, inks, and coatings; and 8) other. For the major products in each market category, see Table 2-14.

The following information about the market sectors is drawn from Chem Systems, 1987. The growth factors mentioned here should be considered in light of the various "solutions" for plastics waste management discussed in Section 5; these solutions may shift growth potential among the market sectors.

2.4.1 Packaging

In the packaging market segment, LDPE is used in the highest volume of any plastic resin (see Figure 2-1). This segment -- already the largest plastics market -- will continue to grow if traditional materials are replaced with plastics as well as if new packaging products are developed from plastics. Demographic shifts in the United States, including smaller family size, an aging population, and the employment of more American adults, are proving catalysts for the increased use of plastics in packaging. Manufacturers continue to find plastics to be attractive, low-cost materials that can be adapted to their diverse packaging and product presentation needs. Supporting trends include (Chem Systems, 1987):

Table 2-13

CONSUMPTION OF LARGE-VOLUME ADDITIVES
(Millions of Pounds, 1986)

Additive	Consumption	Additive	Consumption
FILLERS (a)		REINFORCING AGENTS	
Inorganics		Fiberglass	780.0
Minerals		Asbestos	90.0
Calcium carbonate	1,700.0	Cellulose	84.0
Kaolin & other	105.0	Carbon & other high performance	7.0
Talc	97.0	TOTAL	961.0
Mica	11.0		
Other minerals	225.0	COLORANTS	
Other inorganic		Inorganics	
Glass spheres	18.0	Titanium dioxide	292.0
Natural	132.0	Iron oxides	11.0
TOTAL	2,288.0	Cadmiums	6.0
		Chrome yellows (includes lead)	6.0
PLASTICIZERS		Molybdate orange	4.0
Phthalates		Others	4.0
Dioctyl (DOP)	290.0	Total – Inorganics	323.0
Diisodecyl	153.0	Organic pigments	
Dibutyl	22.0	Carbon black	86.0
Ditridecyl	21.0	Phthalo blues	3.0
Diethyl	18.0	Organic reds	3.0
Dimethyl	9.0	Organic yellows	1.0
Others	671.0	Phthalo greens	1.0
Total – Phthalates	1,184.0	Others	1.0
Epoxidized oils		Total – Organics	95.0
Soya oil	120.0	Dyes	
Others	16.0	Nigrosines	3.0
Total – Epoxidized oils	136.0	Oil solubles	1.0
Phosphates	50.0	Anthroquinones	0.5
Polymerics	49.0	Others	0.6
Dialkyl adipates	133.0	Total – Dyes	5.1
Trimellates	62.0	TOTAL	423.1
Others			
Oleates	13.0	CHEMICAL BLOWING AGENTS	
Palmitates	4.0	Azodicarbonides	11.3
Stearates	10.0	Oxbissulfonylhydrazide	0.5
All others	168.0	High temperature CBA's	0.4
Total – Others	195.0	Inorganic	0.4
TOTAL	1,809.0	TOTAL	12.6

(Cont.)

Table 2-13 (Cont.)

CONSUMPTION OF LARGE-VOLUME ADDITIVES
(Millions of Pounds, 1986)

Additive	Consumption	Additive	Consumption
FLAME RETARDANTS		**UV STABILIZERS**	
Additive Flame Retardants		Benzotriazoles	-
Aluminum trihydrate	218.0	Benzophenes	-
Phosphorous compounds	60.0	Salicylate esters	-
Antimony oxide	36.0	Cyanoacrylates	-
Bromine compounds	36.0	Malonates	-
Chlorinated compounds	33.0	Benzilidenes	-
Boron compounds	11.0	Others	-
Others	19.0	TOTAL (1984)	5.5
Total – Additive Flame Retardants	413.0		
Reactive Flame Retardants		**IMPACT MODIFIERS**	
Epoxy reactive	28.0	Acrylics	-
Polyester	12.0	MBS	-
Urethanes	12.0	ABS	-
Polycarbonate	8.0	CPE	-
Others	10.0	Ethylene–vinyl acetate copolymers	-
Total – Reactive Flame Retardants	70.0	Others	-
TOTAL	483.0	TOTAL	135.0
LUBRICANTS			
Metallic stearates	37.0	**ANTISTATIC AGENTS**	
Fatty acid amides	21.0	Quaternary ammonium compounds	-
Petroleum waxes	18.0	Fatty acid amides & amines	-
Fatty acid esters	13.0	Phosphate esters	-
Polyethylene waxes	6.0	Fatty acid ester derivatives	-
TOTAL	95.0	Others	-
		TOTAL	6.5
HEAT STABILIZERS			
Barium–cadmium	35.0	**ANTIOXIDANTS**	
Tin	25.0	Hindred phenols	-
Lead	23.0	Others	-
Calcium–zinc	5.0	TOTAL (1984)	35.0
Antimony	1.0		
TOTAL	89.0		

(a) Data presented are not fully consistent with estimates of filler consumption given in Table 2-11 because of differences in the definition of categories.

Notes: Data are for 1986 unless otherwise indicated at TOTAL.

Data for free radical initiators, antimicrobials, catalysts and curing agents not available.

"-" means not separately available.

Source: Modern Plastics and Rauch Associates; as cited in Rauch Associates, 1986.

Table 2-14

MAJOR PRODUCT AREAS IN THE
PLASTICS MARKET CATEGORIES

Market Category/ Product Area	Product as Percentage of Category	Category as Percentage of U.S. Sales
PACKAGING		
Flexible packaging except household & inst. refuse bags & film	24.1 (a)	
All other categories, except those below	23.9 (a)	
Bottles, jars, and vials	18.8 (a)	
Food containers (Excl. disp. cups)	17.0 (a)	
Household & inst. refuse bags and film	16.1 (a)	
Packaging, Total	100.0	33.5
BUILDING AND CONSTRUCTION		
Pipe, conduit and fittings	39.8	
Siding (incl. accessories and structural panels)	11.5	
Insulation materials	11.1	
Flooring	8.4	
All other	29.2	
Building and Construction, Total	100.0	24.8
CONSUMER AND INSTITUTIONAL PRODUCTS		
All categories, except those below	42.0	
Disposable food serviceware (incl. disp. cups)	nd	
Health care and medical products	nd	
Toys and sporting goods	9.6	
Hobby and graphic arts supplies	nd	
Consumer and Inst. Products, Total	100.0	11.1

(Cont.)

Table 2–14 (Cont.)

MAJOR PRODUCT AREAS IN THE
PLASTICS MARKET CATEGORIES

Market Category/ Product Area	Product as Percentage of Category	Category as Percentage of U.S. Sales
ELECTRICAL AND ELECTRONIC		
Home and industrial appliances	30.8	
Electric equip. combined		
with electronic components	26.4	
Wire and cable	nd	
Storage batteries	nd	
Communications equip.	3.4	
Electrical and Electronic, Total	100.0	6.1
FURNITURE AND FURNISHINGS		
Carpet and components	nd	
Textiles and furnishings, nec	28.3	
Rigid furniture	10.6	
Flexible furniture	nd	
Furniture and Furnishings, Total	100.0	4.9
TRANSPORTATION		
Motor vehicles and parts	77.7	
Ships, boats and recr. vehicles	19.0	
All other trans. equip.	3.3	
Transportation, Total	100.0	4.5
ADHESIVES, INKS AND COATINGS		
Inks and coatings, nec	67.1	
Adhesives and sealants	32.9	
Adhesives, Inks and Coatings, Total	100.0	4.0
ALL OTHER	100.0	11.0
TOTAL		100.0

Note: Market shares are calculated based on product sales and captive use by weight.
nd – Not disclosed by source.   nec – Not elsewhere classified.
(a) The market share estimates do not include a small residual of product sales. The unallocated
residual sales, however, represent only 0.1% of the packaging market and have been ignored.

Source: The Society of the Plastics Industry, 1988. The source utilized 1987 data.

Figure 2-1
PLASTIC RESINS IN PACKAGING USES

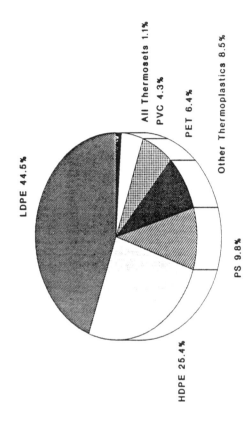

Source: Chem Systems, 1987; based on 1985 data.

- Decreasing time devoted to food and beverage preparation in the home, resulting in a demand for products in convenient, single-service packages such as microwavable prepared-entree trays and single-serving juice boxes

- Efforts by fast food outlets to convert paper wraps and boxes to disposable plastic containers; increased bulk food distribution for the increasing restaurant and institutional demand is also contributing to development of new products (e.g., sauce canisters)

- Increasing substitution of plastic shopping bags for paper; plastic bags are forecast to capture 75% of the market by 2000

- Increasing use of composites (several resins combined in one product) as well as development of high-barrier polymers and the technology to combine dissimilar polymers; these changes in the rigid packaging market are creating the opportunity for plastics to replace other materials in products that require oxygen, carbon dioxide, flavor, odor, and solvent permeation protection

- Increasing combination of traditional packaging materials with plastics to meet product-specific package-performance requirements (i.e., aseptic box packages combining paperboard, metal foil, and various resins)

The last two of these growth factors, the use of composites and the combination of plastics with traditional packaging materials, are of particular interest for environmental analysis. Both of these types of packaging tend to limit recycling options available (a topic discussed extensively in Section 5) because they make it difficult to separate different plastic resins or to separate resins from other materials (as needed for reprocessing).

Plastic composites are used for "high-barrier" packaging, that is, packaging that provides sufficient barriers against gas or moisture permeation to allow it to compete with traditional materials. Table 2-15 presents a 1985 forecast of the expected growth of high-barrier plastics as a share of the food and beverage packaging market. This market share is forecast to grow from negligible in 1983 to 7.8% (representing 14.5 billion containers) in 1993 (Agoos, 1985). Another forecast, published in 1986, estimated a 15.1% market share for high-barrier plastics in 1995 (Prepared Foods, 1986). This market share would represent 29 billion containers.

2.4.2 Building and Construction

Most end products used in building and construction can be made from commodity resins, though specialty resins are needed for small, functional parts such as casters, pulleys, and latches. The largest demand for thermoplastics in this market sector is for pipes and conduits; PVC, HDPE, LDPE, and polypropylene are used, for example, in potable water pipe and gas pipe. Polystyrene is also used for various building and construction needs, such as light fixtures and ornamental profiles. Thermoset resins, on the other hand, are used for the bonding and laminating of plywood, wood products, and protective coatings. For the volumes of each plastic used in this market (by percentage of the total market), see Figure 2-2.

Table 2-15

PROJECTED GROWTH OF
PLASTICS USE IN FOOD AND
BEVERAGE CONTAINERS
(Billions of Containers)

Material	1983		1988		1993	
	No.	As % of Total	No.	As % of Total	No.	As % of Total
Aluminum	58.08	35.8	68.87	39.5	74.30	40.2
Steel	34.32	21.1	26.42	15.2	18.56	10.0
Glass	42.12	25.9	39.91	22.9	35.64	19.3
Plastic – Commodity	20.63	12.7	28.02	16.1	32.68	17.7
High–barrier	0.01	0.0	2.00	1.1	14.50	7.8
Paper/foil combinations	7.27	4.5	8.99	5.2	9.10	4.9
Total	162.43	100.0	174.21	100.0	184.78	100.0

Source: Agoos, 1985.

Figure 2-2
PLASTIC RESINS IN BUILDING AND
CONSTRUCTION USES

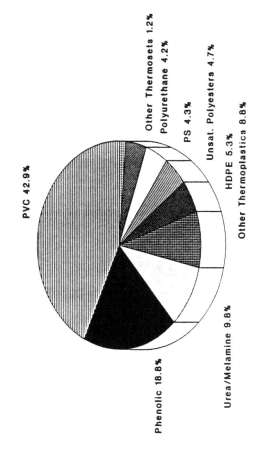

PVC 42.9%

Other Thermosets 1.2%
Polyurethane 4.2%
PS 4.3%
Unsat. Polyesters 4.7%
HDPE 5.3%
Other Thermoplastics 8.8%

Phenolic 18.8%

Urea/Melamine 9.8%

Source: Chem Systems, 1987; based on 1985 data.

The following trends support the increased use of plastics in this market sector:

■ Increasing numbers of smaller, multi-family dwellings, in which many types of plastics will be used to give design functionality at reduced cost

■ Increasing refurbishments of older homes rather than new construction; plastics will be used in advanced wiring systems, expanded attics, and finished basements

■ Increasing replacement by plastics of wood, metal, and glass in windows

■ Development of new polymers that offer product design economies for insulation, decorative moldings, wall coverings, roofing materials, and weight-supporting structural applications (e.g., beams of glass and resin rather than metal in buildings containing sensitive electronics equipment)

2.4.3 Consumer and Institutional Products

This market segment is defined as including such products as disposable food serviceware (including disposable cups), dinner and kitchenware, toys, sporting goods, health and medical care products, hobby and graphic arts supplies, and luggage. In this segment, polystyrene (PS) is used in the greatest volume (see Figure 2-3). Performance improvements and parts consolidation have been the driving forces behind the increased use of PS and other plastics in this market segment. Key areas for growth include:

■ The medical market, where medical gowns, operating table covers, and other fabrics can be replaced by single-use plastic films

■ The toy market, where electronic toys and action figures are becoming more popular

■ The household market, where dual-ovenable, disposable, and reusable food trays are an increasingly important application for plastics

■ The office supply market, where plastics can replace metals in such products as tape dispensers, stapler bodies, and desk organizers

2.4.4 Electrical and Electronics

This market includes home and industrial appliances, electrical and industrial equipment, components, computers and peripherals, records and batteries. In electrical and electronic applications, no one resin has cornered more than a quarter of the market (see Figure 2-4) -- in contrast to the packaging sector, for example, where LDPE represents 44.5% by volume of resins used. The fastest growing applications for plastics lie in the appliance and computer/peripheral areas; these trends include:

Figure 2-3
PLASTIC RESINS IN CONSUMER
PRODUCT USES

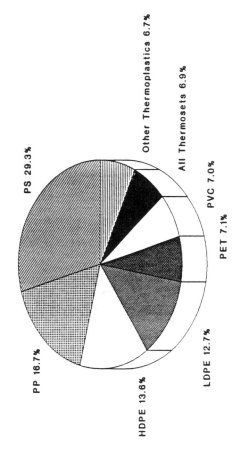

Source: Chem Systems, 1987; based on 1985 data.

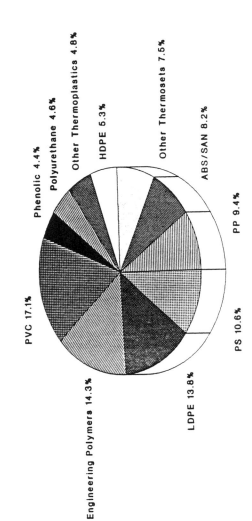

Figure 2-4
PLASTIC RESINS IN ELECTRICAL AND
ELECTRONIC USES

Source: Chem Systems, 1987; based on 1985 data.

- Increasing use of specialty plastics in small appliances (e.g., lawn mowers and power tools) traditionally made of metal

- Increasing use of plastics in large appliances, especially for housings, due to increased efficiency and design flexibility

- Increasing factory automation, resulting in a demand for plastics in such components as control panels, sensors, and printed wiring boards

- Increasing residential automation, in which microcomputers are used to control lighting, security, and appliances

- Increasing acceptance of high-reliability batteries containing inherently conductive polymers for medical and other fault-intolerant equipment

2.4.5 Furniture and Furnishings

This market (Figure 2-5) is dominated by polyurethane foams and polypropylene, which are used largely in upholstery and carpets. The following trends will influence the use of plastics in this segment:

- Increasing plastics substitution for glass because of economic factors and breakage resistance

- Continuing demand for ease of installation, decorating and color options, and ease of care, which supports the use of plastics in such applications as carpeting, flooring, and cabinets

- Increasing demand for relatively inexpensive plastic materials (e.g., polyethylene and polypropylene) at the expense of natural products such as jute

2.4.6 Transportation

The transportation industry's components are automotive, other land-based vehicles (including trailers), mass transit, airplanes/aerospace, marine, and military. Polyurethane is the most important resin in this market by volume used (see Figure 2-6). The following trends are creating significant opportunities for plastics use in this market segment:

- Manufacture in United States of automobiles by Japanese companies, a change that favors domestic consumption of plastics

- Increasing use of polymer systems in cars at the expense of metal, glass, and rubber (e.g., for weight reduction)

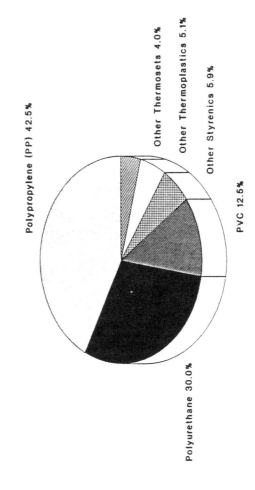

Figure 2-5
PLASTIC RESINS IN FURNITURE
AND FURNISHINGS USE

Source: Chem Systems, 1987; based on 1985 data.

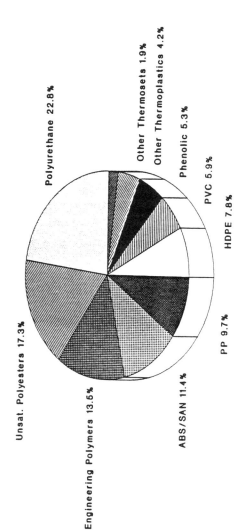

Figure 2-6
PLASTIC RESINS IN TRANSPORTATION USES

Polyurethane 22.8%

Other Thermosets 1.9%

Other Thermoplastics 4.2%

Phenolic 5.3%

PVC 5.9%

HDPE 7.8%

PP 9.7%

ABS/SAN 11.4%

Engineering Polymers 13.5%

Unsat. Polyesters 17.3%

Source: Chem Systems, 1987; based on 1985 data.

■ Reduction of the 5 mph bumper impact standard to 2.5 mph; a number of plastics can now meet Federal regulations (by 2000, plastics are forecast to capture 70% of the market)

■ Development of polymeric alloys and blends (e.g., nylon, polyester) specifically tailored for automotive exterior parts such as body panels and bumpers; advantages include lighter weight, resistance to salt corrosion, and economics of production

■ Increasing military spending on advanced systems such as stealth aircraft

2.4.7 Adhesives, Inks, and Coatings

This market is dominated by thermosets (e.g., urea and melamine), styrenics (e.g., styrene-butadiene), and vinyls (e.g., polyvinyl acetate) (see Figure 2-7). The following trends will influence the growth of this segment:

■ Increasing use of adhesives to replace mechanical fasteners in automotive, aerospace, and other structural applications

■ Increasing use of multilayer constructions of noncompatible materials for packaging, which will require adhesives to bond dissimilar materials

2.4.8 Other

This segment consists primarily of sales to resellers, compounders, and distributors. Often, this material does not meet the primary suppliers' intended product specifications and is thus relegated to less demanding applications.

2.5 DISPOSITION OF PLASTICS INTO THE SOLID WASTE STREAM

Plastic end products and materials eventually contribute to the solid waste stream. This section characterizes the plastics share of general waste volumes and, to the extent possible, the types of plastics included in these waste materials. The solid waste management methods used to handle these wastes are described in Sections 3 and 4 of this study.

As plastics are discarded or lost, they contribute to various waste streams. The stream on which this report focuses is MSW, i.e., the waste generated by households, institutions, and commercial establishments and managed by community services. In addition, some plastic wastes are:

■ Discarded, discharged, or lost to inland water bodies or the ocean

■ Improperly disposed of as litter; these wastes may be eventually added to MSW or remain uncollected indefinitely

Figure 2-7
PLASTIC RESINS IN ADHESIVES, INKS,
AND COATINGS USES

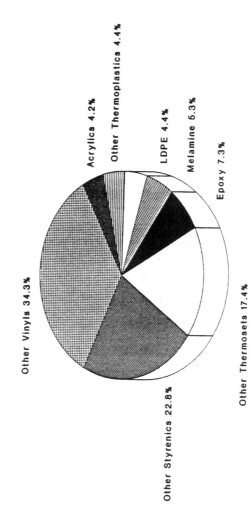

Acrylics 4.2%

Other Thermoplastics 4.4%

LDPE 4.4%

Melamine 5.3%

Epoxy 7.3%

Other Vinyls 34.3%

Other Styrenics 22.8%

Other Thermosets 17.4%

Source: Chem Systems, 1987; based on 1985 data.

■ Disposed of as building and construction wastes, which are often sent to different landfills than MSW

■ "Disposed of" in automobile salvage yards and then discarded or recycled (e.g., plastic used on the dashboard of an automobile)

Industrial waste streams are not considered a component of MSW (as defined by EPA); thus, these streams, with one exception, do not fall within the scope of this analysis of post-consumer waste. That exception -- the stream of plastic resin pellets apparently released to the marine environment in the chain from plastics manufacture to transportation to processing -- is considered here because the impact of these pellets on the marine environment is an issue of increasing concern. Available information on building/construction wastes and plastic automobile waste is also included. In addition, litter and materials discarded in the marine environment are also post-consumer waste and thus are discussed in this report.

The following sections describe the contribution of plastics to the solid waste streams addressed here. The more comprehensive information covers the components of MSW. Separate analyses of plastics in building and construction wastes were not located.

2.5.1 Plastics in Municipal Solid Waste

Information on the composition of the plastic waste stream is extremely important for analyzing waste management options. The studies described below offer limited information regarding the amounts and types of plastics in the municipal solid waste stream.

The best data available for characterizing discarded plastics are those developed in studies of MSW. Because the primary focus of this study is post-consumer waste, the discussion here appropriately concentrates on household, institutional, and commercial wastes -- the primary constituents of MSW.

Table 2-16 presents a characterization of MSW as developed for EPA (Franklin Associates, 1988). The data show the contribution of plastic wastes, by weight, to the municipal solid waste stream for 1986 (the most recent year for which data have been prepared) and for the years 1970 and 2000. The data were generated using a "materials-flow" methodology, which relies on published data series on production or consumption of materials and products that enter the municipal solid waste stream. The researchers also made adjustments to the data to reflect materials or energy recovery.

Franklin Associates estimated that plastics represented 7.3% of MSW by weight in 1986. Paper and paperboard (35.6%) and yard waste (20.1%) combine for over one-half of the total. Metals (8.9%) and glass (8.4%), two materials that compete with plastics in many product applications, contribute slightly more weight to the MSW total than do the plastic wastes. (Analyses of relative volumes for waste materials are made in Section 4, where landfill issues are discussed.) The aggregate quantity of the plastic waste in MSW was estimated at 20.6 billion pounds, or 10.3 million tons. Plastics are predicted to increase to 31.2 billion pounds in Table 2-16 the year 2000 as both glass and metals decrease in their contribution to the waste stream. The Franklin Associates research does not include certain wastes that are not

Table 2-16

TYPES OF MATERIALS DISCARDED INTO THE MUNICIPAL
WASTE STREAM (a) AND THEIR SHARE OF THE TOTAL WASTE STREAM

Material	1970		1986		2000	
	lb (b)	%	lb (b)	%	lb (b)	%
Paper and paperboard	73.0	32.4	100.2	35.6	132.0	39.1
Glass	25.0	11.1	23.6	8.4	24.0	7.1
Metals	27.0	12.0	25.2	8.9	28.8	8.5
Plastics	6.0	2.7	20.6	7.3	31.2	9.2
Rubber and leather	6.0	2.7	7.8	2.8	7.6	2.3
Textiles	4.0	1.8	5.6	2.0	6.6	2.0
Wood	8.0	3.6	11.6	4.1	12.2	3.6
Other	0.2	–	0.2	–	0.2	–
Food waste	25.6	11.4	25.0	8.9	24.6	7.3
Yard waste	46.2	20.5	56.6	20.1	64.0	19.0
Miscellaneous organics	3.8	1.7	5.2	1.8	6.4	1.9
TOTAL	225.0	100.0	281.6	100.0	337.6	100.0

Source: Franklin Associates, 1988a.

– Negligible.

Notes: (a) Wastes discarded after materials recovery and before energy recovery.
Details may not add due to rounding.
(b) Expressed in billions of pounds.

considered municipal solid waste, such as building and construction wastes and automobile bodies and scrap. Further, this study considers the net import balance in product flows, but it does not include the packaging materials for imported products.

The same research group estimated the aggregate quantity of MSW in 1986 at 281.6 billion pounds, or 140.8 million tons. This estimate corresponds to a rate of MSW generation of approximately 3 pounds per capita per day.

Evidence from direct excavations of municipal landfills also provides information on the constituents of MSW. Researchers from the University of Arizona have measured the constituents of a number of landfills (Rathje et al., 1988). They reported that plastic wastes represented 7.4% by weight of MSW materials excavated from three landfills in geographically dispersed parts of the United States (see Table 2-17). The wastes exhumed in this study were first landfilled between 1977 and 1985. The researchers did not include in their total any contribution of plastics to "mixed wastes" in the landfills, including textiles, fast food packaging, and diapers (all products that could include plastics). Thus, foamed polystyrene ("clam-shell" type) fast-food containers are not included in the plastics total.

These excavation results are not directly comparable to the Franklin Associates studies. For instance, the excavation studies attempt to characterize the historical content of MSW as it is represented by wastes in a landfill, which would contain both recently discarded and older wastes. The Franklin Associates data only attempts to characterize the current flow of wastes.

Limited research has been performed on the types of plastics identifiable in MSW, making it difficult to complete the understanding of the flow of plastic materials through the economy and into the waste stream. A few elements of this work can be summarized here.

Franklin Associates research provides the best indications of the types of plastic materials entering the municipal solid waste stream. Table 2-18 presents a breakdown of MSW into product categories. In the container and packaging category, for example, Franklin Associates estimated 2.8 million tons of plastic containers and 2.8 million tons of "other [plastic] packaging." The remaining plastic wastes (out of the 10.3 million ton total) are included in the totals for durable and nondurable goods. The specific breakdown between these product categories is not given. It has been noted, however, that plastic products have been growing as a share of wastes in the nondurable goods category. This category captures most consumer goods. The durability of a plastic product (as well as numerous other characteristics) is of interest because of its potential impact on the selection of management options for the eventual plastic waste (e.g., is recycling as useful a management option for durable plastic goods). Further, plastics found in some durable goods, such as appliances or as parts of building and construction materials, may not be disposed with MSW but processed by scrap metal recyclers or disposed in separate landfills for demolition wastes.

To extend the analysis of plastics in the waste stream, it is useful to introduce estimates of the lifetime of plastic products. Table 2-19 presents the lifetimes for various plastic products that were used in the Franklin Associates data or extrapolated from that study (extrapolations were developed for certain product categories not covered in the source). Packaging materials are estimated to stay in use for less then one year. The product lifetimes estimates represent a connection between the production statistics for plastic products and waste statistics. Thus,

Table 2-17

WEIGHT OF LANDFILL CONSTITUENTS

Landfill Constituent	Weight (lb)	As % of Total MSW	As % of Excavated Material
BIODEGRADABLE			
Organic			
Yard	255.9	5.2	2.9
Food	59.7	1.2	0.7
Wood	266.5	5.4	3.0
Paper			
Newsprint	790.8	16.0	8.9
Packaging	699.1	14.1	7.9
Non-packaging	486.4	9.8	5.5
Corrugated	251.3	5.1	2.8
Magazines	112.2	2.3	1.3
Ferrous metal	399.3	8.1	4.5
BIODEGRADABLE - TOTAL	3,311.2	66.8	37.2
NONBIODEGRADABLE			
Plastics	367.2	7.4	4.1
Rubber	30.3	0.6	0.3
Aluminum	60.3	1.2	0.7
Glass	187.0	3.8	2.1
NONBIODEGRADABLE - TOTAL	644.8	13.0	7.2
MIXED MATERIALS	999.1	20.2	11.2
Unidentified	744.9	15.0	8.4
Textiles	171.0	3.5	1.9
Diapers	66.3	1.3	0.7
Fast food packaging	16.9	0.3	0.2
MIXED MATERIALS - TOTAL	999.1	20.2	11.2
TOTAL MSW	4,955.1	100.0	55.6
MATRIX MATERIAL			
Fines	2,987.1	—	33.5
Other (mostly clay)	670.3	—	7.5
Rock	293.0	—	3.3
MATRIX MATERIAL - TOTAL	3,950.4	—	44.4
TOTAL SAMPLE	8,905.5	—	100.0

Source: Rathje, 1988.

Table 2-18

NATURE AND DURABILITY OF PRODUCTS
DISCARDED INTO THE
MUNICIPAL SOLID WASTE STREAM

Product Classification	1970		1986		2000	
	Tons	%	Tons	%	Tons	%
Durable goods	13.9	12.4	19.2	13.6	23.0	13.6
Nondurable goods (a)	21.4	19.0	35.4	25.1	47.5	28.1
Containers and Packaging	39.3	34.9	42.7	30.3	50.7	30.0
Other Wastes (b)	37.8	33.6	43.4	30.8	47.5	28.1
TOTAL	112.4	100.0	140.7	100.0	168.7	100.0

Totals may not add due to rounding.

(a) Includes paper products such as newspapers, office papers, and paper towels; also apparel, footware, and miscellaneous nondurables (especially many small plastic products).

(b) Includes yard and food wastes and miscellaneous inorganic wastes.

Source: Franklin Associates, 1988a.

Table 2-19

ESTIMATED LIFETIMES FOR
PLASTIC PRODUCTS

Market Category/ Product Area	Product Lifetimes	Market Category/ Product Area	Product Lifetimes
PACKAGING		ELECTRICAL AND ELECTRONIC	
Flexible packaging except household & inst. refuse bags & film	<1 yr	Home and industrial appliances	10 yr
		Electric equip. combined with electronic components	10 yr
All other packaging	<1 yr	Wire and cable	> 10 yr (a)
Bottles, jars and vials	<1 yr	Storage batteries	> 10 yr (a)
Food containers (Excl. disp. cups)	<1 yr	Communications equip.	> 10 yr (a)
Household & inst. refuse bags and film	<1 yr	FURNITURE AND FURNISHINGS	
		Carpet and components	10 yr
		Textiles and furnishings, nec	10 yr
BUILDING AND CONSTRUCTION		Rigid furniture	10 yr
Pipe, conduit and fittings	NA	Flexible furniture	10 yr
Siding (incl. accessories and structural panels)	NA	TRANSPORTATION	
Insulation materials	NA	Motor vehicles and parts	NA
Flooring	NA	Ships, boats and recr. vehicles	NA
		All other trans. equip.	NA
CONSUMER AND INST. PRODUCTS			
All categories, exc. others	NE	ADHESIVES, INKS AND COATINGS	
Disposable food serviceware (incl. disp. cups)	<1 yr	Inks and coatings, nec	<1 yr (a)
Health care and medical products	<1 yr (a)	Adhesives and sealants	<1 yr (a)
Toys and sporting goods	5 yr		
Hobby and graphic arts supplies	<1 yr (a)		

NA – Not applicable; Category of waste is not normally included in MSW.
NE – Not estimated
nec – Not elsewhere classifiable.

(a) Estimated by ERG.

Source: Franklin Associates, 1988b.

packaging waste represents the current year's production of these materials, whereas consumer durable discards represent production from a decade ago.

With the product lifetimes data, it is possible to return to the estimates of packaging and containers in the solid waste stream to delineate the waste characteristics more clearly. Table 2-20 presents: 1) a distribution of packaging and container wastes by material; 2) distribution according to type of packaging; and 3) distribution by resin. The second and third parts of the table are based entirely on production data for packaging and for resins (as shown earlier in Figure 2-1 and Table 2-12); thus, the production data can be used to indicate the characteristics of the waste. The data indicate that most packaging waste consists of flexible packaging, most made from polyethylene (low- and high-density) plastics.

Other researchers have estimated explicitly the distribution of resins in plastic waste. One study combined assumptions about the approximate life of plastic articles and resin production and end use statistics to calculate the distribution of resins in plastic solid waste. That study generated results for four major resins, as follows: polyethylene (65.3% of the waste), polystyrene (17.1%, includes ABS and other copolymers), polypropylene (8.5%), and polyvinyl chloride (9.1%) (Alter, 1986). These estimates should be considered approximations. Some of the underlying assumptions were developed in the 1970's and were not updated for this report. Further, the estimates do not consider the full range of produced resins, thereby excluding some production.

The University of Arizona researchers have not performed a study on the constituents of plastic waste found in their excavations. Their recent publications, however, refer to methodological issues for measurement of PET soda bottles and of plastic film bags such as cleaner bags, grocery bags, and garbage bags. No quantitative evidence is available, however, about plastics found in the excavation studies.

In conclusion, it should be noted that the Franklin Associates methodology is valuable for clarifying the flow of plastic materials through the economy. In lieu of data about the composition of the aggregate solid waste stream, Franklin Associates developed estimates from the production statistics themselves. Most of the plastic materials produced eventually reach the municipal solid waste stream. One major difference between production and disposal statistics is the lag in disposal for certain plastic products. (Several other adjustments are also needed in order to consider production losses and to adjust for imports and exports.)

The implication of the certainty of eventual disposal is that production statistics are, with certain caveats, the best first-order approximation of the composition of discarded plastic materials. Their estimates do not address comprehensively the research interests of this study because they do not provide specific estimates for all plastic waste sectors (e.g., durable plastic products) and because their estimates cannot describe the composition of construction wastes.

They also do not separately address the wastes disposed to inland waters or the ocean. Nevertheless, the "materials flow" methodology correctly focuses on the aggregate flow of materials through the economy and into the waste stream.

Table 2-20

COMPONENTS OF PACKAGING
AND CONTAINER WASTE STREAM

Part I – All MSW, by Type of Material

Material	Million tons	Percent of Total
Glass	10.7	25.1
Steel	2.7	6.3
Aluminum	1.0	2.3
Paper and paperboard	20.4	47.8
Plastics	5.6	13.1
Wood	2.1	4.9
Other Misc.	0.2	0.5
TOTAL	42.7	100.0

Part II – Plastic Packaging and Containers, by Type of Item

Item	Million tons	Percent of Total
Flexible packaging	3.1	40.3
– Household & Institutional refuse bags & films	1.2	16.2
– All other flexible packaging	1.8	24.2
Bottles, jars, & vials	1.4	18.8
Food containers (excl. disposable cups)	1.3	17.0
All other packaging	1.8	23.9
TOTAL	7.6	100.0

Part III – Plastic Packaging and Containers, by Type of Resin

	Million tons	Percent of Total
LD polyethylene	2.5	44.6
HD polyethylene	1.4	25.0
Polystyrene	0.5	8.9
Other thermo-plastics	0.5	8.9
PET	0.4	7.1
PVC	0.2	3.6
All thermosets	0.1	1.8
TOTAL	5.6	100.0

Note: Assumes annual production for packaging is entirely discarded within one year, thus the production breakdown also represents the breakdown of the waste stream.

Source: Part I – Franklin Associates (1988a); Part II – SPI (1988); Part III – Chem Systems (1987). The Franklin data estimates 1986 waste disposal, the SPI data covers 1987 production and the Chem Systems data covers 1985 production. Consistency in reporting years could not be achieved.

Such information on waste quantities and characteristics as that given above is necessary to draw connections between plastic resins, plastic products, and specific components of the municipal solid waste streams. See Section 4 for an analysis of the effects of plastic wastes on management of municipal solid wastes, including landfilling and incineration.

2.5.2 Plastics in Building and Construction Wastes

The major components of building and construction wastes are mixed lumber, roofing and sheeting scraps, broken concrete asphalt, brick, stone, plaster, wallboard, glass, and piping. Plastics are used in piping, siding, insulation, and flooring as well as in other items. The exact characteristics of building and construction waste vary by location depending on the type of construction and the age of the housing and building infrastructure. No studies were identified that describe a quantitative description of the components of building and construction wastes.

Researchers at the Massachusetts Institute of Technology have produced the most recent estimates of building and construction waste quantities. A 1979 publication estimated the national quantity of building and construction wastes at 33.5 million tons. This estimate was based on observations of demolition waste quantities in selected cities during 1974 to 1976, with results then extrapolated to the national level. Sixty-six percent of the demolition debris generated, by weight, consists of concrete, with 20% wood, 15% brick and clay, under 2% steel and iron, and less than 1% each for aluminum, copper, lead, glass and plastics. Plastic wastes were estimated to total only 1,000 tons per year from this data (Wilson et al., 1979). As previously shown in the sales data, an increasing volume of plastics is being used in the building and construction markets; this trend suggests that the plastic share of building wastes is more than that found by Wilson.

2.5.3 Plastics in Automobile Salvage Residue

Automobiles represent one of the major end markets for plastic products; The transportation sector consumes approximately 4.5% of U.S. sales of plastics. This section examines the final disposition of plastics that are part of automobile scrap.

An estimated 10.8 million vehicles were retired from use in 1986 (the estimate was based on the number of automobiles deregistered that year). Most of these automobiles (92%) were sent to automobile dismantling yards and then to salvage dealers. The remainder of the waste automobiles were abandoned or were driven illegally without registration (U.S. EPA, 1988).

Automobile dismantling yards remove usable parts from automobiles. The auto body hulk is then shipped to a salvage processor. Scrap automobiles consist of ferrous and nonferrous metals, glass, plastic, and other materials. The salvage processor generates revenues by removing the scrap metals that can be resold in international metal markets. This is done by sending the cars through heavy-duty shredding equipment that smashes the auto bodies into small pieces. Magnetic separation equipment then divides the pieces into several types of saleable metals and residues.

Automobile scrap residue (ASR) consists primarily of waste glass, plastic, and dirt. The exact percentage of plastics is not known. ASR is also referred to as "fluff." An estimated 500 to 850 pounds of fluff is generated for each car that is shredded (U.S. EPA, 1988). If 10 million cars are sent to automobile dismantling yards per year, the total quantity of residue created could be estimated at 2.5 to 4.25 million tons per year. These estimates of scrap quantities are not necessarily accurate, however, partly due to uncertainties about data on car deregistrations and abandonments. A representative for the industry trade association estimated that their membership processes 6 to 8 million cars per year (Siler, 1989).

Automobile fluff may contain hazardous materials. Numerous automobile parts -- e.g., batteries, used oil, solvents, mufflers and catalytic converters, paints and coatings, and brake drums -- can contain hazardous chemicals or substances. An industry group, the Institute of the Scrap Recycling Industries (ISRI), has developed guidelines for shredders to help them comply with the environmental requirements that govern the eventual disposal of the fluff. Shredders are urged to require their car suppliers to remove those items from the car body that contain the main hazardous constituents or to refuse the shipment.

ASR is virtually always landfilled for final disposal. The shredder normally must pay for this waste disposal. In selected instances, however, the shredder may obtain a disposal cost discount because the fluff is useful for the landfill operator as daily cover material (Siler, 1989).

2.5.4 Plastics in Litter

Many recent studies of litter have attempted to analyze beach litter as a means of assessing the marine debris problem. The studies of wastes on beaches, however, cannot differentiate between litter left by public beachgoers and that which washes up on the shore. With that caveat in mind, some information about the share of plastics in beach waste and the composition of that litter can be presented.

Table 2-21 reproduces compilations of wastes found in Texas beach cleanup efforts. For 158 miles of Texas beaches, researchers tallied the type and number of items found by cleanup workers. As the table indicates, the cleanup workers collected nearly 400,000 individual items of trash. Plastic wastes represented nearly two-thirds of the items collected (66%). Metal (13%) and glass (11%) were much less prevalent. As noted, several of the items (e.g., fishing nets) originate in the marine environment rather than from land-based littering. Many other items, however, including numerous sorts of plastic packaging and containers, could originate from either marine or land sources. This information is thus insufficient to differentiate between the two categories.

The Texas beach results reflect the unique combinations of industrial activity and ocean current found in that area. Other beach debris studies, however, have also found large proportions of plastic materials (Vauk and Schrey, 1987; Dixon and Dixon, 1983).

Section 3 addresses plastic wastes in the marine environment. Section 4 addresses the issue of plastic wastes in litter and includes information to characterize these wastes.

Table 2-21

COMPOSITION OF MATERIALS FOUND IN TEXAS COASTAL CLEAN-UP
(1987)

Material	Number of Items	Material	Number of Items
PLASTICS		PLASTICS (Cont.)	
Bags	31,773	Toys	2,820
Caps/lids	28,540	Straws	2,639
Misc. pieces	21,619	Lighters	2,429
Rope	18,878	Computer read/write rings	2,337
Bottles – other	16,784	Vegetable sacks	2,023
Beer rings (6-pack yokes)	15,631	Diapers	1,914
Cups/utensils	12,486	Shoes/sandals	1,750
Milk jugs	7,460	Fish nets	1,719
Bottles – green	7,170	Buckets	1,703
Bottles – soda	6,341	Tampon applicators	1,040
Strapping bands	4,933	Syringes	930
Large sheeting	4,817	Hardhats	225
Fish lines	4,225	Misc. foamed polystyrene pieces	22,609
Light sticks	4,179	Foamed polystrene cups	14,998
Gloves	4,127	Foamed polystyrene buoys	1,048
Egg cartons	3,417		
TOTAL – PLASTICS			252,569
METAL		PAPER	
Beverage cans	20,580	Misc. pieces	12,292
Pull tabs	8,925	Cups	4,511
Bottle caps	8,273	Bags	4,428
Other cans	4,469	Cartons	4,073
Misc. pieces	3,658	Newspaper	1,415
Wire	2,807		
Large containers	1,105	TOTAL – PAPER	26,719
Drums – rusted	268		
Drums – new	225		
		TIRES	546
TOTAL – METAL	50,310		
WOOD		GLASS	
Misc. pieces	9,386	Misc. pieces	21,214
Pallets	605	Bottles	17,902
Crates	292	Light bulbs	2,327
		Fluorescent tubes	1,088
TOTAL – WOOD	10,203		
		TOTAL – GLASS	42,531
TOTAL – ALL MATERIALS			382,878

Source: Interagency Task Force on Persistent Marine Debris (1988).

2.5.5 Plastics in Marine Debris

Marine wastes include wastes generated from vessels or offshore platforms and wastes deposited from land sources. The major vessel categories include merchant marine vessels (including commercial ocean liners and smaller passenger vessels), fishing vessels, recreational boats, offshore oil and gas platforms, and miscellaneous research, educational, and industrial work vessels. Wastes from vessels may be further classified as:

- Wastes from the galley and crew or "hotel" areas of a vessel

- Wastes generated from vessel operations, such as containers from engine room supplies

- Wastes generated as part of the commercial operations, such as fishing gear wastes

Wastes from land sources include:

- Wind-blown or lost debris from municipal solid waste management facilities, including solid waste transfer stations

- Wastes released from sewage treatment facilities or due to combined sewer overflows

- Stormwater runoff and other nonpoint sources

- Beach use and resuspension of beach litter

- Plastic pellets (to the extent they are from plastic manufacturing and processing facilities)

In general, vessel wastes share many of the components of municipal solid wastes. Substantial portions of vessel wastes include food wastes and paper and plastic products. Only the additional commercial wastes are unique to this sector. Wastes from land sources bear some similarity to litter because such wastes would be litter if they remained on land.

See Section 3 for a full characterization of marine wastes. The quantity and characteristics of plastic wastes generated must be addressed separately from each of these sources.

REFERENCES

Agoos, A. 1985. Serving up a better package for foods. Chemical Week. Oct 16, 1985. p. 100.

Alter, H. 1986. Disposal and Reuse of Plastics. In: Encyclopedia of Polymer Science and Engineering. John Wiley & Sons. New York, NY.

Chemical and Engineering News. 1988. Plastics additives: Less performing better. 66:35-57. Jun 13, 1988.

Chem Systems. 1987. Plastics: A.D. 2000 - Production and Use Through the Turn of the Century. Prepared for The Society of the Plastics Industry, Inc. Washington, DC.

Curlee, T.R. 1986. The Economic Feasibility of Recycling: A Case Study of Plastic Wastes. Praeger. New York, NY.

Dixon, T.J. and T.R. Dixon. 1983. Marine litter distribution and composition in the North Sea. Marine Pollution Research. 14:145-148.

Franklin Associates. 1988a. Characterization of Municipal Solid Waste in the United States, 1960 to 2000 (update 1988). Prepared for U. S. Environmental Protection Agency. Contract No. 68-01-7310. Franklin Associates, Ltd. Prairie Village, KS. Mar 30, 1988.

Franklin Associates. 1988b. Characterization of Products Containing Lead and Cadmium in Municipal Solid Waste in the United States, 1970 to 2000. Prepared for U.S. Environmental Protection Agency. Franklin Associates, Ltd. Prairie Village, KS.

Interagency Task Force on Persistent Marine Debris. 1988. Report. Chair: Department of Commerce, National Oceanic and Atmospheric Administration. May 1988.

Kresta, J.E. (ed). 1982. International Symposium on Polymer Additives (Las Vegas, NV). Plenum Press. New York, NY.

Modern Plastics Encyclopedia. 1988. McGraw-Hill. New York, NY.

Modern Plastics. 1989. Resin Report. Jan 1989. McGraw-Hill.

Prepared Foods. 1986. High-barrier Coex: 100-fold increase by 1995. Prepared Foods (155:98).

Radian Corp. 1987. Chemical Additives for the Plastics Industry: Properties, Applications, Toxicologies. Noyes Data Corp. Park Ridge, NJ.

Rathje, W.L., W.W. Hughes, G. Archer, and D.C. Archer. 1988. Source Reduction and Landfill Myths. Le Projet du Garbage. Dept. of Anthropology, University of Arizona. Paper presented at Forum of the Association of State and Territorial Solid Waste Management Officials on Integrated Municipal Waste Management, July 17-20, 1988.

Rauch Associates, Inc. 1987. The Rauch Guide to the U.S. Plastics Industry.

Seymour, R.B. (ed). 1978. Additives for Plastics - Volume 1 - State of the Art. Academic Press.

Siler, D. 1989. Telephone communication between Eastern Research Group, Inc. and Duane Siler, Counsel, Institute for Scrap Recycling Industries, Inc., Washington, DC. March 17.

SPI. 1988. Society of the Plastics Industry. Facts and Figures of the U.S. Plastics Industry. Washington, DC.

Stepek, J. and H. Daoust. 1983. Additives for Plastics. Springer-Verlag.

U.S. Bureau of the Census. 1985. 1982 Census of Manufacturers. U.S. Department of Commerce. As cited in Wirka, 1988.

U.S. Bureau of the Census. 1988. County Business Patterns. U.S. Department of Commerce. Washington, DC.

U.S. EPA. 1988. U.S. Environmental Protection Agency. The Solid Waste Dilemma: An Agenda for Action. Municipal Solid Waste Task Force, Office of Solid Waste. Draft Report. Sep 1988. EPA/530-SW-88-052. Washington, DC.

Vauk, J.M.G. and E. Schrey. 1987. Litter pollution from ships in the German Bight. Marine Pollution Bulletin. 18:316-319.

Wilson, D., T. Davidson, and H.T.S. Ng. 1979. Demolition Wastes: Data Collection and Separation Studies. Prepared under National Science Foundation Grant Number 76-22048 AER. Massachusetts Institute of Technology.

Wirka, J. 1988a. Wrapped in Plastics: The Environmental Case for Reducing Plastics Packaging. Environmental Action Foundation. Washington, DC.

3. Impacts of Plastic Debris on the Marine Environment

Persistent marine debris encompasses a wide assortment of natural and synthetic wastes, particularly plastic materials, that float or are suspended in the water and may eventually be deposited on shorelines and beaches. Either afloat, submerged, or stranded on shores, plastic debris may endanger marine life, pose risks to public safety, impact the economics of coastal communities and generally degrade the quality of the environment.

This chapter provides a review of the types, sources, and quantities of plastic entering the marine environment, the physical and chemical fate of such debris, and the impacts of plastic debris on marine wildlife, beach aesthetics, vessels, and human health and safety.

3.1 SUMMARY OF KEY FINDINGS

Following are the key findings of this section:

- EPA identified several plastic items that are of concern due to the risks they pose to marine life or human safety, or due to the aesthetic or economic damages they produce. The "Articles of Concern" are beverage ring carrier devices, tampon applicators, condoms, syringes (either whole or in pieces), plastic pellets and spherules, foamed polystyrene spheres, plastic bags and sheeting, uncut strapping bands, fishing nets and traps, and monofilament lines and rope.

- Persistent marine debris encompasses a wide assortment of plastic wastes that float or are suspended in the water and may cause harm to marine wildlife, pose risks to public safety or eventually be deposited on shorelines and beaches.

- Marine plastic debris has a number of sources, both land- and marine-based. Land-based sources include solid waste disposal activities, sewage treatment overflows, stormwater runoff, beach litter, and plastics manufacturers or transporters (for pellets). Marine sources include overboard disposal from commercial, military, and recreational vessels operating in marine waters, as well as from offshore oil and gas structures.

- An EPA sampling study of debris in harbors found that most floatable debris consisted of plastic items, that plastic pellets and spherules were ubiquitous and sewage-related items were more prominent in East Coast harbors.

- Improper disposal of plastic materials creates environmental problems including entanglement of marine animals, particularly by derelict fishing gear, ingestion of plastic wastes by wildlife and the aesthetic losses caused by litter deposited on public beaches.

70

- Among marine wildlife, greatest concern has focused on entanglement effects on the northern fur seal, and ingestion of plastic wastes by several endangered or threatened species of turtles.

- Various economic losses occur due to marine debris including loss of tourist revenues in beach communities, depletion of fishing resources, and entanglement and loss of fishing gear and fouling of vessel propellers.

3.2 TYPES AND SOURCES OF PLASTIC DEBRIS

For descriptive purposes, plastic debris may be classified as either raw materials or manufactured products. Plastic raw materials or pellets, in the form of small spherules, disks, and cylindrical nibs, are the least conspicuous and, therefore, most-often-overlooked components of plastic debris. In the marine environment, the most common types of plastic raw materials are polyethylene or polypropylene pellets, from which larger, molded plastic items are made, and polystyrene spherules or beads, the basic structural units of polystyrene products.

Polyethylene and polypropylene pellets, 1-5 mm in diameter, are most often colorless, white, or amber, although black, green, red, blue, and other colors are also produced. Unfoamed polystyrene spherules are generally smaller, 0.1-3 mm in diameter, and are usually white, opaque, or colorless.

The more visible and familiar plastic debris consists of products of sundry sizes, shapes, and composition manufactured from the raw pellets and spherules. Manufactured items that often contribute to plastic debris include containers and packaging materials, fishing gear, disposable dishware, toys, and sanitary sewage-related products. These items are found in the marine environment either intact or as variously sized pieces and fragments.

Plastic debris enters our oceans and estuaries from a number of both land-based sources and marine activities, and for a wide variety of reasons. A large amount of the material drifting at sea or stranded on shores is not easily traced back to its source. Some items, such as derelict fishing equipment, are easily associated with a single source (i.e., the fishing industry), but other items, plastic bags for example, may originate from any number of land-based or marine sources. The types and amounts of plastic debris that end up in the marine environment are greatly influenced by local or regional factors such as climate, physical oceanographic characteristics, uses of the marine environment, and uses of the adjacent land.

EPA and others have characterized the plastic materials littering the marine environment through systematic beach cleanups (CEE, 1989; CEE, 1987a) and surface water observations or net tows (Battelle, 1989; Dahlberg and Day, 1985). Efforts to characterize beach debris have, in recent years, been coordinated by CEE (now called the Center for Marine Conservation, or CMC). In 1988, CEE organized a national beach cleanup and data collection effort that was sponsored by EPA. Results of these studies are presented in Section 3.2.1.5.

EPA has also recently made surface collections of floatable debris, including floating plastic materials, in the harbors of nine coastal U.S. cities: New York, Boston, Philadelphia, Baltimore, Miami, Tacoma, Seattle, Oakland, and San Francisco. These surveys were designed to qualitatively assess debris in the harbors of these cities. Because a unique sampling design was implemented for each specific location and because this design varied among the harbors sampled, the absolute numbers of items collected at each location are not directly comparable. However, comparisons of debris types can be made on the basis of percent of total items collected at each location.

Sampling for the harbor studies was conducted over two or three consecutive days during ebb tide conditions. Samples of debris were collected with a 0.3-mm neuston net towed through surface slicks, areas in which floating debris accumulates. Debris from the tows was then identified, sorted, counted, recorded, and entered into a database. Results of the harbor surveys indicated that 1) 70-90% of the total number of floatable debris items collected was composed of plastic items; 2) plastic pellets/spherules were ubiquitous; and 3) sewage-related items were more prominent in East coast cities. Table 3-1 lists the number of debris items collected and the percentage of items in each debris categories for the harbors surveyed. Medical-type debris included syringes, needle covers, blood vials, pill vials, and similar material. Sewage-related debris referred to condoms, tampons, tampon applicators, grease balls, crack vials, cotton swabs, and similar material that enters the sewage waste stream. Plastic pellets and spheres or styrofoam pieces were the most common debris items encountered in all but one case. The relative abundance of plastic pellets and spheres (percentage of the total items collected at a given harbor) was particularly variable among the different harbors. Plastic pellets and spheres were most prominent in debris collected in Tacoma Harbor, located at the southern end of Puget Sound. The majority of the pellets collected in Tacoma Harbor, however, came from two discrete samples only.

In the following discussion, sources or potential sources for plastic debris have been grouped into three categories: land-based sources, marine sources, and illegal disposal activities.

3.2.1   Land-Based Sources

Plastic debris from land-based sources includes materials that are used and/or disposed of on land but subsequently are washed out, blown out, or discharged into rivers, estuaries, or oceans. Plastic manufacturing and fabricating plants and related transportation activities, facilities for handling solid waste, combined wastewater/stormwater sewer systems, nonpoint-sources runoff, and recreational beach use are all potential land-based sources of plastic debris found in the marine environment.

Table 3-1

SUMMARY OF FLOATABLE DEBRIS COLLECTED DURING EPA'S HARBOR STUDIES PROGRAM

Harbor	Total Items Collected	As % of Total Items Collected						
		Medical-Type Debris[a]	Sewage-Related Debris[b]	Plastic Pellets & Spherule	Misc. Plastic Pieces	Styrofoam Pieces	Plastic Sheeting	All Other Items
New York	13,955	0.3	17.0	19	21	10	4	29[c,e,g]
Boston	9,315	0.2	3.6	30	16	18	1	31[d,h]
Philadelphia	2,835	0.1	7.5	34	5	24	5	24[c,e,f]
Baltimore	4,363	0.8	1.5	19	5	25	13	36[f]
Miami	2,965	0.1	1.5	24	7	37	14	16[g]
Seattle	709	0.3	2.5	16	6	44	6	25[g]
Tacoma	4,935	0.1	1.4	82	3	11	2	<1
Oakland	1,432	0.2	0.3	32	10	36	6	15[b]
San Francisco	3,388	0.4	0.4	16	11	46	5	21[g]

[a]Syringes, needle covers, blood vials, pill vials, etc.
[b]Condoms, tampons and applicators, grease balls, crack vials, cotton swabs, etc.
[c]Includes tar balls, fishing line.
[d]Includes slag.
[e]Includes cigarette butts.
[f]Includes plastic food ware/wraps.
[g]Includes wood.
[h]Includes polyurethane foam.

Source: Battelle, 1989.

3.2.1.1   Plastic Manufacturing and Fabricating Facilities and Related Transportation Activities

Plastic manufacturing, processing, and associated transportation activities represent important potential sources of plastic pellets and spherules found in the marine environment. These raw materials are synthesized at petrochemical plants and are transported in bulk quantities to manufacturing and processing facilities, where they are melted down and fabricated into products. The raw material plastic pellets do not, however, include bits of foamed polystyrene (e.g., Styrofoam) that may result from the physical breakup of food containers, floats, buoys, and various other products.

At both the manufacturing and the fabricating facilities, raw plastic materials can enter the wastewater stream either accidentally or intentionally. Once in the wastewater stream, these materials can be transported to the ocean via inland waterways, directly from industrial outfalls, or indirectly through municipal sewage systems. Plastic pellets may also be released during transport at sea or on land, and accidental spills that occur during loading and unloading at port facilities, (e.g., spillage from bulk containers or rips in smaller paper containers). Any losses of these types could then be washed through storm drains and discharged.

Although plastic pellets are the least noticeable form of plastic pollution, they remain ubiquitous in the oceans and on beaches (Interagency Task Force, 1988; CEE, 1987b; Wilber, 1987). Their overall distribution in the sea tends to parallel the distribution of plastic debris in general (Battelle, 1989; Wilber, 1987). Plastic pellets have been collected in neuston or ichthyoplankton nets in both the Atlantic Ocean (Wilber, 1987; Morris, 1980; Colton, 1974; Carpenter and Smith, 1972) and in the Pacific (Day and Shaw, 1987; Dahlberg and Day, 1985; Wong et al., 1974). Little is known about the sources and distribution of raw plastic materials in the Gulf of Mexico region.

Net tows conducted by Wilber (1987) indicated that polyethylene pellets were present throughout the western North Atlantic; the highest concentrations occurred in the Sargasso Sea where up to 4900 resin pellets per square kilometer of ocean surface were collected. In the same study, it was found that unfoamed polystyrene spherules, up to 8000 per square kilometer, were commonly collected in North Atlantic shelf waters but were rare in the open ocean. Since Carpenter and Smith (1972) first collected plastic pellets in neuston nets towed through the Sargasso Sea 15 years ago, these materials have increased in number nearly two-fold (Wilber, 1987).

During surveys conducted in the Atlantic by Colton et al. (1974), polystyrene spherules were collected in neuston nets only in waters north of Florida, with the highest concentrations occurring in coastal waters south of Rhode Island and south of eastern Long Island. Although these surveys extended to waters off central Florida, only off southern New England and Long Island (Figures 3-1a through Figure 3-1c) were these particles collected at most inshore stations.

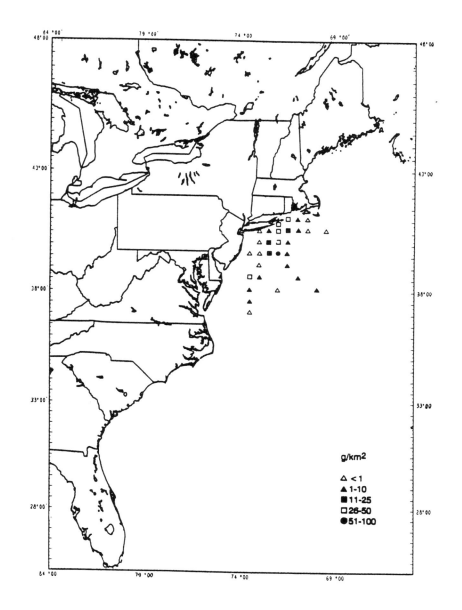

FIGURE 3-1a.   DISTRIBUTION OF OPAQUE POLYSTYRENE SPHERULES IN THE ATLANTIC
OCEAN (adapted from Colton et al., 1974)

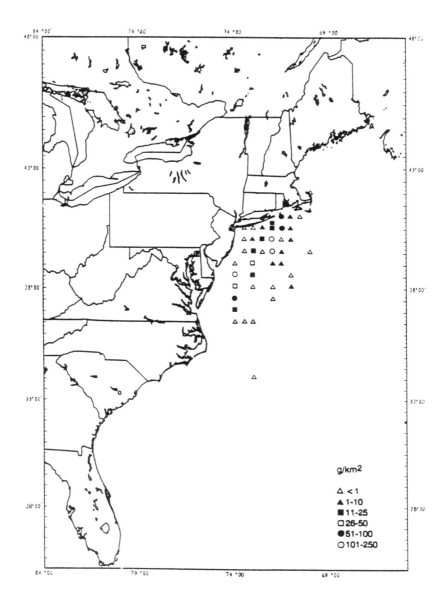

FIGURE 3-1b.   DISTRIBUTION OF CLEAR POLYSTYRENE SPHERULES IN THE ATLANTIC
OCEAN (adapted from Colton et al., 1974)

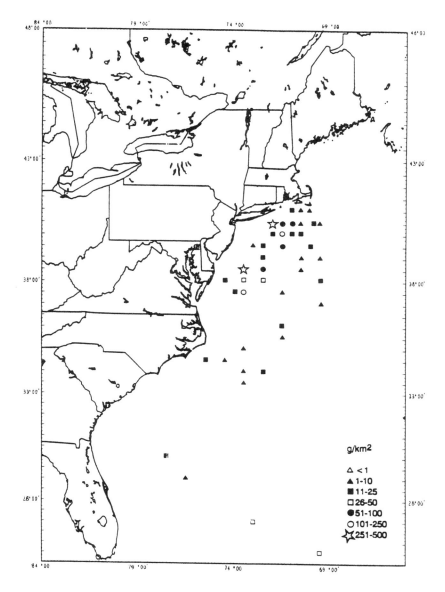

FIGURE 3-1c.   DISTRIBUTION OF POLYETHYLENE CYLINDERS IN THE ATLANTIC OCEAN
(adapted from Colton et al., 1974)

Plastic pellets have also been collected in neuston net tows in the Pacific Ocean (Wong et al., 1974) from California to Japan and northward to the Canadian border, and in the North Pacific and Bering Sea (Day and Shaw, 1987). Wong et al. (1974) reported up to 34,000 plastic pellets per square kilometer in certain locations of the North Pacific Ocean.

During harbor surveys, conducted in New York and Boston (Battelle, 1989), plastic pellets comprised 30 and 40%, respectively, of the plastic items collected, and 20 and 30%, respectively, of all debris items collected. In Tacoma Harbor, located in the southern portion of Puget Sound, plastic pellets comprised 84% of the total debris items collected. The majority of these pellets, however, were found in only two of the samples taken. In Seattle, located further north in the Sound, plastic pellets comprised only 16% of all debris items collected (Battelle, 1989).

Most plastic pellets found in marine waters have been identified as polyethylene, polypropylene, or polystyrene (CEE, 1987b; Hays and Cormons, 1974). Because all these raw materials are shipped worldwide, the specific origin of pellets found in the oceans is difficult to assess. Resin pellets have been collected in coastal areas, near major shipping lanes, and in the vicinity of coastal industrial sources (Morris, 1980). Based on the incidence of pellet ingestion by seabirds in California (Baltz and Morejohn, 1976) compared to the same bird species in Alaska (Day, 1980), Day et al. (1985) suggested that resin pellets are more abundant in waters adjacent to major industrial centers than in areas of the ocean remote from such facilities.

Colton et al. (1974) proposed that the widespread distribution of these materials in rivers, estuaries, and coastal waters of the United States indicated that improper wastewater disposal was a common practice in the plastics industry at the time. Polyethylene cylinders and polystyrene spherules have been found at outfalls from plastics manufacturing plants in New Jersey, Massachusetts, and Connecticut, and downstream of plants in New York and New Jersey (Colton, 1974; Hays and Cormons, 1974). In Massachusetts alone, there are nearly 600 plastics manufacturing and processing companies. Since these studies, National Permit Discharge Elimination System (NPDES) permits have placed stricter requirements on these releases. Nevertheless, Coleman and Wehle (1984) also stated that plastic pellets and particles enter coastal waterways and the ocean from point-source outfalls at plastic manufacturing plants. From studies conducted in the Mediterranean, Shiber (1979) reported that many plastics industries release their wastes directly into the sea. Studies on the West Coast have suggested similar relationships between industrial regions and the distribution of resin pellets in parts of the Pacific (Day et al., 1985).

The widespread occurrence of relatively unweathered resin pellets in oceanic waters south of Cape Hatteras and in the Caribbean Sea indicated to Colton (1974) that, in addition to plastic manufacturing and fabricating plants, resin pellets must also originate from other sources. Although there are no specific data implicating additional sources, foreign and domestic transportation of raw plastic materials by commercial vessels and the loading and unloading operations at port facilities are probably responsible for a certain amount of cargo spillage into both coastal and open-ocean waters. Similarly, transportation on land can result in spillage of pellets that may subsequently be carried to water bodies via stormwater runoff. Pruter (1987) and Day et al. (1985) also reported that resin pellets and spherules may be used on the decks of commercial ships to facilitate moving of cargo containers or other large heavy objects. Such commercial uses increase the potential for plastic pellets to enter the marine environment.

In EPA's recent harbor studies of floatable debris, including plastic pellets, regional differences were found in the composition of debris slicks (Battelle, 1989). Figure 3-2 indicates the percentages of the total items collected that were plastic items and the percentages of total items that were plastic pellets and spheres. Unlike the wastes in other harbors, the majority of those pellets collected in Tacoma Harbor were all of the same size and shape.

3.2.1.2   Municipal Solid Waste Disposal Activities

In coastal regions, municipal solid waste disposal practices can serve as sources of marine debris (Interagency Task Force, 1988; Swanson et al., 1978). Solid waste handling facilities include landfills, incinerators, and transfer stations. Debris from these facilities consists of a diverse assortment of domestic and commercial wastes, some of which is plastic. According to Franklin Associates (1988), the United States annually produces 141 million tons of municipal solid waste. The same researchers have estimated that 7.3% of this solid waste is represented by plastic materials (see Section 2.5; Franklin Associates, 1988). Because of their relatively low density, plastics represent a larger proportion, approximately 15-25%, of the volume of municipal solid waste (see Section 2.5).

In regions of the country where sanitary landfills, marine transfer facilities, and municipal waste incinerators are located in coastal environments, light-weight debris from these facilities may be blown into adjacent waterways and transported out to sea. Persistent materials can inadvertently be released to waterways during solid waste transfer operations, particularly overwater transport of refuse by barges. In the metropolitan New York/New Jersey area, much of the municipal solid waste is transported by barges along coastal waterways to landfill sites. This kind of disposal operation involves a number of marine transfer stations (dock facilities where the barges are loaded and unloaded). Figure 3-3 shows the locations of solid waste handling facilities in the greater New York area.

A recent qualitative survey of waterfront waste-handling facilities within the New York Harbor Complex indicated that such facilities contribute various quantities of debris to the waterways and shores of the harbor complex (U.S. EPA, 1988). The study indicated that winds blow light-weight litter from the open barges and from landfill sites. The EPA report noted that the Fresh Kills Landfill, located on the waterfront on Staten Island (Figure 3-3), may be a significant source of persistent debris within the harbor and in the New York Bight. This large facility receives approximately 28,000 tons of trash per day, of which approximately 50% is transported by barge. Shorelines in the vicinity of the Fresh Kills Landfill have been reported to be heavily littered with municipal waste typically disposed of at the site (U.S. EPA, 1988).

In the State of New York, municipalities have initiated steps to reduce the amount of debris escaping into waterways (U.S. EPA, 1988; Swanson et al., 1978). The City of New York and the State of New Jersey have recently entered into a judicial consent decree which directs waste handling activities. The consent decree, aimed at reducing the amount of debris entering the

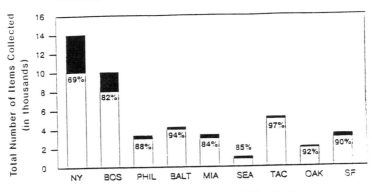

Total Items Collected and Plastic Items as
a Percent of Total Items

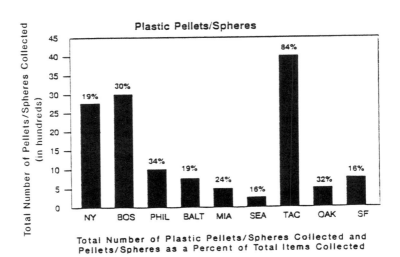

Total Number of Plastic Pellets/Spheres Collected and
Pellets/Spheres as a Percent of Total Items Collected

Figure 3-2.   Total Items Collected and Percent Plastic Items (top);
Plastic Pellets/Spheres (bottom) (Battelle, 1989)

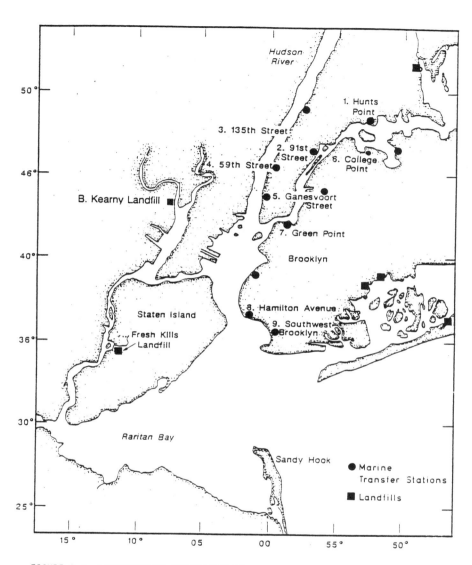

FIGURE 3-3   LOCATIONS OF MARINE TRANSFER STATIONS AND LANDFILLS IN THE
GREATER NEW YORK METROPOLITAN AREA (U.S. Environmental
Protection Agency, 1989a)

marine environment, includes strict waste handling protocols, use of containment booms around loading and unloading facilities, use of barge covers, and frequent removal of floating debris within the containment booms.

The potential contribution of municipal solid waste handling facilities to marine debris has not been quantified on a national level.

3.2.1.3   Sewage Treatment Plants and Combined Sewer Overflows

A significant amount of plastic debris in the marine environment is attributed to inadequate treatment of sanitary sewage and combined wastewater/stormwater sewer systems. The outfall pipes from these systems provide point sources of plastic debris to the environment. Of the more than 15,000 publicly owned treatment works (POTWs) in the United States, approximately 2,000 are located in coastal communities (Interagency Task Force, 1988). Most of these facilities discharge treated effluent into streams and rivers. Nearly 600 of these POTWs, however, discharge effluent directly into estuaries and coastal waters (OTA, 1987).

If properly operated, POTWs should not discharge plastic debris into the marine environment. However, under some circumstances, plastic materials associated with POTWs can enter the marine waters. Plastic debris can be discharged from POTWs to receiving waters for three major reasons (Interagency Task Force, 1988):

- At POTWs that cannot treat the capacity of normal "dry-weather flow," untreated sewage may bypass the system and be released directly into the environment.

- During periods of "down time," when a POTW is not operating because of malfunctions or breakdowns, influent may bypass the treatment system and be released into receiving waters.

- In a community where both sewage and stormwater runoff are combined into one system and the volume of stormwater exceeds a treatment plant's capacity (e.g., during heavy rain), both untreated sewage and stormwater are discharged directly into receiving waters.

Most POTWs are designed to handle the volumes of domestic and industrial wastes generated by municipalities. Even with minimal primary treatment, variously sized screen courses and skimming operations remove most floatable materials from incoming wastewater. These materials are generally disposed of in landfills or at municipal incinerators and the treated effluent is released into local receiving waters. Settled solids are disposed of on land or at sea. Disposal of these settled solids at sea may be through an outfall (such as in Boston) or by direct disposal. Currently, the only example of direct disposal is the 106-Mile Deepwater Municipal Sludge Site used by New York and New Jersey municipalities.

When the volumes of incoming waste are larger than the treatment capacity of the POTW facility or portions of its collection system, untreated sewage bypasses the plant and is released directly into the environment. Similar releases of untreated wastes can occur when a facility is malfunctioning or undergoing maintenance. Under both of these conditions, the untreated waste that is discharged may contain various amounts of plastic debris that generally would be removed by skimmers, screens, and separators during treatment.

Many coastal communities do not have separate sewer systems for domestic/ industrial wastewater and for stormwater. Some older sewer networks transport both sanitary wastes and stormwater to sewage treatment facilities where floatable materials are subsequently removed. Under normal dry-weather conditions, untreated wastewater is carried to the treatment facility by the sewer system. However, during periods of heavy rain, flow of domestic and industrial wastewater and stormwater through the combined sewers may exceed the capacity of the system. When this occurs, portions of the wastewater/stormwater flow are diverted and discharged directly into the receiving water body. These discharges, or combined sewer overflows (CSOs), can occur at various locations throughout the collection system. Of the more than 2,000 POTWs in U.S. coastal communities, 135 have one or more CSOs (Interagency Task Force, 1988). In addition to the impacts of untreated sewage on water quality, CSOs also contain various kinds of sewage-associated plastic debris (e.g., disposable diapers, tampon applicators, condoms, and other disposable sanitary items) as well as street litter collected by stormwater runoff. Additionally, CSOs can contribute syringes to marine wastes. In one study, New York City officials captured an average of 30 syringes per day (with needle intact) from the materials captured in screens and skimmers from 14 wastewater treatment plants (New York DEP, 1989).

Of particular concern are cities with outdated systems, such as in the greater metropolitan areas of New York and Boston. Of the country's 100 largest (on a volume basis) sewage treatment facilities, 36 have collection systems with CSOs. Thirty of these systems are on the U.S. east coast, with twelve located in New York City (NOAA, 1987). Approximately 70% of New York City's sewer systems have CSOs. The locations and numbers of CSOs in the greater New York metropolitan area are shown in Figure 3-4. A more recent study has identified 680 CSOs in the Interstate Sanitation District which encompasses areas in New York, New Jersey, and Connecticut that affect the New York Bight and Long Island Sound (Interstate Sanitation Commission, 1988).

EPA currently is conducting a CSO/storm sewer sampling program in Boston and Philadelphia. These studies will provide data to supplement available information on CSOs and storm sewers as potential sources of plastic debris to the marine environment. EPA is also sampling floatable wastes throughout the POTW systems in these cities in order to determine the potential waste releases such as could occur during heavy rains or in periods when the wastewater system is not operating.

3.2.1.4   Stormwater Runoff/Nonpoint Sources

In addition to the land-based point sources discussed above, many other sources that are nonspecific in nature also contribute to plastic debris in the marine environment. During heavy rains, stormwater runoff, which carries various kinds and amounts of debris that has accumulated during dry periods, enters storm sewers, streams, rivers, bays, and ultimately the ocean. The state of New Jersey, for example, has nearly 5,000 stormwater pipes that discharge

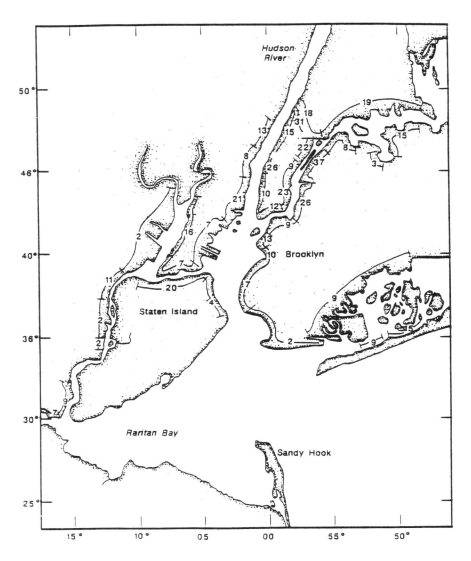

FIGURE 3-4.   LOCATIONS AND NUMBERS OF COMBINED SEWER OVERFLOW SYSTEMS IN
THE GREATER NEW YORK METROPOLITAN AREA (SAIC/Battelle, 1987)

directly into coastal waters. The majority of floating litter that washes up on New Jersey's beaches originates from stormwater runoff and flushing of stormwater pipes after heavy rainfalls (New Jersey DEP, 1988).

Plastic debris originating from stormwater runoff is not found only in coastal environments, however. Floatable wastes including plastic debris can be carried in runoff from inland areas via streams and rivers that empty into the sea. Because of the varied nature of the sources, debris carried by stormwater runoff is difficult to characterize. It can include any and all types of domestic wastes that litter urban and suburban streets, parking lots, and recreational areas. Industrially generated wastes, resulting from spills at storage or transfer facilities and during transportation, can also be collected in stormwater that ultimately is transported to the ocean.

Debris suspended and carried by stormwater flow is ubiquitous throughout the United States. The methods for collecting and transporting stormwater flow may vary from one municipality to another, but along coastal states, the majority of this debris is transported to estuaries and coastal waters. As previously discussed, in many older metropolitan areas that have combined stormwater/wastewater sewer systems, debris is released into coastal waters through CSOs. In other areas, stormwater flow and debris are discharged directly into the marine environment.

3.2.1.5   Beach Use and Resuspension of Beach Litter

The amount of litter observed along our shorelines is one reasonable indicator of the severity of the persistent marine debris problem. Waste materials found on beaches and along shorelines include not only plastic debris left by beach users but also debris that washes ashore from vessels and sea-based commercial activities and from other improper disposal of land-based waste. Because it is impossible to trace the source of many floatable wastes, the relative contributions to the beach litter from beach users and from materials washed ashore are difficult to distinguish. The majority of waste left on beaches by recreational users is floatable debris, consisting primarily of food and beverage containers, six-pack connectors, and other plastic packaging materials. Debris that is washed ashore encompasses a much greater diversity of plastic materials from any number of domestic, commercial, and recreational uses (CEE, 1987a; 1987c).

Using federal and private funds, efforts to characterize beach debris have been coordinated by the Center for Environmental Education in recent years (now called the Center for Marine Conservation, or CMC). Data cards (Figure 3-5) for recording various types of debris were developed by CEE for distribution to beach cleanup volunteers. Analysis of the data from these beach surveys will provide information on the types of debris important in different regions of the United States. The data cards were designed for ease of data collection; in reporting the beach cleanup survey results below, the Styrofoam® (and other foamed plastics) are included in the plastic totals and are not separately reported.

One of the first organized beach debris data collection efforts was carried out in Texas. In this state-wide cleanup campaign conducted in September 1986, an estimated 124 tons of debris were collected from approximately 122 miles of coastline. Of the 171,000 individual pieces of debris recorded on the data cards, 67% were plastic (including foamed plastics such as Styrofoam) (CEE, 1987a). In contrast, paper and wood debris constituted only 8% of all litter

BEACH CLEANUP DATA CARD

Thank you for completing this data card. Answer the questions and return to your area coordinator or to the address at the bottom of this card. This information will be used in the Center for Environmental Education's National Marine Debris Data Base and Report to help develop solutions to stopping marine debris.

Name _____ Affiliation _____

Address _____ Occupation _____ Phone (____) _____

City _____ State _____ Zip _____ M ____ F ____ Age _____

Today's Date   Month _____ Day _____ Year _____ Name of Coordinator _____

Location of beach cleaned _____ Nearest city _____

How did you hear about the cleanup? _____

```
                        SAFETY TIPS

              1.  Do not go near any large drums.
              2.  Be careful with sharp objects.
              3.  Wear gloves.
              4.  Stay out of the dune areas.
              5.  Watch out for snakes.
              6.  Don't lift anything too heavy.

              WE WANT YOU TO BE SAFE
```

Number of people working together on this data card _____ Estimated distance of beach cleaned _____ Number of bags filled _____

SOURCES OF FOREIGN DEBRIS. Please list all items that have foreign labels.

Country	Item Found
Example: *Mexico*	*plastic bottle - "Clarisol"*

STRANDED AND/OR ENTANGLED ANIMALS (Please describe type of animal and type of entangling debris. Be as specific as you can.)

_____
_____
_____
_____

What was the most peculiar item you collected? _____

Comments _____
_____

Thank you!

PLEASE RETURN THIS CARD TO
YOUR AREA COORDINATOR
OR MAIL IT TO:

Center for Environmental Education
1725 DeSales Street, NW
Washington, DC 20036

A Membership Organization

Center for
Environmental
Education

EPA
United States
Environmental Protection
Agency

noaa

FIGURE 3-5.   BEACH SURVEY DATA CARD DEVELOPED BY THE CENTER FOR ENVIRONMENTAL EDUCATION FOR RECORDING AND QUANTIFYING MARINE DEBRIS

(continued)

ITEMS COLLECTED

You may find it helpful to work with a buddy as you clean the beach, one of you picking up trash and the other taking notes. An easy way to keep track of the items you find is by making tick marks. The box is for total items, see sample below

egg cartons ___ ⌐⌐ ⌐⌐ ⌐⌐ ⌐   Total [16]   cups ___ ⌐⌐⌐ ⌐⌐ ⌐⌐ ⌐⌐ ⌐   Total [22]

PLASTIC	Total
bags	
trash ___	☐
salt ___	☐
other ___	☐
bottles	
beverage, soda ___	☐
bleach, cleaner ___	☐
oil lube ___	☐
other ___	☐
buckets ___	☐
caps, lids ___	☐
cups, spoons, forks, straws ___	☐
diapers ___	☐
disposable lighters ___	☐
fishing line ___	☐
fishing net	
longer than 2 feet ___	☐
2 feet or shorter ___	☐
floats & lures ___	☐
hardhats ___	☐
light sticks ___	☐
milk, water gallon jugs ___	☐
pieces ___	☐
pipe thread protector ___	☐
rope	
longer than 2 feet ___	☐
2 feet or shorter ___	☐
sheeting	
longer than 2 feet ___	☐
2 feet or shorter ___	☐
6-pack holders ___	☐
strapping bands ___	☐
syringes ___	☐
tampon applicators ___	☐
toys ___	☐
vegetable sacks ___	☐
"write protection" rings ___	☐
other (specify) ___	☐

GLASS	Total
bottles	
beverage ___	☐
food ___	☐
other (specify) ___	☐
fluorescent light tubes ___	☐
light bulbs ___	☐
pieces ___	☐
other (specify) ___	☐

STYROFOAM® (or other plastic foam)	Total
buoys ___	☐
cups ___	☐
egg cartons ___	☐
fast-food containers ___	☐
meat trays ___	☐
pieces	
larger than a baseball ___	☐
smaller than a baseball ___	☐
other (specify) ___	☐

RUBBER	Total
balloons ___	☐
gloves ___	☐
tires ___	☐
other (specify) ___	☐

METAL	Total
bottle caps ___	☐
cans	
aerosol ___	☐
beverage ___	☐
food ___	☐
other ___	☐
crab/fish traps ___	☐
55 gallon drums	
rusty ___	☐
new ___	☐
pieces ___	☐
pull tabs ___	☐
wire ___	☐
other (specify) ___	☐

PAPER	Total
bags ___	☐
cardboard ___	☐
cartons ___	☐
cups ___	☐
newspaper ___	☐
pieces ___	☐
other (specify) ___	☐

WOOD (leave driftwood on the beach)	Total
crab/lobster traps ___	☐
crates ___	☐
pallets ___	☐
pieces ___	☐
other (specify) ___	☐

CLOTH	Total
clothing/pieces ___	☐

(OVER)

FIGURE 3-5.   (continued)

items collected. The two most abundant items recorded during the single day of Texas beach cleanup activities were plastic bottles and plastic bags. A similar survey was conducted in Texas in September, 1987 (CEE, 1988). The composition of the debris collected in this survey was almost identical to that collected in 1986; 66% of the items was plastic. In 1987, in Mississippi, Louisiana, and North Carolina, plastics represented 52, 64, and 59% of inventoried items, respectively.

The 1987 CEE study also reported the results of a one-day data collection and cleanup event conducted in 19 of 23 marine coastal states. For this effort, approximately 25,000 volunteers collected and inventoried more than 700 tons of debris from 1,800 miles of U.S. coastline. Nationwide, approximately 50% of the number of litter items collected from beaches was persistent synthetic material (CEE, 1987c).

In 1988, volunteers conducted beach cleanups in 24 states, Puerto Rico, and Costa Rica as part of COASTWEEKS '88. The data provided to CEE's National Marine Debris Data Base represent the most comprehensive compilation to date of information regarding beach litter (CMC, 1989). The National Marine Debris Data Base includes data for the following debris types: plastic/Styrofoam® (or other foamed polystyrene), glass, rubber, metal, paper, wood, cloth, fishing gear, sewage-related material, medical items (syringes), balloons, domestic items, beverage six-pack rings, cargo and offshore operations items, strapping bands, and plastic bags/sheeting. Each debris type includes many individual items, which were listed on the data cards used by cleanup participants. Preliminary data, presented in Table 3-2, summarize the quantities and percentages of various types of debris collected by region. The quantity data are not normalized according to either the number of volunteers or the size of the beach area covered so quantities cannot be meaningfully compared among regions. Plastic (including Styrofoam®) was, by far, the most common debris category encountered (Figure 3-6). Paper, metal, and glass were the next most common debris types. Based on the data collected, medical-type debris was among the least common.

Beach litter can also serve as a secondary source of marine debris. Varying oceanographic and meteorological conditions, such as tidal fluctuations, influence the amounts of beach litter that are resuspended from shores and redeposited at other locations.

Data collected by U.S. EPA (1988), as part of the floatable debris investigation in the New York Harbor Complex, indicate that the resuspension of floatable refuse, resulting from above-average tides and/or heavy precipitation, may be a major source of debris slicks in the New York Bight. In August 1987, these two phenomena occurred simultaneously and a 50-mile-long garbage slick formed, leaving debris on beaches between Belmar and Beach Haven, New Jersey. The influence of tides and meteorology on the distribution of floatable wastes on shorelines of 15 New Jersey beaches was recently examined (SAIC/Battelle, 1987). At many locations, total numbers of floatable materials on the beaches were higher during periods of high tides and rain.

The types of debris littering beaches and shorelines of the United States vary geographically. Based on data reviewed to date, derelict synthetic fishing gear appears to be the predominant component of beach debris in the northern region of the North Pacific Ocean (Alaska Sea Grant College Program, 1988). Data from beach cleanup surveys in the Gulf of Mexico (CEE, 1988; CEE, 1987a) suggest that most of the debris on Texas and Louisiana beaches is from

Table 3-2

SUMMARY OF TOTAL ITEMS IN VARIOUS DEBRIS CATEGORIES
COLLECTED DURING COASTWEEKS '88 NATIONAL BEACH CLEANUP
(PERCENT OF TOTAL ITEMS INDICATED IN PARENTHESES)

Debris Type	Northeast Region	Southeast Region	Gulf Coast Region	Southwest Region	Northwest Region
Plastic/ Styrofoam	174,290 (60)	351,504 (58)	448,042 (68)	130,272 (51)	61,006 (65)
Glass	20,865 (7)	52,879 (9)	63,566 (10)	37,844 (15)	7,196 (8)
Rubber	7,940 (3)	9,406 (2)	9,584 (1)	5,533 (2)	1,436 (2)
Metal	35,451 (12)	79,544 (13)	65,500 (10)	30,373 (12)	7,891 (8)
Paper	36,570 (13)	85,496 (14)	48,582 (7)	44,214 (17)	13,458 (14)
Wood	9,871 (3)	20,292 (3)	13,510 (2)	3,490 (1)	1,403 (2)
Cloth	4,092 (1)	7,470 (1)	7,613 (1)	3,026 (1)	843 (1)
Fishing Gear	18,059 (6)	27,382 (5)	34,430 (5)	9,429 (4)	6,669 (7)
Medical Debris	149 (<1)	461 (<1)	785 (<1)	155 (<1)	66 (<1)
Balloons	3,469 (1)	3,471 (1)	1,549 (<1)	1,437 (1)	344 (<1)

(cont.)

Table 3-2 (continued)

Debris Type	Northeast Region	Southeast Region	Gulf Coast Region	Southwest Region	Northwest Region
Domestic Debris	5,674	11,224	25,117	4,834	3,197
	(2)	(2)	(4)	(2)	(3)
Cargo/Offshore	5,427	10,650	18,393	2,636	2,720
Operations	(2)	(2)	(3)	(1)	(3)
Six-Pack Rings	2,781	6,657	15,657	3,507	1,018
	(1)	(1)	(2)	(1)	(1)
Strapping Bands	1,989	3,037	4,198	1,277	1,084
	(1)	(1)	(<1)	(1)	(1)
Plastic Bags/	26,420	44,975	85,435	17,381	10,213
	(9)	(7)	(13)	(7)	(11)
Total Items	289,079	606,591	656,397	254,752	93,233

Source: CMC, 1989.

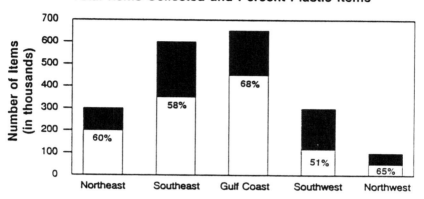

FIGURE 3-6   TOTAL DEBRIS ITEMS BY REGION AND PERCENT PLASTIC ITEMS
(Collected during COASTWEEKS '88 National Beach Cleanup. Adapted
from CMC's National Marine Debris Data Base (Battelle, 1989b)).

offshore sources, primarily commercial shipping and the offshore petroleum industry (CEE, 1988). According to CEE, the debris found on the beaches in Mississippi contained fewer items related to galley wastes and commercial fishing operations than beached debris found in Texas and Louisiana. Data collected in North Carolina appear to identify beach-goers as the major source of debris in that state.

3.2.2 Marine Sources

Marine waste is generated by vessels and by other commercial maritime activities such as offshore oil and gas platforms. The sectors of maritime activity include merchant shipping (including cargo vessels, ocean liners, tug boats, and other vessels), commercial fishing, recreational boaters, military vessels and other government vessels, offshore oil and gas platforms, and miscellaneous (educational, research and industrial vessels). This section looks at the quantities and types of wastes generated from maritime activities.

The disposal of wastes from vessels or other maritime activities has been subject to only limited regulation. Under the Refuse Act of 1899, vessels operating within three miles of shore are prohibited from disposing of wastes that could create hazards to navigation. In actual practice the Refuse Act carries only criminal penalties, making it cumbersome for the Coast Guard to enforce. Waste disposal from offshore oil and gas platforms is regulated separately, as explained further below.

Regulatory coverage for vessels is undergoing significant change, however, with the promulgation of new Coast Guard regulations. These regulations were developed under authority of the Marine Plastic Pollution Research and Control Act. This law directs the Coast Guard to develop regulations implementing the provisions of Annex V of the International Convention for the Prevention of Pollution from Ships (MARPOL) for U.S. vessels and in U.S. waters. The United States ratified this Annex under which each signatory nation prohibits the deliberate disposal of plastic wastes from its vessels and in its waters. Interim final regulations were published by the Coast Guard on April 28, 1989. The Coast Guard regulations prohibit the disposal of plastic wastes from U.S. vessels (regardless of where they operate) and from any vessel operating within 200 miles of the U.S. shoreline. Additionally, the Coast Guard regulations place restrictions on marine disposal of some non-plastic wastes for vessels and platforms operating in near-shore waters. It is important to note that the regulations do not penalize vessel operators for accidental disposal of wastes, such as fishing nets lost during trawling or other normal practices. Nevertheless, the new Coast Guard regulations should substantially reduce the contribution of plastic wastes from vessels and other maritime operations.

The sections below describe some of the wastes generated by the maritime sectors under current operations. For the discussion of vessel waste quantities, wastes are categorized as either domestic or activity-related wastes. The former category captures all of the generic types of wastes generated including galley wastes, wastes from the crew quarters and from any "hotel" areas of the vessels, and normal vessel operating (including engine room) wastes. The latter include any wastes specific to the particular type of commercial vessel activity such as fishing gear wastes, cargo-related wastes, research activity wastes, and so on.

In Section 3.2.2.7 estimates are presented of the pre- and post-MARPOL Annex V marine waste quantities. These estimates were prepared in support of Coast Guard rulemakings under the US law implementing MARPOL Annex V (ERG, 1988, 1989, also Cantin, et. al, 1989). Thus, the estimates of waste quantities cover those sectors that come under Coast Guard responsibility, that is, waste disposal by U.S.-flagged vessels, and by foreign vessels operating in U.S. territorial waters.

3.2.2.1 Merchant Marine Vessels

The merchant marine sector is defined to include ocean-going and domestic cargo vessels, ocean and domestic tugs and barges, ocean liners, and ferries and small charter boat operators. The National Academy of Sciences (NAS, 1975) developed the only near-comprehensive examination of waste disposal from this sector. NAS estimated that domestic waste generation by vessel crew members exceeded 100,000 metric tons annually. Table 3-3 presents the NAS estimates of marine litter. Of this amount, one percent by weight was estimated to be plastic. Since the NAS estimate, crew sizes have declined, but the relative share of plastic waste to shipboard waste has increased. These factors are taken into account in estimates for all sectors that are described below.

Horsman (1982) analyzed merchant marine waste generation by counting the plastic containers that were brought onboard vessels. He estimated that 600,000 plastic containers are discarded at sea by the world merchant fleet.

NAS estimated that 28,000 metric tons of debris are generated each year by cruise ships serving U.S. ports (NAS, 1975). It was estimated at the time that under 2 percent of this material was plastic.

NAS also estimated that cargo-related wastes contributed large amounts to marine debris. Cargo-associated wastes include dunnage (such as wood shoring for cargo compartments), and crates, pallets, wires, plastic sheeting, and strapping bands. NAS calculated, based on a variety of previous international studies, that 5.6 million metric tons per year of cargo-related wastes are discarded.

The NAS estimate is now, however, seriously out-dated. Since the NAS study, world shipping practices have shifted greatly towards containerized cargo. In 1976, U.S. Maritime Administration (MARAD) statistics showed that there were 508 full containerships and 597 partial containerships in the world fleet, and these accounted for 4.7% of the vessel total (including freighters, tankers, bulk carriers, and passenger liners) (MARAD, 1977). By 1988, the world fleet had fallen from 23,586 to 23,307 vessels, but the number of full and partial containerships had risen to 1,097 and 1,720 respectively, and now represent 12.1% of the fleet (MARAD, 1989). With the much greater carrying capacity of the containerized ships, the percentage increase in the cargo carried via containership would be higher still.

Table 3-3

NAS ESTIMATES OF GLOBAL MARINE LITTER

Garbage Types and Sources	Metric Tons/Year	Percent
Regulated Sources under Annex V		
Crew–related wastes		
Merchant marine	11,000	1.8%
Passenger vessels	2,800	0.4%
Commercial fishing	34,000	5.4%
Recreational boats	10,300	1.6%
Military	7,400	1.2%
Oil drilling and platforms	400	0.1%
Commercial wastes		
Merchant cargo wastes or dunnage	<u>560,000</u>	<u>89.5%</u>
Regulated sources– subtotal	625,900	100.0%
Unregulated Sources		
Fishing gear lost	100	1.0%
Loss due to catastrophe(a)	<u>10,000</u>	<u>99.0%</u>
Unregulated sources– subtotal	10,100	100.0%
TOTAL	636,000	100.0%

Note: (a) Debris originating from shipwrecks or due to marine storm damage.

Source: National Academy of Sciences, 1975.

Containerized methods of shipping generate almost no cargo dunnage or other cargo-related debris. As a result, the quantity of cargo-generated waste is much lower than at the time of the NAS research. Estimates of cargo dunnage generation rates for vessels calling at U.S. ports are discussed below (see Section 3.2.2.7).

3.2.2.2 Fishing Vessels

World fishing fleets represent an extremely large number of vessels. The National Marine Fisheries Service estimated that 129,800 U.S.-owned vessels were in operation in 1986. In the past, significant numbers of foreign vessels have also been granted access to U.S. fishing stocks. In recent years, however, the amount of "direct" foreign fishing has declined, as joint ventures between U.S. catcher boats and foreign processing vessels have increased. In 1985, foreign vessels accounted for 41% of the total catch in the U.S. Exclusive Economic Zone (EEZ); by 1987 this had fallen to only 5% of the total (National Marine Fisheries Service, 1988).

The fishing industry, like other sectors, generates domestic wastes and activity-related wastes. Domestic wastes are generated by the substantial population of fishermen onboard these vessels. Using NAS estimates and 1984 data on the number of fishing vessels registered in the U.S., researchers have calculated that more than 92,000 metric tons of galley wastes per year is generated onboard U.S. fishing vessels (CEE, 1987b).

Activity-related wastes consist of fishing nets, floats, lines, traps, and pieces or fragments thereof. Because of its strength, durability, and lower cost, plastic fishing gear materials are employed by virtually all of the world's fleet (Pruter, 1987).

Plastic fishing gear which is lost or discarded at sea becomes a persistent marine pollutant. Normal wear or damage to gear may result in the loss of lines, nets, traps, or buoys. Nets and other gear may be damaged by encounters with marine mammals or predator species, such as sharks. Fishermen may be unable to retrieve submerged nets and traps if marker buoys become lost, severed, or relocated during storms. Operational errors, such as setting traps too deeply, fouling of gear on underwater obstructions, or improper deployment, may result in gear loss. Further, net scraps generated during repair operations have historically been discarded overboard if they cannot be reused; under the new MARPOL Annex V regulations this disposal practice will not be allowed.

Several estimates of the amount of fishing gear lost annually are available. NAS estimated that 13 tons per vessel per year was lost. Data collected in the Bering Sea and Gulf of Alaska indicate that 35 to 65 entire nets or significant pieces were lost annually among approximately 300 trawlers active in the area, between 1980 and 1983 (Low et al., 1985). Merrell (1985) found that commercial fishing operations were a source of 92 percent (by weight) and 75 percent (by number) of plastic debris items categorized on an Aleutian Island beach. Merrell also estimated that more than 1,600 metric tons of plastic debris may be lost or discarded annually from fishing vessels in Alaskan waters.

According to CEE (1987b), accidental or deliberate gear conflicts (i.e., use of two or more types of gear in a fishing area) may also increase the loss of gear. Gear conflicts are common in areas where both fixed gear (e.g., traps, anchored nets) and towed or dragged gear (e.g., trawl nets) are deployed in the same fishing grounds. Commercial fishing conflicts have been especially prevalent off New England and in the Gulf of Mexico (Stevens, 1985; Gulf of Mexico Fishery Management Council, 1984). Fish harvesting policies can also influence the loss of fishing equipment through gear conflicts. In Puget Sound, for example, one type of gill net fishery is active during the same time that Dungeness crabs are harvested, resulting in increased loss of crab pots through entanglement with gill nets (Alaska Sea Grant College Program, 1988).

Two types of fishing nets, drift gill nets and trawl nets, are commonly damaged and lost or discarded at sea (Uchida, 1985). Based on the amount of netting deployed in the North Pacific, the drift gill net is most likely to become derelict. Drift gill nets, some of which are up to 15 miles in length, are generally made of nylon and used to harvest large schools of fish or squid (Interagency Task Force, 1988). Nylon is more dense than seawater and will sink if not buoyed by floats. The nets generally sink if lost or discarded. These nets can last only a few weeks and each vessel can use up to 400 nets in a 4-month season (Parker et al., 1987).

Trawl nets are made in differing mesh sizes depending on the target species. They are generally constructed of nylon or polyethylene in bag-shaped forms that can then be towed at different water depths or along the bottom to harvest a variety of finfish or shellfish species. Bottom trawling can easily damage or entirely detach nets. In the North Pacific, where the trawl net fishery is extensive, trawl net webbing frequently washes ashore on Alaskan beaches (Johnson and Merrell, 1988; Fowler, 1987; Merrell and Johnson, 1987; Merrell, 1985). Derelict gill nets and trawl nets can continue to "ghost fish" for undetermined periods of time.

Several other gear items are also lost at sea. In some U.S. regions, loss of crab and lobster traps can be significant. CEE reports that in New England, lobster traps are lost at a rate of 20 percent annually (CEE, 1987b). Other lost or discarded items include polystyrene buoys and floats, monofilament line and synthetic ropes, and plastic commercial bait, salt, and ice containers.

3.2.2.3   Recreational Boats

Recreational boaters are another source of marine wastes although data on their waste generation rates and disposal habits are extremely limited. This section summarizes the available evidence in this area.

An estimated 16 million recreational boaters use the coastal waters of the U.S. (Interagency Task Force, 1988). The spatial distribution of recreational boats is presented in Figure 3-7. The greatest concentration of boaters is found on the Atlantic Coast. Price and Thomas note that an estimated 160,000 boaters use the waterways of the New York Bight (Price and Thomas, 1987).

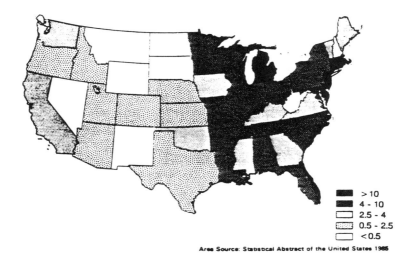

Area Source: Statistical Abstract of the United States 1985

Legend:
- > 10
- 4 - 10
- 2.5 - 4
- 0.5 - 2.5
- < 0.5

FIGURE 3-7 .   DENSITY OF RECREATIONAL VESSELS IN THE UNITED STATES FOR 1984
(number of boats/square mile of land; adapted from CEE,
1987b)

Recreational boaters contribute domestic waste to the marine environment, including food and beverage containers, as well as fishing gear such as nylon monofilament fishing line. CEE and others have estimated the amount of domestic waste generated by recreational boaters. CEE utilized a waste generation rate developed in the NAS study of 0.45 kg per person per day for recreational boaters (CEE, 1987b). This assumption produced an estimate of 51,000 metric tons of trash from U.S. boaters. Researchers working in support of the Coast Guard MARPOL Annex V regulations have estimated recreational waste generation based on the same per capita rate as was applied to the other sectors, which varied between 1.0 and 1.5 kg per person, and assumed that virtually all recreational vessels owned in coastal and Great Lakes states would be used in navigable waters, including marine and inland waterways (ERG, 1988). These estimates produced an aggregate estimate of 636,055 metric tons. Data are inadequate to determine the better estimate of recreational waste generation rates. No recent field studies have been performed, and data have not been developed on the relative amounts of waste disposed of at sea versus that brought back to shore for disposal. In the Coast Guard research it was also estimated that two-thirds of recreational boaters bring wastes ashore, based on conversations with marina operators who noted the frequent tendency of boaters to seek out marina and dockside dumpster facilities.

3.2.2.4 Military and Other Government Vessels

U.S. Navy vessels carry extremely large crew complements, with over 285,000 personnel deployed onboard approximately 600 vessels. Aircraft carriers, the largest vessels in the Navy fleet, carry as many as 5,000 crew at one time (Parker et al., 1987).

The U.S. Navy has performed some of the only quantitative studies of waste disposal at sea. In 1971, a study estimated that Navy ships generated 3.05 pounds of solid waste per person per day, of which only 0.3 percent by weight consisted of plastics. A more recent study, completed in 1987, found that plastics accounted for 7 percent by weight (Figure 3-8) (Schultz and Upton, 1988). Historically, Navy ships have disposed of most garbage overboard. The aggregate rate of plastic waste disposal for the Navy has been estimated at nearly 4 tons per day (Interagency Task Force, 1988).

The U.S. Coast Guard, the National Oceanographic and Atmospheric Administration (NOAA), and the Environmental Protection Agency (EPA) together operate approximately 225 vessels for marine safety, research and other purposes (Interagency Task Force, 1988). These vessels carry approximately 9,000 personnel. Existing Coast Guard policies require ships to dispose of waste onshore (if reasonably possible). No field estimates have been developed of the quantity or manner of waste disposal from the other vessels.

The U.S. Navy and the other government agencies are required to meet the requirements of the MARPOL Annex V within five years of regulatory implementation (1992). An ad-hoc advisory committee has recommended various methods for waste reduction aboard military ships,

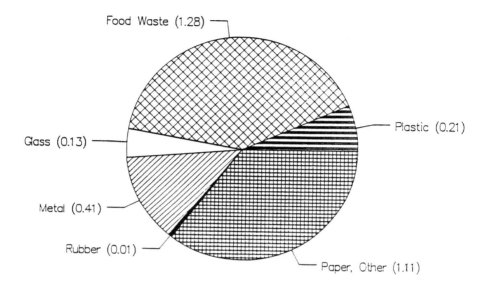

FIGURE 3-8.   QUANTITIES OF SOLID WASTE GENERATED ON U.S. NAVY VESSELS
(pounds/man/day; adapted from Schultz and Upton, 1988)

including source reduction, compaction, thermal destruction, and a technology for melting, compacting, and sterilizing plastic debris (Ad Hoc Advisory Committee on Plastics, 1988). In addition, the Coast Guard, EPA and NOAA have prepared internal operating orders that prohibit disposal of plastic materials from vessels.

3.2.2.5 Miscellaneous Vessels (Educational, Private Research, and Industrial Vessels)

Vessels in the miscellaneous category may generate both domestic and activity-related garbage that may be disposed at sea. This category includes educational vessels (merchant marine training ships), private research vessels (oceanographic research vessels), and industrial vessels (vessels involved in cable-laying operations, work barges, dredges or other vessels engaged in marine construction). Domestic garbage generation is related to the population carried onboard the vessels. Educational vessels (e.g., merchant marine training vessels) can carry large passenger complements on their training cruises. Research and industrial vessels typically have larger crews than cargo vessels, but smaller passenger complements than training ships (ERG, 1988).

Certain vessels within this category can also generate important quantities of activity-related wastes. Research vessels generate substantial plastic wastes from the packaging of research instrumentation and equipment. Wastes from industrial vessels vary with the specific task being performed, and have not been fully characterized in the available literature.

3.2.2.6 Offshore Oil and Gas Platforms

The offshore oil and gas sector includes mobile offshore drilling units (MODUs) used in exploratory drilling, stationary production platforms, which are installed once exploitable reserves of oil and/or gas are located, and a large fleet of support vessels used to transport crew, supplies, and equipment. A recent tally (ERG, 1989) found that there are approximately 200 MODUs, 3,500 production platforms, and over 500 offshore service vessels active in the U.S. offshore petroleum industry. Of the 3,500 platforms, only 779 are manned on a continuous basis. The highest concentrations of platforms are found off the Texas and Louisiana coasts.

Regulations enforced by the Department of the Interior's Minerals Management Service (MMS) prohibit waste disposal from U.S. offshore oil and gas platforms. Nevertheless, some wastes found in beach cleanups in the Gulf of Mexico have included a number of items that may have originated from offshore oilfield operations. Researchers or industry sources have not differentiated between any deliberate or accidental waste disposal.

Offshore oil and gas operations generate domestic waste and a variety of debris from industry activities. CEE used an assumption of 10,000 oilfield personnel working offshore and prepared an estimate of domestic waste quantities. They described their resulting estimate of 1.6 metric tons annually as a conservative estimate of domestic wastes (CEE, 1987b).

Researchers have attributed some of the debris items found in beach cleanups to oilfield operations. Wastes identified in Table 3-4, for example, appear to originate from offshore oilfield operations. For example, computer write-protection rings, which come from magnetic data-recording tapes used in seismic research, and drill pipe protectors are likely oilfield wastes. CEE estimated that 10 percent of the items collected in Texas beach cleanups in 1986 represented oilfield wastes (1987a). They also reported that numerous 30- and 55-gallon drums wash ashore each year. Further, they attribute the large number of milk jugs washing ashore to industry activities as well.

3.2.2.7 Recent Estimates of Plastic Wastes Disposed in U.S. Waters By All Maritime Sectors

Several studies were prepared in the analysis of the impact of the recently-promulgated Coast Guard regulations (ERG, 1988, 1989; Cantin et. al, 1989). These studies provide estimates of the quantity and manner of waste disposal from vessels or offshore structures in all of the sectors for operations within the U.S. EEZ. These estimates cover operations of U.S.-flagged as well as foreign-flagged vessels (including cargo and cruise ships) calling at U.S. ports.

The Coast Guard research utilized estimates of per capita waste generation developed by the International Maritime Organization (IMO). These rates vary depending upon whether the vessel operates over open ocean, coastal, or inland waters. The rates are based on studies of merchant ships, but were assumed to apply also to fishing, recreational and other vessels. Estimates were prepared of the annual-person days of activity for each type of vessel taking into account voyage lengths, crew sizes, passenger-carrying capacities, and vessel utilization rates. The contribution of plastic waste to the total solid waste stream generated was estimated based on the 1987 Navy study (Schultz and Upton, 1988), which found that plastics contribute approximately 7 percent by weight to solid waste. The Navy study is the only direct and recent measurement of this variable for maritime operations.

Table 3-5 presents an example (using the merchant marine sector) of the calculations and forecasts developed for each maritime sector. It should be noted that these estimates may be based in some cases on slightly different underlying data (concerning the number of vessels) than have been reviewed thus far in this section.

For this research, estimates were also developed of the waste disposal practices currently used among the various maritime sectors. The estimates, shown in Table 3-6, were based primarily upon discussions with industry representatives in each of the sectors and indicate that, while much garbage is disposed overboard, vessels operating close to shore bring substantial quantities ashore for disposal. The sectors that bring most of their wastes ashore include commercial passenger vessels, recreational boaters and offshore oil platform operators. Aggregate waste generation was estimated at over 1.2 million metric tons or over 8.3 million cubic meters.

Table 3-4

MARINE DEBRIS ASSOCIATED WITH THE
OFFSHORE PETROLEUM INDUSTRY

- Plastic sheeting
- Computer write-protect rings
- Seismic marker buoys
- Drilling pipe thread protectors
- Diesel oil and air filters
- Hardhats
- Chemical pails
- Plastic and metal drums
- Polypropylene hawsers

Source: Interagency Task Force (1988); CEE (1987a); King (1985).

Table 3-5

QUANTITIES OF DOMESTIC GARBAGE GENERATED PER VOYAGE MERCHANT SHIPPING SECTOR

VESSEL CATEGORY	Voyage Length (days)	Crew Size	Person-Days Per Voyage	Per Capita Generation Rate (kg/day)	Domestic Garbage Generation Per Voyage					
					Total Garbage (kg)	Dry Garbage (kg)	Plastic Garbage (kg)	Total Garbage (cu.m)	Dry Garbage (cu.m)	Plastic Garbage (cu.m)
Foreign Trade										
U.S. Vessels										
Atlantic/Gulf/Pacific	7	25	165	2.0	330.0	196.0	22.1	2.2	2.0	1.4
Non-contiguous – foreign	2	25	53	2.0	105.0	62.4	7.0	0.7	0.6	0.5
Foreign Vessels										
Atlantic/Gulf/Pacific	7	25	173	2.0	345.0	204.9	23.1	2.3	2.1	1.5
Non-contiguous/Great Lakes	2	25	60	2.0	120.0	71.3	8.0	0.8	0.7	0.5
Non-Contiguous Trade (U.S. - domestic)	7	25	175	2.0	350.0	207.9	23.5	2.3	2.1	1.5
Great Lakes (domestic & foreign trade)										
1,000 gross tons & over	2	25	53	1.5	78.8	46.8	5.3	0.5	0.5	0.3
Under 1,000 gross tons	2	25	53	1.5	78.8	46.8	5.3	0.5	0.5	0.3
MSC Charter (U.S.)	7	25	175	2.0	350.0	207.9	23.5	2.3	2.1	1.5
Temp. Inactive Vessels (U.S.)	7	25	175	2.0	350.0	207.9	23.5	2.3	2.1	1.5
Coastal Shipping										
Ships										
1,000 gross tons & Over	5	25	125	1.5	187.5	111.4	12.6	1.3	1.1	0.8
Under 1,000 gross tons	4	25	100	1.5	150.0	89.1	10.1	1.0	0.9	0.6
Tow/Tugboats										
Large (inspected)	4	10	40	1.5	60.0	35.6	4.0	0.4	0.4	0.3
Small	2	6	12	1.5	18.0	10.7	1.2	0.1	0.1	0.1

MSC = Merchant Sealift Command (private ships chartered by the armed services)

Source: Cantin et al., 1989

Table 3-6

FINAL DISPOSITION OF VESSEL-GENERATED DOMESTIC GARBAGE
AGGREGATED SECTOR TOTALS
(PRE-MARPOL ANNEX V; ANNUAL QUANTITIES)

SECTOR	Total Generated Annually (metric tons)	Off-Loaded in Port (metric tons)	(cu.m)	Pre-Annex V Incinerated at Sea (metric tons)	(cu.m)	Dumped Overboard (metric tons)	(cu.m)
Merchant Shipping	30,949	2,097	18,494	1,148	7,684	27,704	179,290
Commercial Passenger Vessels	258,074	232,121	1,553,589	638	4,272	25,315	169,430
Commercial Fishing	233,177	0	0	0	0	233,177	1,560,655
Recreational Boats	636,055	424,036	2,838,081	0	0	212,018	1,419,041
Offshore Oil & Gas	16,710	10,733	102,263	0	0	5,977	9,575
Miscellaneous Sectors	1,637	5	295	0	0	1,633	10,677
U.S. Navy Vessels	57,596	0	0	0	0	57,596	385,493
U.S. Coast Guard Vessels	4,317	2,445	16,366	0	0	1,872	12,527
U.S. Army Vessels	490	0	0	0	0	490	3,279
NOAA Research Vessels	317	99	42	88	588	130	872
TOTALS	1,239,322	671,536	4,529,130	1,874	12,544	565,911	3,750,840

Source: Cantin et al., 1989.

The change in waste disposal patterns that would occur under MARPOL Annex V regulations was also forecast. These forecasts are based on the assumption of full compliance with the new regulations, and include estimates of the likely choices of compliance method among the primary options of 1) bringing wastes ashore (with or without onboard compaction of garbage), 2) incinerating wastes at sea, and 3) continuing overboard disposal (for non-plastic wastes and in authorized areas only). Table 3-7 presents estimates of the final disposition of plastic and non-plastic wastes before and after implementation of MARPOL Annex V. The amount of plastic waste brought ashore by the maritime sectors operating in U.S. waters was estimated to increase from approximately 40,000 tons to almost 90,000 tons. A relatively small number of vessels, consisting primarily of merchant vessels operating over international trade routes and larger research and fishing vessels, were forecast to choose incineration as their disposal option. Such vessels are most likely to find onboard storage of even compacted waste to be disruptive and/or to pose a health risk.

The totals presented here do not include estimates of the quantities of activity-related wastes generated in cargo shipping, commercial fishing, and research sectors (see ERG, 1988). The estimates of activity-related wastes are more speculative, and require considerable additional data and methodological development. In the paragraphs below, an outline of estimates of the main activity-related wastes disposed in U.S. waters is presented (Cantin, et al., 1989).

Dunnage was judged to be generated by general cargo ships only, as cargo carried in containerships does not require the shoring or the construction of separate cargo compartments. Dunnage characteristics and quantities were estimated from discussions with vessel and terminal operators. Most dunnage consists of cardboard and lumber, with only very small amounts of plastics used for special liner requirements. Approximately one-half of the vessels generating such wastes were estimated to dump their dunnage in U.S. waters. The annual quantity of plastic from dunnage disposed in U.S. waters from U.S. and foreign vessels amounts to only 7 cubic meters per year.

Data from observers onboard certain fishing vessels was used to estimate the quantities of fishing gear discarded deliberately at sea. Most of the deliberate discarding is due to the repair of nets that occurs at sea. This occurs relatively infrequently, as much netting is retained for its scrap value. Certain fishing operations, however, generate more substantial waste quantities. Examples include: longline bait fisheries, which generate quantities of packing and strapping materials, and herring fishery vessels, which produce waste salt bags from the salt needed to preserve the catch. The researchers estimated the deliberate at sea disposal of net fragments and other gear at approximately 2,200 metric tons per year.

Finally, this research estimated wastes generated by oceanographic research. These wastes include packing materials from research instruments brought onboard, as well as single-use instruments such as bathometers, which may be cut loose from the vessel once they have transmitted their data to the ship. Research vessels generate 0.1 cubic meters per voyage, for an aggregate total of 70 cubic meters of plastic per year (ERG, 1988).

Table 3-7

DISPOSITION OF GARBAGE GENERATED BY
MARITIME SECTORS -
PRE- AND POST-ANNEX V ESTIMATES

| Disposition | Pre-Annex V | | | | Post-Annex V | | | |
| | Tons (000) | | Cu.Meters (000) | | Tons (000) | | Cu.Meters (000) | |
	Number	Percent	Number	Percent	Number	Percent	Number	Percent
Brought Ashore								
Plastics	40	3.3%	2,583	31.1%	89	7.2%	4,747	61.5%
Other	631	50.9%	1,946	23.5%	833	67.2%	2,136	27.7%
Sub-Total	672	54.2%	4,529	54.6%	922	74.4%	6,884	89.2%
Incinerated	2	0.2%	13	0.2%	10	0.8%	65	0.8%
Dumped	566	45.7%	3,751	45.2%	307	24.8%	770	10.0%
TOTAL	1,239	100.0%	8,293	100.0%	1,239	100.0%	7,718	100.0%

Source: Cantin et al., 1989.

3.2.3 Illegal Disposal of Wastes into the Marine Environment

Although there is a lack of documented information on this topic, it is generally believed that illegal disposal contributes unknown quantities of plastic debris to the marine environment. Illegal disposal of municipal solid wastes, sewage, and medical wastes may represent additional sources of plastic debris to the marine environment. In areas such as New York City, where solid waste disposal involves over-water transport of wastes on barges, there is a potential for accidental spillage of this material into the marine environment. Although the City of New York and the State of New Jersey have, through a U.S. District Court consent decree, established guidelines and protocols governing solid waste handling, light-weight debris from transfer facilities and from loaded barges may illegally enter waterways and be transported out to sea. Noncompliance with the decree requirements, which require adherence to established protocols, use of barge covers and containment booms, and removal of floating debris contained within the barrier booms, may result in illegal disposal of solid waste.

Although assumed to be relatively uncommon, there is potential in all coastal states for garbage trucks to dump their loads from piers or directly into marshes and estuaries. The public may also contribute to illegal disposal of household debris in marshes, estuaries, along shorelines, and on beaches. Evidence of such practices comes from beach survey records reporting debris such as tires, appliances, mattresses, furniture, and other predominantly domestic items.

Sewage sludge from New York and New Jersey municipalities is currently disposed of at the designated 106-Mile Deepwater Municipal Sludge Site located outside of the New York Bight, beyond the continental shelf. The federal ocean dumping regulations strictly prohibit the disposal of "persistent, synthetic or natural materials which may float or remain in suspension." Although the regulations and permits issued for ocean disposal of sludge clearly prohibit disposal of sewage-related plastic and other floatable materials at the site, such materials may potentially enter the ocean illegally if they are not effectively removed at treatment facilities prior to ocean disposal of the sludge (Price and Thomas, 1987). Plastic debris items that may be associated with sewage sludge include tampon applicators, condoms, and disposable diapers.

The recent incidents of medical debris appearing on east coast beaches have caused considerable concern about disposal practices for medical wastes. Because there is no legal pathway for significant quantities of such wastes to enter the marine environment, the occurrence of medical debris on beaches has often been attributed to illegal disposal activities. Rising disposal costs and localized shortages of landfill and incinerator capacity may create an incentive to dump medical wastes illegally. These problems are most severe in the northeast United States. Over the last five years, the cost of disposing of medical wastes has escalated from 17 cents per pound to 50 cents per pound (Boston Globe, 1988). In the New York area, the costs can be as high as 80 cents per pound (Swanson, 1988).

The kinds of medical wastes that have been identified in the marine environment include a wide assortment of syringes, pill vials, surgical gloves, tubing, blood vials, bandages, blood bags, respirators, and specimen cups. Because many medical supplies, originally made of glass and intended for reuse, have been replaced with disposable plastic items, most of these materials can become floatable debris in the marine environment.

In recent years, medical wastes have caused particular alarm in east coast states. New York and New Jersey have each experienced major incidents in 1987 and 1988. Other states, including Massachusetts, Rhode Island, Connecticut, Maryland, and North Carolina, have reported at least one case of beached medical debris. However, the medical waste found on New York and New Jersey beaches represented only 1-10% by volume of the floatable debris that washed ashore on these beaches in the summer of 1988 (New York State DEC, 1988). The additional health risk posed by the medical debris that washed ashore in the Northeast was most likely small, but media attention may have resulted in a large perceived risk. A recent report by U.S. EPA chronologically documented medical waste wash-ups that occurred along the East Coast during the 1988 beach season (U.S. EPA, 1989b). A total of 477 wash-up incidents, in which 3,487 medical waste items were recorded, occurred over a five month period in six East Coast states. Figure 3-9 summarizes these data.

The occurrence of medical wastes in other coastal areas of the country has not been documented as well as along the east coast. However, data collected from beach cleanups held in 1987 indicate that medical debris incidents are not limited to the northeast coast. During the one-day cleanup event in ten coastal states, syringes were collected from Gulf coast beaches in Texas and Mississippi. More than 900 syringes were recorded on Texas shores alone (CEE, 1987c).

Despite the attention given to medical waste found on northeast coast beaches in the summer of 1988, a national beach cleanup effort indicated that the wash-up of medical waste was not unique to the region. The CEE National Marine Debris Data Base includes data from 1988 beach cleanups conducted in 24 states. More than 1100 syringes were reported found during the beach cleanups (CMC, 1989). The region reporting the largest number of syringes was the southeast coast (461), extending from Virginia to Florida and including Puerto Rico. The Gulf coast states reported 303 syringes, followed by the southwest (California and Hawaii), for which 155 syringes were reported. The lowest numbers of syringes, 149 and 66 respectively, were associated with beaches in the Northeast, extending form Maine to Maryland, and in the Northwest (Oregon, Washington and Alaska). The quantity figures, however, have not been normalized to consider the number of volunteers involved in the beach surveys or the miles of beaches covered. Thus, comparisons among regional findings should be made with caution.

In a preliminary study of the 1988 medical waste incidents in the northeast, the New York Bight was identified as the source of much of the medical debris in southern New England and Long Island (Spaulding et al., 1988). The prevailing winds from mid-June to mid-July were identified as the major factor in transporting waste from the New York Bight to southern New England. Swanson (1988) suggests that the winds during this period were from a different direction than the normal summer wind path observed in this region. The Fresh Kills landfill, sewer discharges, CSOs, and marine transfer stations were identified as the major sources of the medical debris that washed ashore on New York beaches this summer (New York State DEC, 1988). Further, during investigations of the medical debris problem, it was found that a laboratory in Brooklyn had illegally disposed of blood vials on the banks of the Hudson River. It is believed that this activity may be responsible for one instance of large numbers of

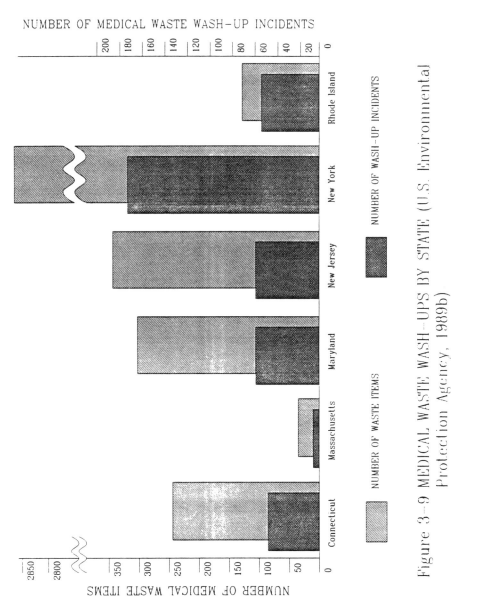

Figure 3-9 MEDICAL WASTE WASH-UPS BY STATE (U.S. Environmental Protection Agency, 1989b)

blood vials on the New Jersey shoreline (New York State DEC, 1988). Finally, the medical waste incidents in North Carolina have been traced to Navy vessels discharging debris offshore. New Jersey and New York have developed state regulations requiring a cradle-to-grave tracking system for certain medical wastes. Other states are also developing manifest systems or have them in place for these materials, while some states have not developed tracking programs at the state level.

Under the Medical Waste Tracking Act of 1988, EPA has promulgated interim final regulations under which generators of more than 50 pounds per month of regulated medical wastes will be required to segregate, package, and label medical waste shipments according 'o the requirements of Code of Federal Regulations Part 259 (54 Federal Register 12326, March 24, 1989). A standardized tracking form will also be attached, which will be signed by both the transporter(s) of the wastes and an individual at the final disposal facility. A copy of the form will then be sent from the disposal facility to the generator to complete the process. Generators of smaller quantities must also segregate, package, and label their wastes, but in some cases they need not complete a tracking form. The regulations currently apply to medical wastes generated in New York, New Jersey, Connecticut, and states bordering the Great Lakes.

3.3   FATE OF PERSISTENT MARINE DEBRIS

The fate of floatable debris may be described in terms of physical transport mechanisms or biological and chemical degradative processes. Several studies have examined transport mechanisms, and the influence of oceanographic and meteorological conditions on the distribution and fate of floatable material (Swanson, 1988; SAIC/Battelle, 1987; Wilber, 1987; Swanson et al., 1978; NOAA/MESA, 1977). Much less information, however, is available on rates and processes of degradation of floating debris.

3.3.1   Physical Fate and Transport Processes

Physical transport mechanisms include high river runoff, winds, and surface currents. In coastal regions impacted by significant discharges from rivers, such as the Hudson River discharge to the New York Bight, transport of marine debris is strongly influenced by the river plumes. The transport and fate of floatable materials in marine waters is also largely influenced by short-term patterns of surface currents (SAIC/Battelle, 1987). Although long-distance transport is influenced by large-scale current systems (Mio and Takehama, 1988), patterns of wind direction and velocity, offshore oceanic circulation, and short-term meteorological and oceanographic events are primarily responsible for strandings of debris on beaches (SAIC/Battelle, 1987).

The initial fate of plastic debris relates to the density of the material, the location of dumping or release, and meteorological and oceanographic conditions. Ultimately, it is a complex interaction of the physical properties of the waste materials and oceanographic and meteorologic conditions that will determine the fate of plastic debris. In general, lightweight floatable debris, such as plastic materials, is confined to surface waters and is transported by the dominant currents. Transport by currents may be modified by wind-driven transport.

In the northern North Pacific Ocean, marine debris is primarily transported by the North Pacific Current that flows east and by seasonal winds (Mio and Takehama, 1988). If material is transported far enough east to reach northern U.S. coastal waters, the California Current continues to transport it south and west, along with debris from U.S. waters, to waters northwest of the Hawaiian Islands where debris converges (Mio and Takehama, 1988). In the northwestern Hawaiian Islands, Henderson (1988) found that trawl web and gill net fragments accumulated on northeast-facing beaches that are exposed to the predominant northeasterly trade winds, but certain promontories on the leeward side of some islands also accumulated debris as a result of inshore currents.

Field sampling conducted by Wilber (1987) in the North Atlantic showed that the distribution of plastic materials in this region is influenced by three major forces (Figure 3-10). The large, clockwise circulating Central Gyre exerts the major initial effect on debris transport. Within the gyre, which is centered north of Bermuda, the fate of plastic debris is controlled by smaller scale rotating features known as eddies or rings that continually traverse the Central Gyre. Lastly, the effect of winds over the ocean surface creates Langmuir cells, which concentrate debris in long linear features known as windrows or more commonly, "slicks." The specific pathway of a plastic item in the North Atlantic is specifically related to where, in relation to the Central Gyre, it enters the marine environment. Debris items may be intra-gyral, originating solely from vessels operating within the gyre, or they may be extra-gyral, originating from terrestrial sources or vessels in near-coastal waters. Extra-gyral debris may be removed before entering the gyre or may become entrained in the gyre with intra-gyral debris. The islands of Bermuda, the Bahamas, and the Florida Keys act as sieves, continuously removing debris entrained in the gyre. This scenario seems to explain the abundance of plastic litter on many remote beaches of these islands.

In regions of the country subject to severe weather (e.g., portions of Alaska), storms may have a major influence on the fate of marine debris. During storms, plastic debris that is buried may be uncovered and debris stranded on shore may either be transported inland or buried (Johnson, 1988).

From several tag-and-recovery studies on Alaskan beaches, Johnson found that 10% of stranded trawl net fragments can move 1 km or more laterally along the beach and that 10% of tagged recoveries were buried and reexposed again within one year. He concluded that in severe climates such as in Alaska, once debris is stranded on shore, most remains there and is not resuspended. The amount of debris visible on beaches in any given year may be dictated largely by storms.

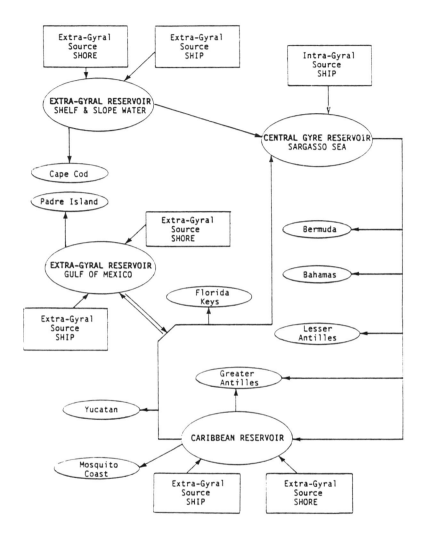

FIGURE 3-10   FLOW DIAGRAM FOR FATE OF PLASTIC DEBRIS IN THE WESTERN NORTH
ATLANTIC OCEAN (adapted from Wilber, 1987)

Floatable debris may alternately become stranded on shores, resuspended from shores, and redeposited elsewhere. Other studies have shown that in nearshore environments, above-average water levels, resulting either from heavy rains, extreme tides, or a combination of these events, resuspend a higher percentage of stranded debris (Swanson, 1988; U.S. EPA, 1988; Swanson et al., 1978). Slicks may form if these phenomena occur simultaneously (U.S. EPA, 1988). The slicks may disperse floatable debris onto shorelines, drift out of harbor areas into the open ocean, or both. In the open ocean, debris slicks may disperse, and plastic items eventually sink or accumulate along windrows, areas where currents converge.

3.3.2   Degradative Processes

Plastic debris is subjected to various physical and chemical processes that combine to weaken the integrity of the material and initiate some degree of physical or biological degradation (Figure 3-11). Most plastic materials are highly persistent in the environment. Their molecular structure and configuration, which generally consist of very densely compressed long-chain molecules (polymers), render these materials recalcitrant to natural processes of decay. If the polymer is fragmented or reduced in size, the plastic material eventually loses strength, becomes brittle, and may fragment.

Because most plastic debris has some degree of buoyancy, it is continually exposed to ultraviolet (UV) radiation from sunlight while floating. UV energy initiates a chemical reaction that leads to the fragmentation of the polymers (photodegradation). This reaction is slow and, for some types of plastic material, it can be many years before the fragmentation begins. Thin plastic sheets and bags are most susceptible to this breakdown process; thicker, denser plastic items are less susceptible. Subsequent physical stress from wind or wave action eventually destroys the integrity of the plastic material and results in fragmentation. The buoyancy of fragmented plastic may change when it is colonized by epifaunal organisms, such as hydroids, barnacles, and bryozoans. These organisms may also shield the plastic from the effects of UV radiation. Increase in density may result in sinking of the fragments. Photodegradation of some types of plastic fragments may continue until the fragments are reduced to such a small size that, under optimal environmental conditions (e.g., nutrients, temperature), microbial degradative processes will become efficient in breaking down the plastic fragments.

While the significance of these processes in the marine environment remains largely unassessed, degradation of plastic materials probably does not reduce the impact of plastics on the marine environment, because the process is too slow. In a study investigating six commercially available plastic materials, Andrady (1988) found that most of the materials degraded much more slowly in seawater than in air. Two types of trawl netting investigated did not degrade significantly in air or water over one year and expanded polystyrene foam degraded more rapidly in seawater than in air. Plastics that are manufactured to enhance degradation are discussed in Chapter 5.

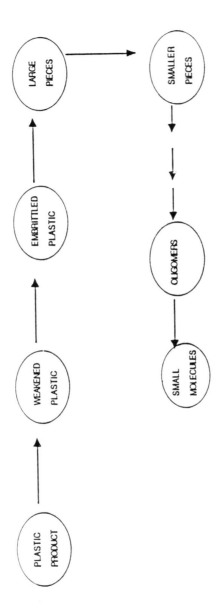

FIGURE 3-11   DEGRADATIVE PROCESSES FOR PLASTIC MATERIALS (adapted from Interagency Task Force, 1988)

3.4   EFFECTS OF PLASTIC DEBRIS

Despite the fact that marine debris has only recently emerged as a serious environmental issue and that the effects of plastic debris in the ocean are far from being completely assessed, numerous studies document the negative effects of this material on the marine environment, and on recreational and commercial uses of marine waters.

Because of their buoyancy, long-term persistence, and ubiquity in the marine environment plastic wastes pose a variety of hazards to marine wildlife. Studies of the impacts of plastic debris on marine animals have been compiled by Shomura and Yoshida (1985). Although additional research is required to completely understand all the biological impacts of plastic debris on marine organisms, the physical effects of entanglement, suffocation, and starvation are often very apparent. In addition to the impacts on marine animals. plastic debris also aesthetically degrades the environment, and has impacts on our economy and on human health and safety.

3.4.1   Impacts on Marine Wildlife

Plastic materials in the marine environment, either as buoyant debris or deposited on the sea bottom, pose a variety of hazards to marine mammals, fish, turtles, crustaceans, and seabirds. The two major mechanisms by which plastic debris is known to impact marine species are entanglement and ingestion.

Entanglement typically refers to the encircling of body parts by various types of plastic debris that ensnare the animal. Plastic litter most often responsible for entanglement of marine life includes fragments of synthetic fish nets (commonly trawl and gill net webbing), monofilament fishing line, ropes, beverage container rings, rings and gaskets, and uncut polyethylene cargo strapping bands. The results of entanglement can be debilitation and death by drowning, loss of limbs through strangulation and infection, starvation, and increased susceptibility to predation. Historically, most of the studies of entanglement of marine organisms have been conducted on northern fur seals. However, an increasing volume of literature describes entanglement of other marine species as well. Entanglement has been reported for marine mammals, sea turtles, seabirds, fish, and crustaceans.

Ingestion refers to the consumption of plastic debris by marine organisms. Common plastic wastes known to be ingested by animals include small polyethylene pellets and polystyrene beads, and larger debris such as bags, balloons, and packaging materials. Ingestion of plastic debris may result in intestinal blockage, nutritional deficiencies due to a false feeling of satiation, suffocation, intestinal ulceration, and intestinal injury. Numerous reports in the literature indicate that a variety of seabirds, marine mammals, turtles, and fish ingest plastic materials.

It is not possible to estimate the threat posed by entanglement and ingestion to the populations of many species because not enough studies have been conducted to date. The problems associated with plastic debris in the marine environment have been recognized only recently and require further study. However, the existing data indicate that plastic debris in the marine environment harms numerous species and that some of the species affected are threatened or endangered. The persistence of plastic debris in the environment guarantees that populations will continue to be affected for a long time. Existing data can be used to identify the species whose populations are most likely to be affected adversely by plastics in the environment.

3.4.1.1 Entanglement

Entanglement of marine animals can result from accidental contact with debris or as a result of normal activities. Marine animals may be attracted to debris because prey species have already been attracted to or entangled in the debris. The animals may then become entangled in attempting to catch prey entrapped in or near debris, by attempting to rest on debris, or by playful contact with debris (of particular concern for juvenile animals) (Laist, 1987). The consequences of entanglement include drowning, reduced ability to catch food, reduced ability to escape predators, wounds and associated infections, or altered behavior patterns. Fishing-related gear poses the greatest threat and, therefore, entanglement may be of particular concern in the North Pacific Ocean where drift-net and trawl fishing is extensively employed (Laist, 1987).

MARINE MAMMALS -- The greatest scientific attention to effects of persistent marine debris has historically been directed toward entanglement of marine mammals. Species of seals, sea lions, and cetaceans are widely reported entangled in debris. The most common entangling debris items are, in decreasing order of importance, fishing nets and net fragments, uncut plastic strapping bands, ropes, and plastic sheeting (Laist, 1987). The fact that most reports of entanglements are from areas where fishing and marine transportation are common activities is consistent with findings that fishing gear and packing straps are the most common entangling materials (Interagency Task Force, 1988). The incidence of entanglement of northern fur seals in the Pacific Ocean has increased since it was first reported in the 1930s, with a noticeable increase in percent entanglement in the late 1960s, when commercial fishing efforts increased and when synthetic materials were commonly employed in the construction of fishing nets (Fowler, 1987).

Entanglement of an individual animal can restrict its normal activities such as feeding and swimming, and require the animal to expend more energy on these activities (CEE, 1987b). Other impacts include starvation if the animal is unable to capture prey, strangulation or severed carotid arteries if an entrapped animal grows into constricting debris, infection of wounds caused by entangling materials, drowning if swimming ability is impaired, increased vulnerability to predation, or a combination of these impacts (Fowler, 1987). Entangled seals spend more time at sea than seals that are not entangled (Fowler, 1987), swim at reduced speeds, and dive for shorter periods of time than nonentangled seals (Yoshida and Baba, 1988). Henderson (1988) reported that 49% of entangled Hawaiian monk seals were able to free themselves of debris; that number might have been higher if some seals that were assisted had been left to free themselves. Entangled northern fur seals were marked and, one year later, were resighted at the same rate (25%) as seals that were not entangled, indicating that entanglement does not increase mortality over a one-year period (Scordino, 1985). Eighteen percent of the entangled seals resighted in that study had freed themselves of the debris during the year.

It is more difficult to determine the effects of entanglement on a whole population of animals. Determining precisely how many animals in a population are or have been entangled is difficult because entangled animals can succumb to predators or drown and sink, making an accurate count impossible (Laist, 1987). In addition, the geographic range of the animals contributes to the difficulty in conducting such a study.

Scars and bruises which are believed to result from entanglement are often observed on marine mammals. These scars, found around the animals' necks and shoulders, are characteristic of encounters with entangling debris (Scordino, 1985). The scarred or bruised animals can be included in population counts as animals that are or have been entangled. Scordino (1985) has shown that scars and bruises often are not visibly apparent on northern fur seals, but become apparent when harvested seal skins are processed, indicating that counts of scarred animals may actually underestimate the number of seals that have been entangled. Even with these limitations, researchers have been able to estimate entanglement rates for some populations. Table 3-8 shows the entanglement rates, the number of animals entangled and the total sample sizes in three study locations and for four pinniped species. However, it is still difficult to determine the significance of entanglement rates on the local population of animals.

The northern fur seal population of the Alaskan Pribilof Islands has been studied relatively extensively, and entanglement has been related to declining numbers of these seals (Fowler, 1987; 1985). The Pribilof Island population has been declining at a rate of approximately 4-8% per year since the 1970s (Fowler, 1985). On St. Paul Island in the Pribilofs, the current incidence of entanglement for subadult male northern fur seals is about 0.4% (as shown in Table 3-8, 101 seals out of a total sample of approximately 25,000), which is two orders of magnitude greater than the rate determined in the 1940s (Fowler, 1987). Trawl webbing made up 62-72% of the entangling debris on St. Paul Island (Scordino, 1985). Fowler (1987; 1985) has related the decline in the seal population, the decline in the number of seal pups, and an unexpected increase in juvenile mortality to entanglement, particularly the entanglement of young seals, although further study is necessary to provide accurate estimates of mortality caused by entanglement.

Entanglement is believed to have an adverse impact on the endangered Hawaiian monk seal in the northwest Hawaiian Islands. The population of these seals has declined from an estimated 1000-1200 seals in the late 1950s to 500-625 in the mid to late 1970s (Kenyon, as cited in Laist, 1987). At least part of this decline is believed to be due to entanglement (Laist, 1987). Henderson (1988) found that Hawaiian monk seal pups became entangled at a higher rate than adult seals, with 41% of observed entanglements involving weaned pups. Because of their small size, young seals can become entrapped in net of smaller mesh sizes than entrap adult seals, making young seals more vulnerable to entrapment in a wider range of net mesh sizes (Merrell and Johnson, 1987). The pup seals' tendency toward exploration and their proximity to shore where debris concentrates may also contribute to elevated rates of entrapment (Henderson, 1988). The evidence that young seals become entangled at higher rates than adults is of

Table 3-8

OBSERVED PERCENT ENTANGLEMENTS FOR VARIOUS PINNIPED SPECIES

Pinniped Species	Location	Percent Entanglement	Number Entangled	Total Sampled
California Sea Lions*	San Nicolas Island, CA	0.14	41	28,919
California Sea Lions*	San Miguel Island, CA	0.22	15	6,905
Northern Elephant Seals*	San Nicolas Island, CA	0.17	18	10,870
Northern Elephant Seals*	San Miguel Island, CA	0.15	10	6,468
Harbor Seals*	San Nicolas Island, CA	0.11	2	1,900
Harbor Seals*	San Miguel Island, CA	0.07	1	1,494
Northern Fur Seals*	San Miguel Island, CA	0	0	826
Northern Fur Seals**	Pribilof Islands, AK	0.4	101	24,932

Sources: *Stewart and Yochem (1987).
            **Fowler (1985).

particular concern in terms of impacts on populations (Fowler, 1987; Henderson, 1988) because increasing rates of entanglement and the tendency for juveniles to become entangled could result in a decline in future birth rates (Stewart and Yochem, 1987).

Entanglement in marine debris has been observed for a number of other pinniped species in a variety of geographic locations. Fowler (1988) summarized available information on entangled species, and identified ten species of otariid seals (fur seals and sea lions) and six species of phocid seals (true seals and elephant seals) that have been reported as entangled (Table 3-9). For some of these species, relatively large numbers of individuals have become entangled. Fewer numbers of species and individuals within a species have been reported for phocids, and these entanglements are not considered to be of great significance. Fowler (1988) also summarizes explanations that have been offered for the different rates of entanglement for the otariid and phocid species; these include differences in body shape, behavior, and location of habitat. Phocids tend to live in high-latitude environments with less developed fisheries and presumably less fishing-related debris. And, because they have more rounded body shapes and larger necks in proportion to the head, entanglement may be minimized. Otariids are generally more playful and curious than phocids and, therefore, may tend to investigate and become entangled in debris at a higher rate.

A variety of explanations have been offered for entanglement of seals with plastic debris. It is possible that debris, which has entangled or attracted fish or other prey organisms, also attracts the seals and they themselves become entangled when attempting to feed on the prey (Laist, 1987). Objects present in the water attract fur seals; they commonly respond by inserting their heads through holes in debris (Fowler, 1987). In a study of captive seals, Yoshida and Baba (1988) found that adult seals often become entangled when they inadvertently swim into debris, but that young seals become entangled as a result of play activities.

Cetaceans primarily become entrapped in gill nets and buoy lines used to mark traps (Interagency Task Force, 1988). Entanglement of cetaceans in nets and trap lines usually involves active fishing gear (CEE, 1987b). Off the coast of New England, scars, presumed to be from entanglement, have been identified on 56% of photographed right whales and on 40% of humpback whales (Weirich, as cited in Interagency Task Force, 1988). Off the coast of the northeastern United States, 20 humpback, 15 minke, and 10 right whales were observed entangled in gill net or lobster pot lines between 1975 and 1986 (Laist, 1987). Along the Oregon coast, gray whales, about 16,000 of which migrate along the coast twice each year, have become entangled with fishing gear (Mate, 1985). In particular, these whales become entangled with crab pot lines; an average of two gray whales is reported entangled this way each year off the Oregon coast, and others have been reported entangled in their winter calving area off Baja, Mexico (Mate, 1985). Although much of this entanglement results from contact with active fishing gear, it is reasonable to conclude that lost and abandoned gear also entangles cetaceans.

Table 3-9

OTARIID AND PHOCID PINNIPED SPECIES OBSERVED ENTANGLED IN
PLASTIC MARINE DEBRIS

OTARIID SPECIES

Arctocephalis australis	South American fur seal
Arctocephalis forsteri	New Zealand fur seal
Arctocephalis gazella	Antarctic fur seal
Arctocephalis phillippi	Juan Fernandez fur seal
Arctocephalis pusillis	Cape or South African fur seal
Callorhinus ursinus	Northern fur seal
Eumetopias jubatus	Northern sea lion
Otaria flavescens	South American sea lion
Phocarctos hookeri	Hooker's sea lion
Zalophus califorianus	California sea lion

PHOCID SPECIES

Halichoerus grypus	Grey seal
Mirounga angustirostris	Northern elephant seal
Mirounga leonina	Southern elephant seal
Monachus schauinslandi	Hawaiian monk seal
Phoca groenlandica	Harp seal
Phoca vitulina	Harbor seal

Source: Fowler (1988).

Entanglements have additionally been reported for other marine mammals. Crab-pot lines entangle the West Indian manatee, an endangered species (Wallace, 1985). Approximately 5,000 Dall's porpoises are entangled and die each year in drift nets of the Japanese salmon fleet (Eisenbud, 1985). Eisenbud (1985) estimates that 0.06% or 639 miles of drift net are lost each year by the Japanese driftnet fishery, and it is likely that porpoises and other marine mammals become entangled in the debris.

TURTLES -- Marine turtles can become entangled in various types of plastic debris. Entanglement has been reported for each of the five species of sea turtles that inhabit U.S. waters. All of these species are listed as either threatened or endangered (Interagency Task Force, 1988). In Balazs's (1985) review of reported cases of turtle entanglement throughout the world, monofilament fishing line was the most common entangling debris item. Rope, trawl webbing, and monofilament net were other types of entangling debris (Table 3-10). A total of 68% of reported entanglements involved materials associated with the fishing industry. Turtles may be attracted to floating masses of net for shelter and concentrated food, as they are attracted to sargassum mats, increasing the probability of entanglement as the turtles swim near the netting (Balazs, 1985). Most entangled turtles are not able to function normally and suffer a variety of effects including drowning, reduced swimming efficiency, reduced ability to escape from predators, lacerated appendages, and limb necrosis (Interagency Task Force, 1988; Balazs, 1985).

Balazs's review identified 60 reports of turtle entanglements worldwide, 95% of which occurred after 1970. This statistic may correspond to the introduction of synthetically constructed fishing nets, which were in common use by the early 1970s (Fowler, 1987; Pruter, 1987). The green turtle was involved in 42% of the entanglements reported in Balazs' review, and the general trend was for immature turtles to be more frequently involved than adults (Table 3-11). Most of the reported turtle entanglement cases in the United States have been on the eastern, southeastern, and Gulf coasts and in Hawaii. During the 1986 and 1987 beach surveys, 25 entangled turtles were observed (CEE, 1988). Figure 3-12 shows the locations of reported incidents of entanglement in the United States. (The pattern observed is partly due to the limited available data. A more accurate picture of turtle entanglement will emerge as more studies are conducted.)

Recent evidence has indicated that juvenile turtles may spend from three to five years in a pelagic stage, during which they drift in surface waters (Carr, 1987). The young turtles prefer to concentrate along areas of current convergence or gyres, in which high concentrations of plastic debris, including floating nets and lines, accumulate (Interagency Task Force, 1988). The evidence that young turtles may drift along areas of current convergences for extended periods of time, thus increasing their likelihood of contacting debris, heightens concerns about the potential impacts of debris on turtle populations (Carr, 1987).

BIRDS -- There are three types of plastic debris in which seabirds become entangled. Trash and net fragments have openings that can trap the bird's head, feet, and wings; lengths of monofilament line and string can wrap around the wings, beak, and feet; and large pieces of netting can entangle the bird, causing immediate drowning (Wallace, 1985). Entanglement in

Table 3-10

PERCENT OCCURRENCE OF TYPES OF DEBRIS FOUND
ENTANGLED ON MARINE TURTLES

Type of Debris	Percent Entanglement*
Monofilament fishing line	33.3
Rope	23.3
Trawl net	20.0
Monofilament net	13.3
Plastic sheets or bags	3.3
Plastic objects	1.7
Line with hook	1.7
Cloth	1.7
Parachute anchor	1.7

Source: Balazs (1985).

* Sample size = 60

Table 3-11

AGE DISTRIBUTION OF MARINE TURTLES BECOMING
ENTANGLED IN MARINE DEBRIS

Species	Percent of Cases		Sample Size
	Adult	Immature	
Green turtle	42	58	24
Loggerhead	0	100	4
Hawksbill	11	89	9
Olive ridley	50	50	4
Leatherback	100	0	7
All species	42	58	48

Source: Balazs (1985).

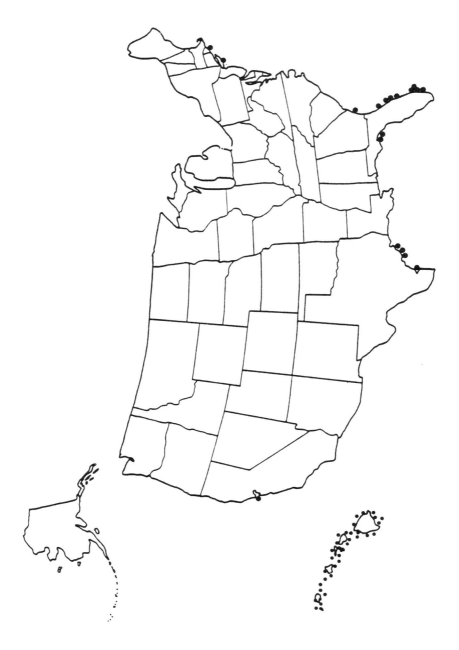

FIGURE 3-12   LOCATIONS OF REPORTED MARINE TURTLE ENTANGLEMENTS (adapted from Balazs, 1985)

derelict fishing nets is a major concern for seabirds. Studies have shown that large numbers of seabirds are caught and killed in active fish nets, and derelict nets continue to entrap seabirds as they drift (Eisenbud, 1985). Nets can be lost, abandoned, or discarded in the marine environment, where they continue to catch fish, the phenomenon called "ghost-fishing." Birds are attracted to the entrapped fish as prey, and the birds themselves can then also become entangled in the nets. Seabirds are also entrapped in monofilament fishing line, beverage container rings, and pieces of net. Entangled bird carcasses washed up on beaches are frequently noted by researchers and during beach cleanup efforts (CEE, 1987a; Piatt and Nettleship, 1987).

It is difficult to assess the impacts of bird deaths by entanglement on the population of a particular species. Populations of certain seabirds, including gannets, razorbills, and common guillemots, have been shown to be significantly impacted by entanglement in active fishing nets, whereas other populations, such as cormorants and puffins, are not significantly impacted (Piatt and Nettleship, 1987). A study of gannets in Germany estimated that at least 2.6% of the population was entangled but still able to fly, and that approximately 13-29% of observed gannet mortality was due to entanglement, although it is not believed that the population is being significantly affected by entanglement (Schrey and Vauk, 1987). The brown pelican, an endangered species, is significantly impacted by entanglement in monofilament fishing line (Wallace, 1985). Further information and studies on the impacts of entanglement on seabirds must be assessed in order to determine the effects of entanglement on seabird populations.

FISH AND CRUSTACEANS -- The phenomenon known as "ghost fishing" is responsible for the entrapment and death of large numbers of finfish and shellfish. Ghost fishing occurs when lost or discarded fishing gear continues to fish until the gear deteriorates or is rendered ineffective. If lost or abandoned, surface and bottom-set gill nets and traps can all continue to ghost fish. The gear can entrap fish or crustaceans, which in turn attract and entangle other wildlife. Other forms of entanglement of fish and crustaceans can occur, such as the entanglement of manta rays in monofilament line (Wallace, 1985), but these impacts are considered minor relative to the impacts of ghost fishing.

It is difficult to quantify the impacts of ghost fishing, but large quantities of fishing gear are known to be lost each year (Uchida, 1985; CEE, 1987b). High (1985) reports that Alaskan fisherman lose approximately 10% of their crab pots each year and estimates that 30,000 derelict pots may continue to be operating in Alaskan fishing grounds. Experiments show that about 20% of legal-size king crabs and 8% of sublegal-size crabs can eventually escape from the traps (High, 1985). However, given the numbers of pots estimated to be lost, a significant number of crabs may still be taken. There is evidence that crabs confined in pots, for periods of at least ten days before escape, experience increased mortality (High, 1985). In 1978, more than 500,000 lobster traps were reported lost in New England. These traps have the capacity to trap over one million pounds of lobster in a single year (CEE, 1987b).

Fragments of derelict gill nets have been reported to contain dead fish, sometimes in large numbers, and other marine organisms (CEE, 1987b; High, 1985; Wallace, 1985). As an estimate of the potential magnitude of the lost gill net problem, Eisenbud (1985) reported that the Japanese pelagic drift-net salmon fishery loses an average of 12 miles of net per night.

Examination of a portion of net 1500 m long showed more than 200 entangled salmon (Wallace, 1985). Abandoned nets and traps can continue to ghost fish for many years. In a study of derelict salmon nets in Puget Sound, High (1985) found that fish continued to be caught for more than 3 years and that crabs were still being entangled after six years.

The quantities and types of debris, in relation to the geographic distribution of various species, determine which species are affected in various regions of the United States. Marine mammals become entangled in areas where fishing and marine transportation are common. A high rate of entanglement is reported for seals in Alaska, where trawl-net fishing is widespread. Whales become entangled in gill nets and lobster-pot lines in the northeastern United States, and in crab-pot lines and fishing gear in the Pacific northwest. Most reports of turtle entanglements come from the southeast Atlantic and the Gulf coasts as well as from Hawaii. Ghost fishing affects crustaceans, fish, and birds, which become entrapped in derelict fishing nets. Concern over entrapment of crabs, fish, and lobsters in these nets has focused on the northeastern and northwestern United States, including Alaska, regions of intense fishing activity. As more and more studies are conducted and as data are compiled, a clearer picture of the regional importance of entanglement and the respective susceptible species will emerge.

In summary, entanglement of wildlife in persistent marine debris has been reported on all three coasts of the contiguous United States, and in Alaska and Hawaii. The greatest threat for most species is posed by fishing-related debris, including nets, lines, and traps. As a result, areas of concentrated fishing activity are of particular concern for wildlife entanglement.

3.4.1.2 Ingestion

It is likely that ingestion of persistent marine debris is closely related to feeding behavior of animals. Debris may resemble natural prey, or may be covered with organisms that result in the animal misidentifying the debris as natural material. Some animals may inadvertently ingest debris while feeding on other materials, as may occur with a filter-feeding whale, for example.

Plastic pellets and beads, small fragments of plastic bags and sheeting, and other forms of debris are ingested by marine wildlife. The consequences of debris ingestion can be quite severe, and include inadequate nutrition, internal injury or blockage, and suffocation. Because these materials are ubiquitous in distribution, ingestion of plastic debris is of concern throughout the marine environment.

MARINE MAMMALS -- Ingestion of plastic debris by marine mammals has been documented in only a relatively small number of cases. The deaths of one elephant seal and one Steller sea lion due to choking on foamed polystyrene have been reported off the coast of Oregon (Mate, 1985). In an examination of the stomachs of 38 sperm whales stranded on the Oregon coast, one was found to contain about 1 liter of trawl net (Harvey, as cited in Mate, 1985). The stomachs of 1500 pelagic cetaceans, including porpoises, dolphins, and whales, were examined and none were found to contain plastic (Walker, as cited in Interagency Task Force, 1988). There are reports from the Atlantic, Pacific, and Gulf coasts of the United States, of individuals of a variety of cetacean species that had ingested plastic debris (CEE, 1987b). In 1985, the death of a sperm whale in New Jersey was attributed to a mylar balloon blocking its intestinal tract (Audubon, 1988). The death of a Minke whale off the coast of Texas may have been caused by plastic sheeting found in its digestive tract (Sport Fishing Institute, 1988). The autopsy of an infant pygmy sperm whale, found orphaned off the coast of Texas, showed that the number of plastic bags filling its stomach resulted in starvation (CEE, 1987a). Laist (1987) describes a report of two endangered West Indian manatees that had died from ingestion of debris.

TURTLES -- Marine turtles ingest a variety of plastic materials. Balazs (1985) reported 79 cases worldwide of persistent marine debris in the digestive tracts of turtles. Plastic bags or sheets of plastic represented 32%, tar balls represented 21%, and plastic particles accounted for 19% of the reported cases of plastic ingestion. During the 1986 and 1987 Texas beach surveys, 35 turtles were found with persistent marine debris in either their mouths, throats, stomachs, or intestines (CEE, 1988). Plastic bags were the most common type of debris ingested.

Turtles are believed to ingest plastic materials as part of their feeding behavior. The debris may resemble food in size, shape, and movement, and is often covered with natural growth that may attract the turtles while disguising the nature of the plastic (Balazs, 1985). Various types of debris covered with fish eggs and mussels have been reported in the stomach of turtles (Balazs, 1985). Leatherback turtles ingest sheets of plastic film and plastic bags that are mistaken for jellyfish, a primary source of food for turtles (Carr, 1987). That the turtles mistake plastic materials for jellyfish is supported by a study in which the alimentary canal and feces of loggerhead turtles captured in the Mediterranean Sea were examined and found to contain only translucent white plastic pieces, the color of jellyfish, even though various colors of plastic materials are available to turtles in the environment (Gramentz, 1988). Of the cases of ingestion of debris documented by Balazs (1985), the green turtle was most commonly involved, followed by, in decreasing order, loggerhead, leatherback, hawksbill, and a few Kemp's ridley turtles. As is the case for entanglement, there is a higher frequency of involvement for immature turtles for most species, with the exception of the leatherback turtle (Table 3-12).

There have been many reports of ingestion of persistent marine debris by turtles along the coasts of the continental United States (Balazs, 1985). Figure 3-13 maps the locations of such reports. As with entanglement, the observed pattern is partly due to the limited available data.

Young sea turtles may spend from three to five years in an epipelagic period, during which they drift with currents and feed at the surface (Carr, 1987). The young turtles migrate to areas where currents and marine debris converge. Drifting plastic debris, particularly small plastic pellets, resembles the sargassum floats that are a food source for young turtles. The stomachs of young loggerhead turtles washed ashore in Florida contained plastic beads similar in size and shape to sargassum floats (Carr, 1987). Because young turtles are attracted to areas with high concentrations of plastic pellets and debris, there are concerns about the impacts of ingestion of this debris on declining turtle populations.

Table 3-12

AGE DISTRIBUTION OF MARINE TURTLES INGESTING
ENTANGLED IN MARINE DEBRIS

Species	Percent of Cases		Sample Size
	Adult	Immature	
Green turtle	19	81	21
Loggerhead	19	81	15
Hawksbill	9	91	11
Olive ridley	100	0	11
Leatherback	0	100	3
All species	31	69	62

Source: Balazs (1985).

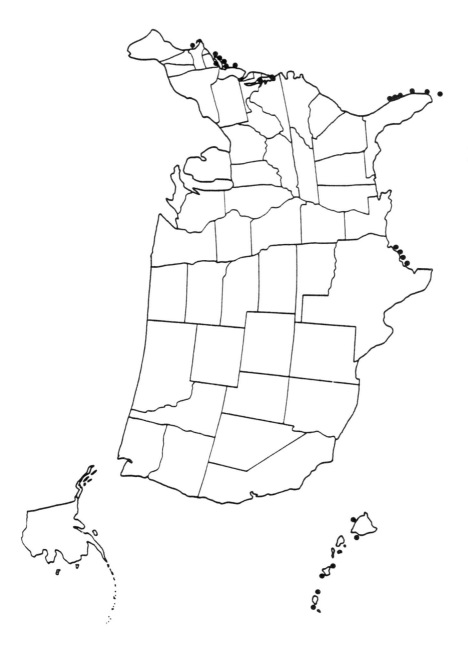

FIGURE 3-13   LOCATIONS OF REPORTED INGESTION OF PLASTIC MATERIALS BY
MARINE TURTLES (adapted from Balazs, 1985)

The extent of occurrence and impacts of ingestion of plastic debris by young and adult turtles is not clear. The National Marine Fisheries Service reported that one-third to one-half of necropsied turtles contained plastic products (Interagency Task Force, 1988). It is not known if the ingested plastic was the cause of death. There is some evidence that ingested plastic materials do not always completely obstruct the digestive tract, but may be voided naturally by the turtles. Balazs (1985) reported two cases of living turtles with plastic sheeting protruding from their cloacas. However, the presence of plastic debris in the digestive system can result in lost nutrition, reduced ability to absorb nutrients, the possibility of absorption of toxic compounds present in plastic materials, and reduced ability to dive due to the buoyancy of plastic materials (Balazs, 1985; Bauer, 1986). To date, there has been one reported turtle death resulting directly from the ingestion of plastic material (Interagency Task Force, 1988).

The issue of ingestion of balloons by marine turtles has become one of great concern recently. Releases of large quantities of helium-filled balloons have long been an attractive, crowd-pleasing spectacle whose consequences have been considered only recently. The released balloons can be transported great distances and eventually fall to the earth as litter; those released near the coast can land in the ocean. Most balloons are made of latex rubber; some balloons, however, are made of plastic (such as Mylar). In the marine environment, plastic and rubber balloons pose many of the same threats to turtles that other plastic debris does. The death of a leatherback turtle in 1988 in New Jersey was linked to a balloon and 3 feet of ribbon found blocking the animal's digestive system (Smith, 1988). The neck of a latex balloon was part of the 1 lb of plastic debris found in the intestinal tract of an 11-lb hawksbill turtle in Hawaii (Bauer, 1986). Some of the turtles found during the Texas beach surveys had pieces of balloons in their digestive tracts (CEE, 1988).

The public is becoming aware of the dangers that balloons present to marine animals. Due to pressure from environmental groups, citing the potential harm to marine organisms, some organizations, such as the Triangle Coalition for Science and Technology Education and the Arthritis Foundation of Hawaii have cancelled scheduled balloon releases. Some states are considering legislation that would prohibit the release of large quantities of balloons; New Jersey has a bill pending and Connecticut and Massachusetts are considering submitting bills. However, releases of large quantities of balloons continue to be a part of many outdoor activities such as football games, business openings, and amusement park entertainments.

BIRDS -- A great deal of attention has been focused on the ingestion of plastic materials by seabirds. This phenomenon has been reported in 50 species of seabirds worldwide (Table 3-13; Day et al., 1985). For some bird populations, a large percentage of examined birds had ingested plastic debris. For example, 87% of fulmars examined on the Dutch coast in 1983 and 60% of shearwaters examined in Hawaii in 1982 and 1983 contained plastic materials (Fry et al., 1987; Van Franeker, 1985).

The ingestion of plastic debris by seabirds is related to their feeding behavior. The most commonly ingested types of debris are small, floating plastic pellets. Birds that feed by seizing food on the water surface or by pursuit-diving have the highest rates of plastic ingestion, and are believed to mistake plastic pellets for food sources such as pelagic eggs, the eyes of squid

Table 3-13

MARINE BIRD SPECIES RECORDED AS HAVING INGESTED PLASTIC

Wandering albatross	Leach's storm petrel
Royal albatross	Sooty storm petrel
Black-footed albatross	Fork-tailed storm petrel
Laysan albatross	Blue-footed booby
Gray-headed albatross	Red-necked phalarope
Northern fulmar	Red phalarope
Great-winged petrel	Laughing gull
Kerguelen petrel	Heerman's gull
Bonin petrel	Mew gull
Cook's petrel	Herring gul
Blue petrel	Western gull
Broad-billed prion	Glaucous-winged gull
Salvin's prion	Glaucous gull
Antarctic prion	Great black-backed gull
Fairy prion	Black-legged kittiwake
Bulwer's petrel	Red-legged kittiwake
White-chinned petrel	"Terns"
Parkinson's petrel	Dovekie
Pink-footed shearwater	Thick-billed murre
Greater shearwater	Cassin's auklet
Sooty shearwater	Parakeet auklet
Short-tailed shearwater	Least auklet
Manx shearwater	Rhinoceros auklet
White-faced storm petrel	Tufted puffin
British storm petrel	Horned puffin

Source: Day et al. (1985).

and fish, or the bodies of larval fish (Day et al., 1985). Twenty-six percent of Alaskan birds that feed by pursuit-diving, 16% that feed by surface-seizing, 9% that feed by dipping, and none that feed by plunging or piracy contained plastic debris (Day et al., 1985). Day et al. also found that birds that feed primarily on crustaceans or cephalopods ingest more plastic debris than birds that feed primarily on fish. In addition, certain shapes and colors of plastic materials, presumably those that resemble food items, are ingested at higher rates by certain species of birds.

Some species of birds, such as gulls and terns that regurgitate food, are able to clear themselves of debris (Wallace, 1985). The problems are more significant for those species not able to rid themselves of the plastic materials they have ingested. Also, reports indicate that three species of birds that prey on seabirds, the bald eagle, the Antarctic skua, and the short-eared owl, have ingested plastic materials by feeding on prey containing plastic debris (Day et al., 1985).

There are both direct and indirect effects of ingestion of plastic waste materials (Day et al., 1985). Direct effects include starvation resulting from decreased feeding activity associated with stomach distension, intestinal blockage, and ulceration or internal injury to the digestive tract. Indirect effects include decreased reproductive or physical health of the bird due to the presence of plastic materials or pollutants associated with the debris. A controlled study of the effects of plastic ingestion on chickens showed that birds fed plastic materials ate less than those not fed plastic, probably because of reduced gizzard volume (Ryan, 1988). The reduced consumption of food may limit the ability of the bird to store fat and, thereby, reduce its ability to survive and reproduce.

The feeding of regurgitated plastic materials to young seabirds by adult birds can result in adverse impacts on the young. Ninety percent of Laysan albatross chicks examined in Hawaii in 1982 and 1983 had plastic pellets in their upper digestive tracts; these pellets caused obstruction of the gut, ulceration, and starvation (Table 3-14; Fry et al., 1987). Higher rates of plastic debris consumption by young birds has been noted for several species (Day et al., 1985).

Concern has been raised about ingestion of plastic debris and the potential toxicity of chemicals such as polychlorinated biphenyls (PCBs) and other organochlorine compounds that are either used in the manufacture of plastic polymers or adsorb to plastic materials (Ryan et al., 1988; Van Franeker 1985). Hydrocarbons are suspected of adversely affecting reproduction in birds, and ingestion of hydrocarbons associated with plastic materials may, therefore, have an impact on reproductive success (Day et al., 1985). Fry et al. (1987) suggested that obstruction and impaction of the bird's gut is of much greater concern than the toxicity of plastic materials, but additional research is needed to better define potential adverse impacts.

Ingestion of plastic debris may be higher in regions of plastic production. Day et al. (1985) compared data for bird species from Alaska, an area of low plastic production, with the same bird species in California, an area of high plastic production (Table 3-15). His study indicated that fewer of the Alaskan species contained ingested plastic debris and at lower concentrations than found in the California birds. However, because plastic debris can be transported by various mechanisms once it enters the marine environment, ingestion is not only of concern in areas where plastic is produced (Day et al., 1985).

Table 3-14

INCIDENCE OF PLASTIC INGESTION BY LAYSAN ALBATROSS
CHICKS, NORTH WESTERN HAWAIIAN ISLANDS

| Date | Chicks with Ingested Plastic | | Sample Size |
	Number	Percent	
August, 1982	75	3	4
April, 1983	94	16	17
May, 1983	100	5	5
July, 1983	87	21	24
Average	90	45	50

Source: Fry et al. (1987).

Table 3-15

COMPARISON OF PLASTIC INGESTION IN SEABIRD
SPECIES EXAMINED IN CALIFORNIA AND ALASKA

Species	Alaska		California	
	% with Ingested Plastics	Sample Size	% with Ingested Plastics	Sample Size
Northern fulmar	38	58	3	100
Sooty shearwater	76	43	21	43-67
Short-tailed shearwater	200	84	6	100
Mew gull	10	0	4	25
Glaucous-winged gull	63	0	8	13
Black-legged kittiwake	188	5	8	13-25
Rhinoceros auklet	20	0	26	4

Source: Day et al. (1985).

FISH -- Plastic materials have been reported in the digestive tracts of a variety of fish (CEE, 1987b; Wallace, 1985), although the reports tend to be anecdotal. However, there are no data to indicate that significant harm or mortality occurs as a result of ingestion of plastic debris by fish.

In summary, ingestion of plastic materials has been reported on all coasts of the U.S. and Alaska and Hawaii. Marine turtles ingest plastic bags or sheets of plastic, plastic pellets and balloons. Birds ingest plastic pellets or pieces most frequently. Ingestion is a threat when plastic materials are present where marine animals eat and particularly when the debris has a similar appearance to the animals' food.

3.4.2   Aesthetic and Economic Effects

The most noticeable impacts of plastic debris on the environment are degraded aesthetics of the coastal waterways and shorelines. Floating debris, either in massive slicks or as dispersed items, is visually unappealing and poses marine safety threats. Similarly, debris stranded on beaches and shorelines seriously degrades the coastal environment, resulting in economic losses due to the decline in tourism. Littered beaches along the Atlantic coast have, in the past, been closed solely because of objectionable aesthetics.

Plastic litter in the water and on shorelines can cause serious negative impacts on both commercial and recreational activities, including fishing and fishing resources, vessel operation, and beach use. The magnitude of these impacts is often difficult to quantify. The consequences of aesthetic deterioration of one area are borne by the entire regional population.

The loss of fishing gear has economic impacts on the fishing industry. It is difficult to estimate the value of lost gear because accurate records of gear losses are not available. In the Gulf of Maine, where conflicts between recreational and commercial fishing interests are intense, an estimated $50,000 worth of equipment and $1,000,000 in operating expenses are lost each year by party-boat operators because gear becomes entangled in monofilament gill nets and lost (CEE, 1987b).

Lost fishing gear also impacts fishery resources by continuing to ghost fish for many years after it is lost. Both nets and traps can continue to indiscriminately ghost fish, and commercially important species are removed from the total stock available for commercial catches. The economic impact of lost lobster traps in New England was estimated in 1978 to be almost $250 million, representing 1.5 million pounds of lobster in lost or abandoned traps (Smolowitz, as cited in CEE, 1987b).

Wallace (1985) suggested that plastic debris can also have impacts on activities involving birds. The debris can impact recreational activities such as birdwatching. Birdwatching in the Alaskan Aleut communities of St. Paul and St. George Islands generates hundreds of thousands of dollars in revenue (Wallace, 1985). The effects of plastic debris on the bird populations and on the aesthetics of birdwatching could have an adverse impact on this tourist industry.

Plastic debris can interfere with normal operations of military, commercial, and recreational vessels. A variety of plastic materials, including gill nets, garbage bags, plastic sheeting, and monofilament line, can foul propellers and clog cooling water intake systems (CEE, 1987b). Costs associated with these types of problems include both the cost of repairing the damage to the vessel and the costs associated with loss of operating time. Although there are no records to document the frequency of vessel damage by floating debris, the fact that some boat builders are now installing devices on propellers to alleviate the problem may suggest that such incidents occur often enough to be a source of concern (CEE, 1987b).

Plastic debris on beaches is aesthetically unpleasing and can result in significant economic impacts on local businesses, communities, and governmental budgets. (While the tourist dollars not spent at seaside resorts are spent elsewhere, in this report the economic effects on the coastal communities alone are considered.) Many communities spend money to routinely clean debris from their beaches. These efforts are not without substantial cost. Padre Island, Texas spends over $10,000 per year on beach cleanup efforts (CEE, 1987b); 64% of the litter items collected during a beach clean-up were plastic (CEE, 1987b). New Jersey currently collects an estimated 26,000 cubic yards of trash per season from state beaches at a total cost of approximately $2 million to the coastal communities. In some areas, officers have been hired to patrol beaches in an effort to reduce disposal of plastic debris on the beach (CEE, 1987b).

Discretionary beach closings are sometimes necessary due to the presence of floating or stranded litter. In 1976, most of Long Island's public beaches were closed for varying periods because of floatable trash (NOAA/MESA, 1977). More recently, well-publicized incidents of floating and beached hospital waste along the east coast have increased public awareness of the severity of the debris problem and have prompted governmental efforts at all levels to mitigate the floating waste problems.

Preliminary studies have estimated economic losses due to the debris incidents of 1987 and 1988 along the Atlantic coast. One study reports that an estimated $1 billion were lost over the last two summers because of decreased tourism along the Jersey shore (R.L. Associates, 1988). In Ocean Grove, New Jersey, summer beach attendance declined from 1200 people per day in 1987 to 120 people per day in 1988 (Swanson, 1988). Overall, in a comparison of the 1987 and 1988 tourist seasons, it was found that 22% fewer persons travelled to the New Jersey shore and spent 24% fewer days there in 1988 (R.L. Associates, 1988). Also, total expenditures were down 9% in the 1988 tourist season. In Seaside Heights, New Jersey, property taxes were increased 15% to make up for anticipated revenues not generated because of decreased beach use in 1987 (Swanson, 1988). At the two most popular beaches on Long Island attendance between July 7-17 was down 50% in 1988 from that in 1987 (Swanson, 1988). Inns and restaurants in this area were experiencing losses of 50% of their business.

Data quantifying economic impacts of floatable debris pollution were also collected during the 1976 Long Island incident. Debris deposited from the floating slick during one month alone was removed at a cost of $100,000. Long Island businesses affected by decreased tourism suffered major losses; restaurants, bait and tackle shops, and the pier fishing business reported 20-30% declines. Public beach attendance decreased 30-50% during and after the incident. The total economic loss to business was estimated at $30 million (Squires, 1982).

3.4.3   Effects on Human Health and Safety

Floatable waste not only results in loss of aesthetic qualities, decreased recreational opportunities, and adverse economic impacts, but may also pose threats to human safety in the marine environment. An obvious hazard of marine debris to both commercial and recreational activities is the potential of collision with large floating objects. Because such incidents are largely unreported, frequencies are impossible to assess. Vessel disablement by floating debris, for example, may endanger human safety if power or steering control is lost. It is believed that some loss of human lives during storms in the Bering Sea resulted from loss of ship engine power or maneuvering ability due to fouling of propellers, shafts, or intakes of vessels (Wallace. 1985). Submarines are susceptible to entanglement in marine debris, particularly in gill nets. endangering the lives of the crew (CEE, 1987b). Encounters of gill nets with research and military vessels have also been reported (Evans, 1971).

Entanglement of divers in marine debris may also result in injuries or fatalities. Monofilament line and nets can entrap recreational and professional divers. Recreational divers are often not adequately equipped to free themselves and entanglement poses a danger even to divers trained in escape procedures (High, 1985).

3.5   SUMMARY

This chapter has identified the major sources of plastic debris in the marine environment and described the effects of the debris. Based on the types of debris found in the environment. important land-based sources appear to include operations associated with the (1) disposal of solid waste and sewage generated on land (e.g., from CSOs) and (2) plastic manufacturing. fabricating and related transportation activities. Important marine sources include fishing gear from commercial fishing operations and domestic waste generated by all vessels. The presence of certain types of marine waste in the marine environment, such as medical wastes, indicates that illegal disposal of waste is also occurring.

From the various types of plastic waste found in the marine environment, EPA identified several Articles of Concern. These articles are those plastic wastes that pose the greatest threat to human safety, marine wildlife, or aesthetics or economics. The articles selected include beverage ring carrier devices, tampon applicators, condoms, syringes (either whole or pieces), plastic pellets and spherules, foamed polystyrene spheres, plastic bags and sheeting, uncut strapping bands, fishing nets and traps, and monofilament lines and rope. While these articles are among the most evident plastic wastes that contribute to marine pollution, the findings of this chapter suggest that the entire range of plastic wastes found in marine waters are of concern.

The fate of plastic debris in the marine environment is primarily dependent on oceanographic and meteorologic conditions. Degradation does not have a significant effect on the quantity of plastic wastes in the marine environment.

The most important effects of plastic debris are the hazards presented to marine wildlife and the economic losses incurred due to debris on public beaches. Entanglement of marine animals in discarded fishing gear is another important problem. The potential threat to northern fur seals has been well-studied. The actual loss of fish and crustacean resources to derelict fishing gear must be studied. Other species whose populations may be affected by entanglement are cetaceans and various species of seabirds. Ingestion of plastic material appears to present the biggest threat to turtles and seabirds. The harm suffered by turtles due to ingestion of plastics, coupled with the threats from entanglement, are troublesome because all species of turtles in North America are threatened or endangered.

There are two major economic effects of plastic debris in the marine environment: (1) the loss of fish and crustacean resources to ghost fishing, and (2) the losses resulting from aesthetic degradation of public beaches due to the presence of plastic debris. Losses from ghost fishing have not yet been quantified. Losses of tourist revenues due to the aesthetic degradation of beaches have been estimated and found to be significant to the coastal communities.

REFERENCES

Ad Hoc Advisory Committee on Plastics. 1988. Reducing Navy Marine Plastic Pollution. Report to the Assistant Secretary of the Navy for Shipbuilding and Logistics. June 28, 1988. 54 p.

Alaska Sea Grant College Program. 1988. Oceans of Plastic--A Workshop on Fisheries Generated Marine Debris and Derelict Fishing Gear, Portland, OR, February 9-11, 1988. 615 p.

Andrady, A.L. 1988. Experimental demonstration of controlled photodegradation of relevant plastic compositions under marine environmental conditions. Report prepared for the U.S. Department of Commerce, National Oceanic and Atmospheric Administration, Northwest and Alaskan Fisheries Center. Seattle, WA. 88-19. 68 p.

Audubon. 1988. Halftime balloons: A pretty problem. September 1988. p. 18.

Balazs, G.H. 1985. Impact of ocean debris on marine turtles: Entanglement and ingestion. In: R.S. Shomura and H.O. Yoshida (eds). Proceedings of the Workshop on Fate and Impact of Marine Debris. Honolulu, HI. Nov 27-29, 1984. pp. 387-429

Baltz, D.M. and G.V. Morejohn. 1976. Evidence from seabirds of plastic particle pollution off central California. Western Birds 7:111-112.

Battelle. 1989. Study of floatable debris in U.S. harbors. In preparation.

Bauer, D. 1986. Plastic pollution: A persistent problem. Makai 8. Newsletter of University of Hawaii Sea Grant College Program.

Boston Globe. 1988. Medical wastes: Following the trail (Part 1). Volume 234, Number 65. September 4, 1988.

Cantin, J., J. Eyraud, and C. Fenton. 1989. Quantitative Estimates of Garbage Generation and Disposal in the U.S. Maritime Sectors Before and After MARPOL Annex V. Second International Conference on Marine Debris. Honolulu, Hawaii. April, 1989.

Carpenter, E.J. and K.L. Smith, Jr. 1972. Plastics on the Sargasso Sea surface. Science 175:1240-1241.

Carr, A. 1987. Impact of nondegradable marine debris on the ecology and survival outlook of sea turtles. Marine Pollution Bulletin 18(6B):352-356.

CEE. 1987a. Center for Environmental Education. 1986 Texas Coastal Cleanup Report. 52 pages.

CEE. 1987b. Center for Environmental Education. Plastics in the Ocean: More than a Litter Problem. Report prepared for the U.S. Environmental Protection Agency under Contract No. 68-02-4228. 128 p.

CEE. 1987c. Center for Environmental Education. A Review of Data Collected during 1987 Beach Cleanups. Draft Report. 11 p. with appendices.

CEE. 1988. Center for Environmental Education. Texas Coastal Cleanup Report. 105 p.

CMC. 1989. Center for Marine Conservation. Personal communication between Margarete Steinhauer of Battelle Ocean Sciences and Kathryn O'Hara, CMC, regarding preliminary data from CEE's National Marine Data Base. 1989.

Coleman, F.C. and D.H.S. Wehle. 1984. Plastic pollution: A worldwide oceanic problem. Parks 9:9-12.

Colton, J.B. 1974. Plastics in the ocean. Oceanus 18(1):61-64.

Colton, J.B., F.D. Knapp, and B.R. Burns. 1974. Plastic particles in surface water of the northwestern Atlantic. Science 185:491-497.

Dahlberg, M.L. and R.H. Day. 1985. Observations of man-made objects on the surface of the north Pacific Ocean. In: R.S. Shomura and H.O. Yoshida (eds). Proceedings of the Workshop on Fate and Impact of Marine Debris. Honolulu, HI. Nov 27-29, 1984. pp. 198-212

Day, R.H. 1980. The Occurrence and Characteristics of Plastic Pollution in Alaska's Marine Birds. M.S. Thesis, University of Alaska, Fairbanks, AK. 111 p.

Day, R.H., D.H.S. Wehle, and F.C. Coleman. 1985. Ingestion of plastic pollutants by marine birds. In: R.S. Shomura and H.O. Yoshida (eds). Proceedings of the Workshop on Fate and Impact of Marine Debris. Honolulu, HI. Nov 27-29, 1984. pp. 344-386

Day, R.H. and D.G. Shaw. 1987. Patterns in the abundance of pelagic plastic and tar in the north Pacific Ocean, 1976-1985. Marine Pollution Bulletin 18(6B):311-316.

ERG. 1988. Eastern Research Group, Inc. An Economic Evaluation and Environmental Assessment of Regulations Implementing Annex V to MARPOL 73/78. Under contract to the U.S. Coast Guard. Dec 1988.

ERG. 1989. Eastern Research Group, Inc. Development of Estimates of Garbage Disposal in the Maritime Sectors. Prepared for the Transportation Systems Center, Research and Special Programs Administration. U.S. Department of Transportation. March 3, 1989.

Eisenbud, R. 1985. Problems and prospects for the pelagic driftnet. Environmental Affairs 12:473-490.

Evans, W.E. 1971. Potential hazards of non-degradable materials as an environmental pollutant. In: Proceedings of the Naval Underwater Center Symposium on Environmental Preservation. San Diego, CA. May 20-21, 1970. pp. 125-130.

Fowler, C.W. 1985. An evaluation of the role of entanglement in the population dynamics of northern fur seals on the Pribilof Islands. In: R.S. Shomura and H.O. Yoshida (eds). Proceedings of the Workshop on Fate and Impact of Marine Debris. Honolulu, HI. Nov 27-29, 1984. pp. 291-307.

Fowler, C.W. 1987. Marine debris and northern fur seals: A case study. Marine Pollution Bulletin 18(6B):326-335.

Fowler, C.W. 1988. A review of seal and sea lion entanglement in marine fishing debris. Pages 16-63 In: D.L. Alverson and J.A. June (eds). Proceedings of the North Pacific Rim Fishermen's Conference on Marine Debris. Kailua-Kona, HI. Oct 13-16, 1987. pp. 16-63.

Franklin Associates. 1988. Characterization of Municipal Solid Waste in the United States, 1960 to 2000 (update 1988). Prepared for U. S. Environmental Protection Agency. Contract No. 68-01-7310. Franklin Associates, Ltd. Prairie Village, KS. Mar 30, 1988.

Fry, D.M., S.I. Fefer, and L. Sileo. 1987. Ingestion of plastic debris by laysan albatrosses and wedge-tailed shearwaters in the Hawaiian Islands. Marine Pollution Bulletin 18(6B):339-343.

Gramentz, D. 1988. Involvement of loggerhead turtle with the plastic, metal, and hydrocarbon pollution in the central Mediterranean. Marine Pollution Bulletin 19(1):11-13.

Gulf of Mexico Fishery Management Council. 1984. Amendment Number 2 to the Fishery Management Plan for the stone crab fishery of the Gulf of Mexico and Amendment Number 3 to the Fishery Management Plan for the shrimp fishery of the Gulf of Mexico. Tampa, FL. Mar 1984.

Hays, H. and G. Cormans. 1974. Plastic particles found in tern pellets, on coastal beaches, and at factory sites. Marine Pollution Bulletin 5:44-46.

Henderson, J.R. 1988. Marine debris in Hawaii. In: Alverson and June (eds). Proceedings of the North Pacific Rim Fishermen's Conference on Marine Debris. Kailua-Kona, HI. Oct 13-16, 1987. pp. 189-206.

High, W.L. 1985. Some consequences of lost fishing gear. In: R.S. Shomura and H.O. Yoshida (eds). Proceedings of the Workshop on Fate and Impact of Marine Debris. Honolulu, HI. Nov 27-29, 1984. pp. 430-437.

Horsman, P.V. 1982. The amount of garbage pollution from merchant ships. Marine Pollution Bulletin 13:167-169.

Interagency Task Force on Persistent Marine Debris. 1988. Report. Chair, Dept. of Commerce, National Oceanic and Atmospheric Administration. May, 1988. 170 p.

Interstate Sanitation Commission. 1988. Combined Sewer Outfalls in the Interstate Sanitation District: New York, New Jersey, Connecticut. 199 p.

Johnson, S.W. 1988. Deposition of entanglement debris on Alaskan beaches. In: Alverson and June (eds). Proceedings of the North Pacific Rim Fishermen's Conference on Marine Debris. Kailua-Kona, HI. Oct 13-16, 1987. pp. 208-223.

Johnson, S.W. and T.R. Merrell. 1988. Entanglement Debris on Alaskan Beaches, 1986. U.S. Department of Commerce, National Oceanic and Atmospheric Administration, National Marine Fisheries Service, NOAA Technical Memorandum, NMFS. F/NWC-126. 26 p.

Laist, D.W. 1987. Overview of the biological effects of lost and discarded plastic debris in the marine environment. Marine Pollution Bulletin 18(6B):319-326.

Low, L.L., R.E. Nelson, Jr., and R.E. Narita. 1985. Net loss from trawl fisheries off Alaska. In: R.S. Shomura and H.O. Yoshida (eds). Proceedings of the Workshop on Fate and Impact of Marine Debris. Honolulu, HI. Nov 27-29, 1984. pp. 130-153.

MARAD. 1977. U.S. Maritime Administration. Merchant Fleets of the World, 1976.

MARAD. 1989. U.S. Maritime Administration. Merchant Fleets of the World, 1988. Apr 1989.

Mate, B.R. 1985. Incidents of marine mammal encounters with debris and active fishing gear. In: R.S. Shomura and H.O. Yoshida (eds). Proceedings of the Workshop on Fate and Impact of Marine Debris. Honolulu, HI. Nov 27-29, 1984. pp. 453-458.

Merrell, T.R., Jr. 1985. Fish nets and other plastic litter on Alaska beaches. In: R.S. Shomura and H.O. Yoshida (eds). Proceedings of the Workshop on Fate and Impact of Marine Debris. Honolulu, HI. Nov 27-29, 1984. pp. 160-182.

Merrell, T.R. and S.W. Johnson. 1987. Survey of Plastic Litter on Alaskan Beaches, 1985. Report prepared for the U.S. Department of Commerce, National Oceanic and Atmospheric Administration, Northwest and Alaskan Fisheries Center, National Marine Fisheries Service. Auke Bay, AK. 21 pp.

Mio, S. and S. Takehama. 1988. Estimation of distribution of marine debris based on the 1986 sighting survey. In: Alverson and June (eds). Proceedings of the North Pacific Rim Fishermen's Conference on Marine Debris. Kailua-Kona, HI. Oct 13-16, 1987. pp. 64-94.

Morris, R.J. 1980. Plastic debris in the surface waters of the south Atlantic. Marine Pollution Bulletin 11:164-166.

NAS. 1975. National Academy of Sciences. Marine litter. In: Assessing Potential Ocean Pollutants. A report to the Study Panel on Assessing Potential Ocean Pollutants, Ocean Affairs Board, Commission on Natural Resources, National Research Council, National Academy of Sciences. Washington, DC. pp. 405-438.

National Marine Fisheries Service. 1988. Fisheries of the U.S., 1987. National Oceanic and Atmospheric Administration, U.S. Department of Commerce. Washington, DC. May 1988.

New Jersey DEP. 1988. New Jersey Department of Environmental Protection. Department of Environmental Protection. New Jersey's Coastal Ocean, Pollution, and People: A Guide to Understanding the Problems, the Issues, and the Long Term Solutions. Report prepared by the Planning Group, Office of the Commissioner, New Jersey Department of Environmental Protection, Trenton, NJ. 29 p.

New York DEP. 1989. New York Department of Environmental Protection. The Medical Waste Study. April. 62 p.

New York State DEC. 1988. New York Department of Environmental Conservation. 1988. Investigation: Sources of beach washups in 1988. Report prepared by New York State Department of Environmental Conservation. December 1988. 58 p.

NOAA. 1987. National Oceanic and Atmospheric Administration. National Coastal Pollutant Discharge Inventory. U.S. Department of Commerce, National Oceanic and Atmospheric Administration, National Ocean Survey, Ocean Assessments Division. Rockville, MD.

NOAA/MESA. 1977. National Oceanic and Atmospheric Administration. Long Island Beach Pollution: June 1976. U.S. Department of Commerce, National Oceanic and Atmospheric Administration, Marine Ecosystems Analysis; U.S. Environmental Protection Agency, Region II; and U.S. Coast Guard, 3rd District, Marine Environmental Protection Branch. NOAA/MESA Special Report. Boulder, CO. 75 p.

OTA. 1987. Office of Technology Assessment. Wastes in the Marine Environment. United States Congress. OTA-0-334. Washington, DC.

Parker, N.R., S.C. Hunter, and R.J. Yang. 1987. Development of Methodology to Reduce the Disposal of Non-degradable Refuse into the Marine Environment. Final Report for the U.S. Department of Commerce, National Oceanic and Atmospheric Administration, under Contract No. 85-ABC-00203. Seattle, WA. Jan 1987. 85 p. with appendix.

Piatt, J.F. and D.N. Nettleship. 1987. Incidental catch of marine birds and mammals in fishing nets off Newfoundland, Canada. Marine Pollution Bulletin 18(6B):334-349.

Price, R.I. and W. Thomas. 1987. Maritime-Originated Solid Waste in New Jersey Coastal Waters: A Study of the Problem. Report prepared for the State of New Jersey, Department of Environmental Protection, Trenton, NJ. 84 p. with appendices.

Pruter, A.T. 1987. Sources, quantities and distribution of persistent plastics in the marine environment. Marine Pollution Bulletin 18(6B):305-310.

R.L. Associates. 1988. The economic impact of visitors to the New Jersey shore the summer of 1988. Report prepared for the New Jersey Division of Travel and Tourism. R.L. Associates. Princeton, NJ. Nov 1988. 16 p.

Ryan, P.G. 1988. Effects of ingested plastic on seabird feeding: Evidence from chickens. Marine Pollution Bulletin 19(3):125-128.

Ryan, P.G., A.D. Connell, and B.D. Gardner. 1988. Plastic ingestion and PCBs in seabirds: Is there a relationship? Marine Pollution Bulletin 19(4):174-176.

SAIC/Battelle. 1987. Science Applications International Corporation/Battelle Ocean Sciences. 1987. New Jersey Floatables Study: Possible Sources, Transport, and Beach Survey Results. Final Report prepared for the New Jersey Department of Environmental Protection, Bureau of Monitoring and Data Management under U.S. Environmental Protection Agency Contract No. 68-03-3319, U.S. Environmental Protection Agency, Region II. New York, NY. 49 p.

Schrey, E. and G.J.M. Vauk. 1987. Records of entangled gannets (Sula bassana) at Helgoland, German Bight. Marine Pollution Bulletin 18(6B): 350-352.

Schultz, J.P. and W.K. Upton, III. 1988. Solid Waste Generation Survey Aboard USS O'Bannon (DD 987). U.S. Navy, David W. Taylor Naval Ship Research and Development Center, DTRC/SME-87/92. Bethesda, MD. 17 p.

Scordino, J. 1985. Studies on fur seal entanglement 1981-1984, St. Paul Island, Alaska. In: R.S. Shomura and H.O. Yoshida (eds). Proceedings of the Workshop on Fate and Impact of Marine Debris. Honolulu, HI. Nov 27-29, 1984. pp. 278-290.

Shiber, J.G. 1979. Plastic pellets on the coast of Lebanon. Marine Pollution Bulletin 10:28-30.

Shomura, R.S. and H.O. Yoshida (eds). 1985. Proceedings of the Workshop on the Fate and Impact of Marine Debris. Honolulu, HI. Nov 27-29, 1984. 580 p.

Smith, G. 1988. Balloons: A surprising new eco-hazard. Earth Island Journal. Greenpeace. Winter 1988.

Spaulding, M., K. Jayko, and W. Knauss. 1988. Hindcast of medical waste trajectories on southern New England waters. Report prepared for Rhode Island Department of Environmental Management by Applied Science Associates, Inc. 27 p.

Sport Fishing Institute. 1988. Plastic may have caused beached whales death. Sport Fishing Institute Bulletin No. 395:7-8.

Squires, D.F. 1982. The Ocean Dumping Quandary: Waste Disposal in the New York Bight. State University of New York Press. Albany, NY.

Stevens, L. 1985. Will tougher licensing ease gear conflicts? Commercial Fishing News 13(6):21.

Stewart, B.S. and P.K. Yochem. 1987. Entanglement of pinnipeds in synthetic debris and fishing net and line fragments at San Nicolas and San Miguel Islands, California, 1978-1986. Marine Pollution Bulletin 18(6B):336-339.

Swanson, R.L. 1988. Washups of floatable material and their impact on New York Bight beaches. Report submitted to the National Oceanic and Atmospheric Administration. Sep 1988.

Swanson, R.L. et al. 1978. June 1976 pollution of Long Island ocean beaches. Journal of the Environmental Engineering Division, ASCE 104(EE6):1067-1085.

Uchida, R.N. 1985. The types and amounts of fish net deployed in the north Pacific. In: R.S. Shomura and H.O. Yoshida (eds). Proceedings of the Workshop on Fate and Impact of Marine Debris. Honolulu, HI. Nov 27-29, 1984. pp. 37-108.

U.S. EPA. 1988. U.S. Environmental Protection Agency. Floatables Investigation. Report prepared by the U.S. EPA, Region II. New York, NY. 11 p. with appendices.

U.S. EPA. 1989a. U.S. Environmental Protection Agency. Report to Congress on the New York Bight Plastics Study. Section 2302, Subtitle C of Marine Plastic Pollution Research Act. Title II of United States-Japan Fishery Agreement Approval Act, PL 100-220. EPA Document No.: EPA - 503/9-89/002. March.

U.S. EPA. 1989b. U.S. Environmental Protection Agency. Inventory of Medical Waste Beach Washups - June - October, 1988. Prepared for the Office of Policy, Planning and Evaluation by ICF, Inc. March 13.

Van Franeker, J.A. 1985. Plastic ingestion in the north Atlantic fulmar. Marine Pollution Bulletin 16(9):367-369.

Walker, W. 1988. Personal communication. Seattle, WA. As cited in Interagency Task Force, 1988.

Wallace, N. 1985. Debris entanglement in the marine environment: A review. In: R.S. Shomura and H.O. Yoshida (eds). Proceedings of the Workshop on Fate and Impact of Marine Debris. Honolulu, HI. Nov 27-29, 1984. pp. 259-277.

Weirich, M.T. 1987. Managing human impacts on whales in Massachusetts Bay: Countering information limitations through the use of field data. Bull. of the Coastal Soc. 10(3):11-13. As cited in Interagency Task Force, 1988.

Wilber, R.J. 1987. Plastic in the north Atlantic. Oceanus 30(3):61-68.

Wong, C.S., D.R. Green, and W.J. Cretney. 1974. Quantitative tar and plastic waste distributions in the Pacific Ocean. Nature 246:30-32.

Yoshida, K. and N. Baba. 1988. Results of research on the effects of marine debris on fur seal stocks and future research programs. In: D.L. Alverson and J.A. June (eds). Proceedings of the North Pacific Rim Fishermen's Conference on Marine Debris. Kailua-Kona, HI. Oct 13-16, 1987. pp. 95-129.

4. Impacts of Post-Consumer Plastics Waste on the Management of Municipal Solid Waste

In 1986, the principal management method for municipal solid waste was landfilling (80%); the rest of the MSW was handled by incineration (10%) and recycling (10%) (U.S. EPA, 1989). This section discusses the impacts of plastics on the management of MSW by landfilling and incineration.

EPA's "The Solid Waste Dilemma: An Agenda for Action," encourages the use of waste management options other than landfilling, as is suggested by that document's hierarchy of waste management/minimization options:

1. Source reduction
2. Recycling
3. Incineration with energy recovery and landfilling

Although source reduction and recycling are the preferred options, landfilling and incineration are essential components of an integrated waste management system. Of the latter two disposal options, EPA does not have a preference. Each community should consider all the options and select a system that can best handle its waste stream. Section 5 analyzes source reduction and recycling options, particularly as they relate to plastic wastes.

For both landfilling and incineration, the impacts of plastics discussed here include those 1) on the operation, or function, of the MSW management system; and 2) on the release of pollutants from those management systems. Impacts of the released pollutants on the environment or on human health are not discussed here, though some comparisons to EPA regulatory or health advisory limits are included. This section also includes a discussion of the impact of discarded plastic materials on litter problems.

4.1 SUMMARY OF KEY FINDINGS

Following are the key findings of this section:

4.1.1 Landfilling

4.1.1.1 Management Issues

- Available capacity for landfilling of MSW is declining. The growth of MSW generation has contributed to this shortfall. Plastic waste, which represents a growing share of total waste volume, contributes to the rate of capacity use.

- Buried plastic wastes compress in landfills to a greater extent than had been understood, reducing the share of landfill capacity relative to that which would be needed were plastics to retain their shape.

147

■ Plastic wastes are very slow to degrade in landfills; but recent data indicate that other wastes, even those considered to be "degradable," such as paper, are also quite slow to degrade. Thus, degradability appears to have little effect on landfill capacity.

■ Plastic wastes have not been shown to undermine the structural integrity of landfills or to create substantial difficulties for landfill operation.

4.1.1.2 Environmental Releases

■ Plastic polymers do not represent a hazard of toxic leachate formation when disposed in landfills.

■ Data are too limited to determine whether additives in plastics add significantly to the toxicity of MSW landfill leachate.

- One laboratory study indicates that plastic wastes containing cadmium-based pigments do not release toxic metals in sufficient quantities to pose an environmental hazard.

- Analysis of leachate from monitoring of MSW landfills has detected organic chemicals such as are used as plasticizers; one widely used plasticizer, di(2-ethylhexyl)phthalate, has been detected in a number of leachate analyses at a range of concentrations. This additive could have originated in discarded plastic products in MSW.

4.1.2 Incineration

■ Plastics contribute significantly to the heating value of MSW during incineration. Although they contribute only about 7% by weight to MSW, they may contribute 15% or more to the total Btu content of MSW.

4.1.2.1 Management Issues

■ Hydrogen chloride (HCl) gas is emitted during combustion of polyvinyl chloride (or other chlorinated polymers), and may result in corrosion of municipal waste combustor internal surfaces. Ongoing research by both EPA and the Food and Drug Administration (FDA) is addressing the extent of potential impacts on incinerator operation and the potential options to address identified impacts.

4.1.2.2 Environmental Releases

■ HALOGENS. HCl emissions from MSW combustion are correlated with the polyvinyl chloride (PVC) content of MSW. PVC and related chlorinated polymers contribute not more than about 1% by weight to the MSW stream, but may nonetheless be one of the major sources, along with paper and food wastes, of HCl in MSW incinerator emissions. Data, however, are quite limited regarding the relative contribution to HCl emissions of plastic and other wastes. EPA and FDA are both conducting further review of the contribution of plastics to HCl emissions from MSW combustion.

- DIOXINS. Although polyvinyl chloride in MSW has been postulated to be a principal chlorine donor in the formation of dioxins and furans in municipal waste combustors, experimental evidence is inconsistent. It remains unclear whether reducing PVC levels in MSW would have any impact on dioxin formation. Both EPA and FDA are conducting further analyses of the potential link between PVC and dioxin formation.

- PRODUCTS OF INCOMPLETE COMBUSTION. Under sub-optimal operating conditions, all organic constituents of MSW (including plastics, wood, paper, food wastes, yard wastes, and others) may release toxic products of incomplete combustion. Proper incinerator operation is far more important to controlling emissions of these compounds than the quantity of plastics or any other MSW constituent.

- VOLUME OF INCINERATOR ASH. Plastics ash contributes proportionately less to the volume of incinerator ash requiring disposal than plastics contribute to the uncombusted MSW waste stream.

- INCINERATOR ASH TOXICITY. Lead- and cadmium-based plastic additives contribute to the heavy metal content of MWC ash. Because they are distributed in a combustible medium, plastics additives tend to contribute proportionately more to fly ash than to bottom ash. Additional investigation is warranted to determine with greater accuracy the impact of plastics additives on MSW fly ash toxicity (i.e., the contribution of plastic additives to leachable lead and cadmium in ash).

4.1.3 Litter

- Some beach areas receive unusually large quantities of marine debris, much of which is plastic. These areas must be cleaned of debris or suffer aesthetic losses and economic damages.

4.2 LANDFILLING

Net MSW discards in the United States currently amount to approximately 140.8 million tons per year, of which plastics are estimated to account for 7.3% by weight. Landfilling has been the predominant disposal method for MSW. The impacts of plastics on the successful management of these landfills is a subject of increasing concern. The United States faces dwindling landfill capacity and an increasing flow of MSW. Plastics represent an increasing share of the total weight of the solid waste stream (see Table 4-1). The areas considered in this section include both landfill management issues and environmental releases.

Table 4-1

GROWTH IN THE CONTRIBUTION OF PLASTIC WASTES
TO THE MUNICIPAL WASTE STREAM
(1960-2000)

Year	Weight (million tons)	As percentage of MSW by weight
1960	0.4	0.5
1965	1.4	1.5
1970	3.0	2.7
1975	4.4	3.8
1980	7.6	5.9
1981	7.8	5.9
1982	8.4	6.5
1983	9.1	6.8
1984	9.6	6.9
1985	9.7	7.1
1986	10.3	7.3
1990	11.8	7.9
1995	13.7	8.6
2000	15.6	9.2

Source: Franklin Associates, 1988a.

4.2.1 Management Issues

The management issues raised by plastics in MSW include impacts on landfill capacity, structural integrity, and daily operations.

4.2.1.1 Landfill Capacity

Plastic wastes add to the volume of MSW generated. This section reviews recent literature findings on each topic and develops conclusions about the impact of plastic waste disposal.

The available capacity of MSW landfills in the United States is decreasing as a result of two related factors:

- Remaining capacity of existing landfills is dwindling rapidly. In the next five to seven years, EPA predicts that 45% of the MSW landfills in the United States will reach capacity (U.S. EPA, 1988a; see Table 4-2). Similarly, a survey by the American Public Works Association indicated that 40% of the responding landfill operators claimed that their community landfill capacity will be depleted within five years (EPA Journal, 1988). Figure 4-1 shows that landfill capacity is most tightly constrained in heavily populated areas, especially the Northeast and Great Lakes region.

- As landfills reach capacity, few are being replaced by new sites. By 1994, for example, the number of operating landfills is predicted to decrease by 83% from 1976 levels (Figure 4-2) (U.S. EPA, 1989). Reasons include increasingly stringent regulations and more public concern regarding potential landfill problems. Tougher environmental regulations may force older landfills to close regardless of their capacity.

As plastics increase in use (see market projections in Section 2), they will require a greater share of remaining landfill capacity. Estimating the size of that share, however, remains difficult for several reasons. First, future recycling rates for plastic are unknown. Second, the amount of plastic that will be incinerated cannot be estimated at this time. Finally, there are unresolved issues in the research on the relationship of the weight of plastics waste to its volume after the waste has been placed in the landfill.

Several studies have considered the volume:weight ratio of plastics. In one study, Jack Schlegel of International Plastics Consulting Corporation calculated the volume of plastic wastes using estimates of the density of various types of plastic wastes (see Table 4-3). These density estimates were derived from industry and literature sources. Overall, this research estimated that plastics volume (as a percentage of MSW volume) is three to five times that of plastics weight. Using Franklin Associates' estimates of the percentage by weight of plastic wastes, Schlegel calculated that in 1984 plastics accounted for 25.4% by volume and 6.8% by weight of MSW (Schlegel, 1989; Schlegel's calculations represent estimates of plastic volumes in MSW at the point of disposal and are not intended as empirical estimates of plastic volumes when compressed in landfills). In addition, estimates have been made that by the year 2000, as plastic packaging continues to increase, the volume of municipal plastic waste could be as high as 40% (Modern Plastics, 1988). If such high volume figures are accurate, plastics waste disposal will consume much of the remaining landfill capacity.

Table 4-2

REMAINING YEARS OF OPERATION OF
MUNICIPAL SOLID WASTE LANDFILLS
(AS OF 1986)

Remaining Years	Number of Landfills	% of Landfills
0	535	8.9
1-5	2,167	35.9
6-10	612	10.1
11-15	1,126	18.7
16-20	360	6.0
> 20	1,234	20.5
All Years	6,034	100.0

Source: U.S. EPA, 1988a.

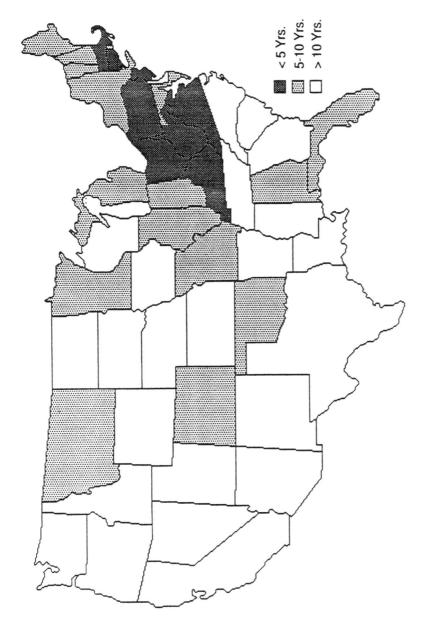

Figure 4-1. Remaining U.S. Landfill Capacity.

Source: National Solid Waste Management Association, 1988.

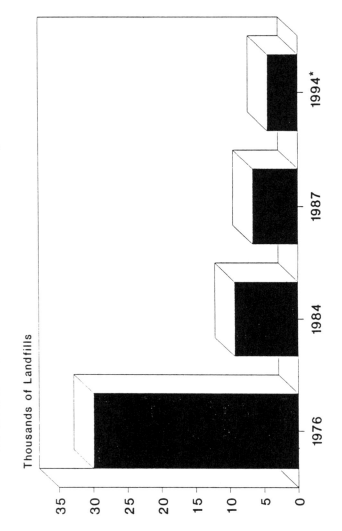

Figure 4-2
OPERATING MUNICIPAL LANDFILLS, 1976–1994

Thousands of Landfills

Source: Forester, 1988; and EPA, 1989b.
* Projected

Table 4-3

ESTIMATES OF PLASTIC WASTE

Information Source	Year	Plastic Component of MSW Weight(%)	Volume(%)	Volume: Weight Ratio	Methodology
International Plastics Consultants Corporation	1984	6.8[a]	25.4	3.7	Calculated from Franklin Associates weight estimates; estimated for current year's waste
Franklin Associates Ltd.	1984	6.8	NE	NE	Materials-flow based on consumption; estimates for current year's waste
Franklin Associates Ltd.	1986	7.3	NE	NE	
W.L. Rathje et al. University of Arizona					
Share of MSW only	1989	7.4[b]	17.9[c]	2.4	Measured from landfill excavations; Average of more than 20 years of waste
Share of entire quantity of excavated material	1989	4.1[b]	15.9[c]	3.9	

[a]IPCC volume estimates are based on 1984 Franklin Associates weight figures.

[b]Plastics represent 4.1 percent of landfilled wastes by weight when matrix materials (i.e. soil, fines) are included and 7.4 percent by weight of MSW alone. The former figure is used here because it is most relevant to the volume discussions.

[c]Dr. Rathje explains that because plastic films expand after being uncovered, this volume estimate is higher than the actual volume of buried plastics.

NE=Not estimated.

Sources: Franklin Associates, Ltd. 1988a.
            Schlegel, 1989.
            Rathje et al, 1988.

Recent field research by W.L. Rathje and others at the University of Arizona, however, suggests that such volume:weight ratios do not accurately reflect plastics compression in a landfill and that plastic wastes do not consume such large percentages of available landfill space. Rathje's study examined wastes excavated from three municipal waste landfills serving major metropolitan areas from 1960 to the present. He noted that virtually all plastic containers were flattened or crushed in the landfill and did not recover their shapes after excavation. Uncovered plastics were found to have a 2 or 3 to 1 ratio (% volume to % weight). Plastic wastes were measured at 17.9% of the volume of all MSW, and 15.9% of all excavated materials including fill, fines, and other materials.

Further, Rathje estimated that even his measured volume:weight ratios overstated true volumes of plastic waste. This measurement discrepancy occurred because Rathje was not able to correct for the tendency of excavated plastic film (e.g., refuse bags) to billow and fill with air. Currently, his group is developing new methods to measure volumes for these plastic wastes that reflect actual volumes under compression in the landfills.

The Rathje et al. data and other estimates must be compared with some care. Rathje's estimate represents an historical average of the share of landfill volume used by plastic waste; the derived estimates of plastics volumes will underestimate the plastics share for the current stream of garbage, as that share increases over time. Nevertheless, these estimates suggest that the landfill volume consumed by plastic wastes is less than had been estimated by Schlegel.

In summary, plastic wastes consume a substantial portion of landfill capacity, but do not represent as large a share as has been estimated in some theoretical studies of waste volumes. Further, plastic wastes are compressed in the landfill and do not consume exceptional capacity because of their resiliency.

4.2.1.2 Landfill Integrity

Some authors have suggested that the low bulk density of plastic can undermine the integrity of landfill design. This characteristic may lead to the upward migration of plastic waste through landfills (Center for Plastics Recycling Research, 1986). Migrating pieces of plastic may create voids or air pockets in their wake, thus weakening the landfill's structural integrity.

Research fails to confirm that upward migration does occur or that landfills suffer a loss of integrity due to the presence of plastics. Again in Rathje's work, in which municipal solid waste was excavated from landfills, no general concentration of plastics could be observed within the excavated cores, such as might be expected if plastics were shifting upward in the landfill (Rathje, 1989). No upward migration of plastics was observable even within the distinct landfill "lifts" (i.e., separate layers of waste) within the landfill.

Further, The National Solid Waste Management Association conducted an informal telephone survey of its membership (i.e., of MSW landfill operators) several years ago to determine whether plastic wastes were affecting landfill integrity. The consensus of membership opinion was that plastic wastes were not a significant threat to landfill integrity (Repa, 1989). Because no organized survey was performed, however, no rigorous findings on this subject were developed.

4.2.1.3 Other Management Issues

Plastic wastes, as well as paper wastes, contribute to the wind-blown debris in MSW landfills. Proper waste management techniques (i.e., covering wastes immediately after deposition) should limit the amount of this blowing debris. Some additional blowing wastes are likely, however, due to the presence of plastics in the waste. Landfill operators can use windbreaks to capture such debris.

Plastic wastes may also slow the rate of degradation of the wastes with which they are disposed. Several researchers have examined the rate of waste degradation in landfills in order to better understand methane gas generation, as well as the potential for uneven settling of wastes from uneven degradation. The degradation rate for wastes determines the rate of methane gas generation in landfills, another concern for landfill management.

One study indicates a potential for plastics to slow decomposition of wastes substantially. EPA-funded research by Kinman et al. (1985) involved the construction and analysis of 19 simulated landfills for a period of nine to ten years. The study was designed to analyze the physical, chemical, and microbial conditions of MSW landfills and to examine the effect of co-disposal of industrial waste with MSW to evaluate effects on the decomposition process.

Kinman et al. discovered well-preserved organic materials within plastic bags or underneath and encompassed by plastic materials throughout the landfills. Many readily biodegradable items, including food waste and fecal matter, were found to be well protected by household garbage bags and by other wrappings after up to a decade in the landfill. Paper bags also substantially protected food wastes from decomposition.

Kinman et al. concluded that plastic and paper bags should be torn open as much as possible to allow readily biodegradable materials to decompose. The group's findings of intact garbage bags and plastic-wrapped materials, however, appear to be exceptional and are due to the nature of the simulated landfill, in which garbage was relatively undisturbed. In normal landfill operations, the bulldozing operation will rip open garbage bags and expose wastes to the landfill element.

Rathje's study provides further information about biodegradation patterns in landfills. He noted that the plastic bags and wrapping in the excavated garbage were torn or shredded, and the materials inside were exposed to landfill elements. Thus, plastic wastes did not appear to inhibit the rate of degradation for nonplastic wastes. Rathje also noted, as did Kinman, that other materials such as paper were quite slow to degrade and could be found largely intact after decades in the landfill.

4.2.2 Environmental Releases

Leaching of any potentially harmful chemical is the primary pathway of concern for environmental releases of plastics or plastics additives from landfills. The other potential environmental release from landfills is methane gas, which forms as the result of bacterial decomposition of organic matter. Because plastic products contribute negligibly to bacterial activity in landfills, gas formation is not attributable to this component of the waste stream.

4.2.2.1 Leaching of Plastic Polymers

Plastic polymers do not contribute to negative environmental releases from landfills. Polymers are largely impervious to various forms of attack, including biodegradation. Nor can plastics be attacked or dissolved by the weak acidity found in landfills. The slow rate of biodegradation is due to the high molecular weight of most plastic polymers. The stability of the carbon-carbon bond in the polymer chain also contributes to the slow rate of degradation. Such molecules cannot be broken down into smaller portions that microorganisms can consume. This high molecular weight is a characteristic of nearly all the major commodity plastic resins. Only cellulosic resins, and a specialty resin, PHBV (poly [3 hydroxybutrate-3 hydroxyvalerate]), are readily biodegradable. Cellulosic resins, however, generally lose the characteristic of biodegradability when they are processed into useful plastic products.

Most studies on contributions to leachate from plastic materials have focused primarily on issues concerning the use of plastics in groundwater monitoring systems. Several studies have raised concerns about the use of plastics in leachate collection systems, because of their tendency to contribute organic chemicals primarily from plasticizer additives to leachate (Barcelona et al., 1985). Nevertheless, these studies focus on the potential for releases of organic materials from plastic additives and not from deterioration of plastic polymers. Curran and Tomson (1983) analyzed groundwater quality monitoring techniques by measuring leaching from different types of plastic pipes. The group found that using rigid PVC pipe (a plastic product with few additives) resulted in insignificant leaching -- i.e., the pipe could be used without measurably altering groundwater test results.

Not only do plastic polymers not generate toxic leachate in landfills, but low leaching and corrosion resistance make some plastics the material of choice in applications in which leachates must be controlled. For example, plastics are used as liners for hazardous waste landfills (Lu et al., 1985) and as casings in groundwater monitoring wells (Sykes et al., 1986).

4.2.2.2 Leaching of Plastics Additives

Numerous chemical additives can be employed in manufacturing to modify the properties of the resins in processing and design. Section 2 outlined over a dozen categories of additives that serve a variety of purposes and encompass a range of chemical properties. This section analyzes the available evidence on indicated or potential leaching of these chemicals in landfills.

Because of the wide variety of chemicals that are used as additives, it is necessary to develop screening methods to select those with the greatest potential for generating environmental concerns. After this screening, several categories of information will also be needed in order to analyze their contribution to potential or actual leachates from landfills. The principal screening of additives is based on 1) the toxicity of some of the major classes of chemicals used in each class of plastics additives, and 2) the levels of production for the additives.

Table 4-4 summarizes the potential toxicity concerns associated with the various additive categories. The discussion is not exhaustive because a wide variety of chemicals are included in some of the additive categories. Nevertheless, the discussion identifies the potentially toxic compounds among the chemicals consumed in substantial quantities. In terms of toxicity, several specific compounds warrant further attention. These compounds include phthalates from plasticizers, metal constituents of colorants, flame retardants and heat stabilizers, metallic stearates from lubricants, and antimicrobial additives.

The toxic additives are examined here in the context of 1) the overall level of production of the additive, and 2) the concentration level of the additive in plastic products. For the purposes of this report, EPA is focusing on those additives that could be present in some quantity in products disposed as MSW. EPA recognizes that other compounds, present in small quantities, may be toxic. Based on a consideration of research priorities, however, this study will not examine these minor additives further.

Several additives are used in significantly larger quantities than others and, therefore, pose a greater possibility of leaching. Antimicrobials and lubricants are used in small amounts and would not be present in a landfill in notable quantities. For additives present in large or moderate volumes, further analysis was performed to assess whether the additive can be leached from the plastic. See Table 4-5 for information concerning the quantity of additives that may be found in landfilled wastes. The phthalate plasticizer is an example of an additive that, while it mixes thoroughly with a polymer, is only weakly bonded to it. Some leaching of the phthalate is therefore possible. Reinforcements and plasticizers are used in the greatest bulk. The tendency of other additives, such as metal-based colorants, to be released from polymers is unknown, although it is probably limited. Metal-based colorants do not react with polymers and are merely embedded in the plastic products; nevertheless, the colorants are not readily released from the polymers (Radian, 1987).

Finally, Table 4-5 presents conclusions regarding the relative concern about leachate that is presented by the various additives disposed in plastic wastes found in landfills. These findings are based on the volume data and additional considerations that affect the likelihood that they would pose a problem in landfills. This analysis was not extended to define an absolute measure of the significance of the potential leaching of the additives; only relative judgments were developed. It should also be noted that plastics represent only a portion of the wastes in a landfill, and that plastics with any specific additive represent a portion of the plastic wastes. Section 2 presents the statistics necessary to place the plastic waste issue in the perspective of total MSW quantities.

Based on this analysis, plasticizers, fillers and reinforcing agents are present in the largest quantities. The latter two additives, however, do not pose a hazard for leaching due to lack of toxicity in leachate, so they are not considered further. Several other additives may be present (marked as uncertain in the table) in sufficient quantities to present a leaching concern. These include colorants, flame retardants, and heat stabilizers. One other additive, impact modifiers, is

Table 4-4

TOXICITY AND POTENTIAL FOR LEACHING OF PLASTICS ADDITIVES

ANTIMICROBIALS - Although the amounts used in plastics are small, antimicrobials are bioactive compounds designed to be toxic to microorganisms and thus could be potentially toxic to larger organisms. Several types used in plastics contain tin or mercury (Radian, 1987). No information is available concerning their leaching properties.

ANTIOXIDANTS - FDA has regulations governing the use of antioxidants in plastics having food-contact applications. These additives are mainly high-molecular weight phenols. These materials do not degrade appreciably and do not represent a notable source for leachates. Antioxidants are the most widely used additives; but because they are used in such small quantities, aggregate production is quite low and concentrations in plastic products are extremely small.

ANTISTATIC AGENTS - The antistatic agents used in largest volume are quaternary ammonium derivatives and these chemicals have some potential toxicity. The FDA regulates the use of antistatic agents in food and medical applications, and has regulated several organic amines under 21 CFR 178-3130.

Antistatic agents by their nature must be at the surface of the polymer and to have only partial compatibility with the polymer. As a result, the antistatic agent can leach from the surface of the plastic.

CATALYSTS - These additives facilitate reactions among other compounds and are not themselves joined into the plastic product. As such, they are present in only residual amounts or to the extent they are not removed from the resulting product. While certain types of catalysts contain toxic metals, these should be present in final products in extremely small amounts.

CHEMICAL BLOWING AGENTS (CBAs) - CBAs are introduced as solids, which decompose to form volatile gases and solid residues. Consequently, both the agents themselves, as well as their decomposition products, must be considered, but the quantities in plastic waste would be extremely small. Among the agents and components, several including benzene, 1,2-dichloroethane, trichloroethylene, and barium are considered priority pollutants or hazardous wastes. Others, such as diazoaminobenzene and tetramethyl-succinonitrile are not listed but are known to be toxic. Residues may be washed from the product following processing, however, and may not be present in the final product (Radian, 1987).

COLORANTS - The colorant in widest use (as measured by production), titanium dioxide, does not present any environmental hazard. Numerous other colorants, however, include heavy metals and thus could pose some concern for disposal. Many of the colorants or their constituents are listed as priority pollutants or as hazardous under the RCRA program by EPA. In particular, lead and cadmium compounds are often singled out because they are widely recognized as toxic (e.g., 40 CFR 261).

The leachability of colorant constituents from plastics is unclear and may be quite limited. The colorants are embedded in inert plastic; they are also chosen for their resistance to migration in the product. This suggests that colorants do not readily release from the plastic.

Table 4-4 (cont.)

FILLERS - Most fillers (e.g., calcium carbonate, clay) are unreactive and insoluble and thus do not present a leachate concern. Asbestos continues to be used as a filler in some applications, and theoretically could present a particulates issue. Asbestos is embedded in the polymer, however, and it is not likely to be released in any quantity from plastic wastes.

FLAME RETARDANTS - A wide variety of organic and inorganic compounds are used in flame retardants. Among the inorganic chemicals, antimony oxide is toxic; it is known to leach from some plastics (SPI, 1985), and it is on the EPA list of hazardous constituents found in waste (40 CFR 261). Some organic compounds were found to be toxic, but were banned from use and are no longer produced. Certain chloride compounds still in use are toxic, but the most widely used chlorine compounds, chlorinated paraffins, are non-toxic. Flame retardants can also be categorized as reactant and nonreactant (or additive). Reactant compounds, as the name suggests, react with the plastic polymer and, because they are bound to the polymer, will not degrade any faster in a landfill than the polymer itself. Most flame retardants, however, are nonreactant: they are encapsulated as small, relatively insoluble particles by the plastic or dissolved in the plastic and also function as plasticizers. These are the flame retardants that are most prone to volatilize and to leach from plastics during use or disposal.

FREE RADICAL INITIATORS - These additives are largely consumed by the polymers in the reactions that they initiate. For this reason they are of relatively little environmental concern. Only residues of the additives are likely to remain in the plastic product, and their potential for leaching is likely to be small.

HEAT STABILIZERS - Many heat stabilizers are organometallic chemicals and are considered toxic. Several of the most effective stabilizers are among the most toxic; less toxic stabilizers such as calcium and zinc are less effective. All PVC plastic requires heat stabilizing additives during processing.

The likelihood of heat stabilizers released from PVC is as yet unknown. Many heat stabilizers are highly compatible with plastic polymers and would not readily be released. Other applications may not call for such compatibility, however, and present some potential for release.

Note that heat stabilizers are a more significant source of lead and cadmium in MSW than are colorants. In discarded plastics they have been shown to contain more than twice the amount of lead and cadmium found in colorants (Franklin Associates, 1988b).

IMPACT MODIFIERS - Impact modifiers are polymers themselves and are generally inert in the environment. No release of these additives is likely.

(cont.)

Table 4-4 (cont.)

LUBRICANTS - Lubricants are used primarily to improve the processing characteristics of the plastic or to improve the characteristics of the plastic in use. In either case, a residue of the lubricant or more substantial quantities of the material could remain with the plastic product until disposal, although quantities of lubricant would be quite small.

Most lubricants are chemically inert, and others are derived from natural sources. This class of additives is not believed to pose an environmental threat. Exceptions include certain "metallic-soap" lubricants used to improve processing (i.e., the metal constituents of these additives can be toxic).

PLASTICIZERS - Numerous phthalate plasticizers are listed as hazardous wastes or as priority pollutants by EPA. Included among these is di(2-ethylhexyl) phthalate, which is used in large quantities in plastic products (Life Systems, 1987). Overall, the toxicity of the plasticizers varies considerably with the specific additive used and its concentration in the plastic.

Plasticizers are somewhat extractable from the polymers into which they are incorporated; therefore, they tend to exude during use or after disposal. After disposal, they may also be extracted by water or by solvents. Water, however, can extract only small amounts of plasticizer.

REINFORCING AGENTS - These agents are similar to fillers except they consist of fibers rather than particle-type additives. Glass fibers, the predominant reinforcing agent, are nontoxic. They do not present a hazard in the landfill. Another reinforcing agent, asbestos, can also be used and could present a particulate hazard if fibers are released in quantity. Since these fibers should be well embedded in the polymer, however, notable releases are unlikely.

UV STABILIZERS - Some benzophenones and benzotriazoles are approved by the FDA for food packaging uses (Radian, 1987). Nickel organic stabilizers, which account for a significant percentage of total consumption, are toxic. Cyanide and zinc may also be used.

U.V. stabilizers are chosen for their nonmigratory properties, since leaching would leave the polymer vulnerable to UV degradation. Thus, leaching of additives is not expected to be a significant problem.

Source: Radian (1987) and data compiled by Eastern Research Group, Inc.

Table 4-5

POTENTIAL FOR PRESENCE OF TOXIC, PLASTICS-DERIVED
ADDITIVES IN LANDFILLS AND IN LANDFILL LEACHATE

Category of Additive/ Primary Toxic Concerns	Additive's Presence in Landfills			Additional Considerations Affecting Leaching Concern
	Volume Use in Plastics in 1986 (million lbs)	Additive to Resin Ratio (a) (lbs additive/ 100 lbs resin)	Is Additive Present in Landfill–Discarded Plastics In Relatively Large Quantities?	
Antimicrobial Agents Various	11.0	<1	No	--
Antioxidants None identified	35.0	<1	No	--
Antistatic Agents Quaternary ammonium derivatives	6.5	<1	No	--
Catalysts None identified	40.0 (b)	<1	No	--
Chemical Blowing Agents None identified	12.6	1–5	No	Generally most of agent evaporates
Colorants	423.0	<1	Uncertain	Colorants used in low concentrations but used in most plastics
Cadmiums	6.0	<1	Uncertain	
Chromium yellow	6.0	<1	Uncertain	
Lead compounds	<1.0	<1	Uncertain	
Fillers None identified	2,288.0	10–50	Yes	Low toxicity so little leaching concern
Flame Retardants	483.0	10–20		Some toxicity, but used mostly in building products that aren't disposed of with MSW
Antimony oxide	36.0	<5	Uncertain	

(cont.)

Table 4–5 (cont.)

POTENTIAL FOR PRESENCE OF TOXIC, PLASTICS-DERIVED ADDITIVES IN LANDFILLS AND IN LANDFILL LEACHATE

Category of Additive/ Primary Toxic Concerns	Additive's Presence in Landfills			Additional Considerations Affecting Leaching Concern
	Volume Use in Plastics (million lbs)	Additive to Resin Ratio (a) (lbs additive/ 100 lbs resin)	Is Additive Present in Landfill-Discarded Plastics in Relatively Large Quantities?	
Free Radical Initiators	NA	<1	No	--
Heat Stabilizers	83.0	0.2–5		Some toxicity, but used mostly in building products that aren't disposed with MSW
Barium–cadmium	35.0	0.2–5	Uncertain	
Lead	23.0	0.2–5	Uncertain	
Impact Modifiers None Identified	135.0	10–20	Uncertain	Polymer itself; Not leachable
Lubricants	95.0	<1	No	--
Metallic stearates	37.0	<1		
Plasticizers	1,809.0	20–60		Liquid; most readily released
Phthalates	1,184.0	20–60	Yes	
Reinforcements	961.0	10–40		Particulate, not leaching concern
Asbestos	90.0	10–40	Yes	
UV Stabilizers None Identified	5.5	<1	No	--

NA = Not Available.

Notes: (a) Ratios shown are for the additive category as a whole; ratios for individual chemicals not separately estimated.
(b) Volume estimate covers only organic peroxide.

Source: Volume data from Rauch, 1987.   Other estimates by Eastern Research Group.

sometimes used in relatively high concentrations but this is a polymer itself and would not be released into leachate.

In the rest of this section, the available laboratory and field data are reviewed to highlight which chemicals have been shown to leach from plastic products. Based on the previous discussion, the following chemicals merit further analysis:

- Phthalate esters, the most widely used class of plasticizers

- Toxic flame retardants, particularly antimony oxide

- Heavy-metal colorants, particularly lead- or cadmium-based (although used in small quantities, their high toxicity could warrant attention)

- Metal-based heat stabilizers, particularly lead- or cadmium-based

The toxicity of metals, particularly lead and cadmium, used in certain of the plastic additives, has received attention in previous research. EPA commissioned a study by Franklin Associates to estimate the aggregate quantity of lead and cadmium that may be present in MSW (Franklin Associates, 1988b). Franklin Associates used a "material flows" methodology (see Section 2) to derive these estimates. Their study was not designed to estimate the potential for leaching of these wastes.

Table 4-6 presents the principal findings of the Franklin Associates study. Plastic additives were found to contribute 3,576 tons of lead and 564 tons of cadmium to MSW. These quantities represent 1.7% of the total lead and 31.5% of the total cadmium present in MSW. The great bulk of the lead and cadmium in MSW is from automobile and household batteries, respectively.

Available research on heavy metals as well as on other additives of concern falls into two categories: 1) laboratory studies of leaching potential, and 2) evidence from monitoring studies of municipal solid waste landfills. The intended scope of both types of studies is to determine if leachate contains the various chemicals that could originate from plastic wastes and, if so, in what quantity.

LABORATORY STUDIES OF LEACHING POTENTIAL -- Limited data were identified on the topics of potential or actual leaching of additive components from plastic materials. The two available laboratory studies examined the leachability of toxic heavy metals used in plastic additives.

In the first of these studies, Wilson et al. (1982) examined leaching of cadmium from pigmented plastics in simulated landfill conditions. That group performed several different leachate tests with plastics (e.g., ABS, polystyrene, high-density polyethylene, PVC) containing cadmium-based pigments. The plastics contained 1% yellow cadmium pigment by weight. This concentration of pigment, as Wilson et al. noted, is the maximum used in commercial practice.

Wilson performed tests using a series of large glass columns filled with mixed pelletized plastic (which increased the possible surface area for leaching) and wet-pulverized refuse. The ratio of

Table 4-6

CONTRIBUTIONS OF LEAD AND
CADMIUM FROM DISCARDED PLASTIC
PRODUCTS IN LANDFILLS
(tons)

Source	Lead	Cadmium
Heat stabilizers	2,586	309
Colorants	990	255
TOTAL PLASTICS	3,576	564
All sources	213,652	1,788
Plastic contribution as % of all sources	1.7	31.5

Source: Franklin Associates, 1988b.

plastics to general refuse was 1:5, a ratio that overestimates the concentration of plastics in a landfill. The columns were irrigated and leachate collected for periods of six months to a year. In one set of tests, wastes were extracted using a solution of 5,000 ppm acetic acid, buffered to pH 5 with sodium hydroxide. Wilson et al. employed distilled water in a second set of tests. Some data interpretations were complicated by possible surface contamination of laboratory equipment; even pristine elements indicated the presence of cadmium.

Wilson et al. found that the cadmium levels in the leachate were not higher for columns with pigmented plastic than for control columns with uncolored plastic. The group also concluded that such leaching as did occur was so minimal that it must have occurred only from the surface of the plastic and not from the bulk. Furthermore, the overall contribution of cadmium leachate from a normal mixture of plastic in a landfill was estimated not to exceed that from trace contaminants of cadmium in paper and food.

A second study performed by the Society of the Plastics Industry used extraction test procedures of EPA and California hazardous waste programs to estimate prospective leachates from wastes. The California Waste Extraction Test (WET), similar in theory to the extraction program employed in the EPA RCRA program, is designed to indicate the nature of leachate that is produced from wastes co-disposed with MSW. The California test, however, requires waste to be milled before the leachate test is performed.

The SPI results from the California test are summarized in Table 4-7. SPI ran tests on four samples of ABS plastic, three samples of polyvinyl chloride (PVC), and three samples of nylon. Metal-based additives, including lead, chromium, zinc, antimony, and molybdenum, were contained in the plastics. The results show leaching of very small percentages of the metals content of the various plastics.

These results are compared in the table with limits defined for the Extraction Procedure toxicity test in the EPA RCRA program. None of the samples exceeded the defined limit for pollutant concentrations. SPI concluded that the California test, while more stringent than the EPA test, did not indicate that plastic wastes would be classified as hazardous wastes. The California requirement to mill the waste was particularly stringent because much more surface area was made available for potential leaching than may be the case in normal landfill disposal of plastic products. When the effect of the milling was measured by re-applying the test in one case (sample 4, ABS) with recombined granules, the amount of metals in the leachate declined by an order of magnitude.

These data suggest that the leaching potential of cadmium from plastic products is low and that environmental risk is low. Leaching of organic constituents of additives, however, such as from plasticizers, was not addressed by any of this research.

LEACHATE OR GROUND-WATER MONITORING AT MUNICIPAL SOLID WASTE FACILITIES -- Leachate and ground-water monitoring results provide additional information about the apparent leachate formation from plastic wastes deposited in municipal landfills. These data represent direct field evidence of leaching of general municipal solid waste. It must be presumed that discarded plastic materials were included in the waste. The exact contribution of plastic wastes to leachate, however, cannot be isolated from these data.

Table 4-7

RESULTS OF THE CALIFORNIA WASTE EXTRACTION TEST APPLIED TO PLASTICS

Plastic Sample	Sample Additives (mg/kg)		Metals in WET Test results (mg/L)[a]	Maximum Concentration-Extraction Toxicity (mg/L)
Acrylonitrile-butadiene-styrene				
Sample 1	Pb	5,000	3.1	5.0
	Cr(VI)	5,000	0.6	5.0
Sample 2	Pb	5,000	1.6	5.0
	Cr(VI)	5,000	0.2	5.0
Sample 3	Cd	10,000	0.3	1.0
Sample 4	Pb	4,300	2.3	5.0
(granules)[b]	Sb	21,400	36.0	n/a
	Cd	5,200	0.02	1.0
	Zn	500	0.5	n/a
	Cr	800	0.4	5.0
	Mo	250	<0.1	n/a
Sample 4	Pb	4,300	0.3	5.0
(in plaque form)	Sb	21,400	1.1	. n/a
	Cd	5,200	<0.1	1.0
	Zn	500	0.02	n/a
	Cr	800	0.03	5.0
	Mo	250	<0.1	n/a
Polyvinyl Chloride				
Sample 1	Pb	5,000	2.6	5.0
	Cr	5,000	0.6	5.0
Sample 2	Pb	5,000	1.5	5.0
	Cr	5,000	0.4	5.0
Sample 3	Cd	10,000	0.07	1.0

(cont.)

Table 4-7 (cont.)

Plastic Sample	Sample Additives (mg/kg)		Metals in Test (WET) results (mg/L)[a]	Maximum Concentration- Extraction Toxicity (mg/L)
Nylon				
Sample 1[c]	Cd	25,000	10.00	1.0
Sample 2	Cd	7,000	0.5	1.0
Sample 3	Cd	4,000	0.05	1.0

[a]The WET test procedure is as follows:
1. Sample is milled to pass through a 2-millimeter sieve.
2. Fifty grams of sample is extracted with 500 milliliters of the deaerated (anaerobic) extractant solution in the range of 20 to 40 degrees Celcius for 48 hours. The extracting solution contains 0.2 molar sodium citrate at pH 5.0, which simulates the MSW landfill environment.
3. The mixture is filtered and analyzed using U.S. EPA methods.

[b]This sample was not prepared with strict controls as to particle size.

[c]This sample was derived from a "masterbatch", which is heavily loaded with cadmium-based pigment. The masterbatch is used to color a large amount of plastic.

Source: SPI, 1985.

The first of the available studies, performed by Dunlap et al. (1976), involved collection and analysis of groundwater from wells near a municipal landfill. This site included an old landfill site that had been used as a dump for 38 years as well as a new landfill site in which disposal operations had begun in 1960. Monitoring wells were placed near both the old and new landfills. The new landfill, as the authors explained, never received appreciable quantities of industrial solid waste; thus, they categorized the groundwater contamination there as originating from consumer products and other MSW. The waste in the new landfill had been placed in unlined cells of 20 feet in thickness.

Chemical analysis of the groundwater from near the landfill indicates the presence of a number of organic compounds that are used as plasticizers. See Table 4-8 for the levels of constituents found, and for information on the commercial uses of the chemicals. (Much of the production data on specific chemicals could be developed only from 1979 data.) The commercial use data are needed to determine whether the chemicals found in the groundwater could have originated from other products. For example, one of the chemicals found, dioctyl phthalate, is used almost solely as a plasticizer (97% of use). That chemical, therefore, most likely originates from plastic products.

The Dunlap study only analyzed organic contaminants; no information about metals in the groundwater are presented. This study also predates the development of the EPA Subtitle D RCRA program for municipal solid waste. More recent studies have developed substantial quantities of leachate monitoring data. Thus the Dunlap study should be considered only one of many on leachate from MSW.

EPA prepared a summary of these landfill leachate data as part of its 1988 Report to Congress on Solid Waste Disposal in the United States (U.S. EPA, 1988a). In that summary, EPA compiled the number of instances in which various chemicals were detected in the monitoring studies and then compared these results with available standards indicating health risk, including priority pollutant limits.

Table 4-9 presents some of these data on a selected set of organic and inorganic chemicals that were identified in some leachates from MSW landfills. Table 4-9 also presents information on specific uses of the chemicals as plastics additives and in other products. The data indicate that several organic chemicals often used in plasticizers were found in MSW leachate. Most notably, concentrations of bis(2-ethylhexyl)phthalate (also called di(2-ethylhexyl)phthalate) were found in a number of investigations. A wide range of concentration levels were found.

Certain caveats must be considered for interpreting the phthalate levels in the leachate data. As was noted in Section 4.2.2.1, research by EPA and others indicates that there is some potential for bias in leachate results from plastic and other materials used in monitoring systems. Plastic materials could increase pollutant levels because additives (particularly plasticizers) are released from piping or other materials used in constructing the leachate collection system. Conversely, plastic materials in some cases could absorb organic chemicals from the leachate, biasing pollutant level measurements downward (Barcelona, 1989).

For the studies compiled by the EPA Report to Congress, it is presumed that careful and appropriate monitoring systems were employed. Nevertheless, some errors in measurements due to plastic materials could have occurred. It is estimated that up to 50 to 60 ppb of phthalate

Table 4-8

ORGANIC CONSTITUENTS OF GROUND WATER IDENTIFIED IN A
STUDY OF AN MSW LANDFILL

Compound found in ground water under landfill[a]	Ground water concentration (ppb)	Product and commercial uses[b]
Diethyl phthalate	4.1	1.0×10^7 kg produced in 1978, used as plasticizer and solvent for cellulose esters, solid rocket propellant and insecticide spray. Cellulose esters are used in acetate fibers, lacquers, protective coatings, photographics film, transparent sheeting, thermoplastic molding composition, cigarette filters, and magnetic tapes.
Diisobutyl phthalate and Di-n-butyl phthalate	0.1	7.7×10^6 kg produced in 1979, 35% used as plasticizer for plastisols, which are dispersions of plastic in a plasticizer used to mold thermoplastics, chiefly polyvinyl chloride; also used in lacquers, elastomers, explosives, nail polish, perfumes, textile lubricants, printing inks, paper coatings, and adhesives.
Butyl benzyl phthalate	---	6.8×10^7 kg produced in 1979, plasticizer for polyvinyl chloride and cellulosic plastics; carrier and dispersing media for pesticides, cosmetics, and colorants.

(cont.)

Table 4-8 (Cont.)

Compound found in ground water under landfill[a]	Ground water concentration (ppb)	Product and commercial uses[b]
Dicyclohexyl phthalate	0.2	Plasticizer for nitrocellulose, ethyl cellulose, chlorinated rubber, polyvinyl acetate, and polyvinyl chloride.
Dioctyl phthalate (di(2-ethylhexyl)phthalate)	2.4	1.4×10^7 kg produced in 1979, plasticizer for polyvinyl chloride (86%), cellulose esters (4%), synthetic elastomers (3%), other vinyl resins (3%), other polymers (1%); nonplasticizer uses (3%).
N-Ethyl-p-toluene-sulfonamide and N-Ethyl-o-toluene-sulfonamide	0.1 ---	Plasticizer in polyurethanes, nylons, polyesters, alkyds, phenolics, epoxides, and amine resins.
Tri-n-butyl phosphate	1.7	2.3×10^8 kg produced in 1979, solvent for nitrocellulose, cellulose acetate; and plasticizer.
Triethyl phosphate	0.3	4.1×10^8 kg produced in 1979, solvent; plasticizer for resins, plastics, and gums; lacquer remover; and flame retardant for polyesters.

[a]Source: Dunlap et al., 1976. ppb is parts per billion; ug/L.
[b]Source: Radian, 1987; Sax and Lewis, 1987.
[c]Detected but not quantitated by Dunlap et al. (1976).

Table 4-9

CONCENTRATIONS OF CHEMICALS USED AS PLASTICS ADDITIVES
FOUND IN LEACHATE FROM MUNICIPAL SOLID WASTE LANDFILLS

Compound	No. of Sites at Which Constituent Was Detected (a)	Concentration Range (ppb)	Median Concentration (ppb)	Use in Plastics Manufacturing	Other Major Uses/ Comments
ORGANIC COMPOUNDS					
Bis-(2-ethylhexyl) phthalate	8	16-750	80	Plasticizer	Limited other uses: Some use as synthetic elastomer (b)
Diethyl phthalate	12	3-330	83	Plasticizers	See Table 4-8; Limited other uses: Some use as a solid rocket propellant
Vinyl chloride (c)	6	8-61	40	Intermediate product in PVC manufacturing	Limited other uses; some exporting
Xylene (c)	6	32-310	71	Polyester resins	Protective coatings. solvents, aviation gas (b)
INORGANIC COMPOUNDS					
Antimony	9	0.0015-47	0.066	Flame retardant in PVC and as a colorant	Limited other uses (b)
Cadmium	31	0.007-0.15	0.0135	Colorants, Stabilizers	See Table 4-6; Use in plastics is 31.5% of all cadmium in MSW
Lead	45	0.005-1.6	0.063	Colorants, Stabilizers	See Table 4-6; Use in plastics is 1.7% of all lead in MSW

(a) Leachate from a total of 51 sites was analyzed for organic constituents and from 62 sites
for inorganic constituents.
(b) From Radian, 1987.
(c) Vinyl chloride and xylene are not additives but are used in plastics manufacturing.
Source: U.S. EPA, 1988.

could have originated from PVC pipe in monitoring systems in which such pipe is used (Barcelona, 1989). If this is the case, then the number of landfills at which phthalates have been detected may overstate the frequency at which these chemicals leach from wastes. If Teflon piping was used for most of the studies, however, virtually no release of phthalates from piping would be expected. No estimate was obtained of the reverse effects of absorption of phthalates from wastes.

Three metals, antimony, lead and cadmium, are included in the table as well. For lead and cadmium, the numerous potential contributors in landfilled wastes make it difficult to interpret the monitoring results. As noted previously, the plastic additives containing lead and cadmium contribute approximately 1.7% and 31.5% of the quantities of these two metals in the landfill. Antimony may have originated in plastic wastes, but it has not received as much attention as lead and cadmium as a potential landfill problem. Other inorganic chemicals besides antimony, lead and cadmium could also be present due to plastic additives.

The table also reports the presence of vinyl chloride monomer in some of the leachate samples, the source of which is uncertain. Some researchers have theorized that biological activity in the landfill can produce vinyl chloride (Murthy et al., 1989). Specifically, biochemical anaerobic reactions involving wastes such as lignin from paper can generate vinyl chlorides. Also, PVC manufactured before 1975 could contribute some vinyl chloride monomer to leachate because the less-complete PVC processing techniques employed until that time left some monomer in the materials (Webster, 1989). Researchers have also noted that vinyl chloride appears in leachate from numerous landfills with no apparent correlation with the types of waste disposed (Webster, 1989).

Two weaknesses of the landfill leachate data are concerns about the range of possible monitoring conditions and the possibility of industrial wastes being included in the landfill. It is also worthwhile, therefore, to consider recent research in which such factors were well controlled. EPA published a study of controlled landfill experiments involving municipal wastewater treatment sludge and municipal solid waste (SCS Engineers, 1989). The study was designed to evaluate sludge landfilling as a disposal option for that waste. As part of their research effort, however, the researchers also evaluated the environmental releases from co-disposed sludge and MSW and from MSW alone. The MSW-alone cells were used as control cells in the experiment.

The SCS research team designed 28 lysimeters which were loaded with various combinations of sludge, sludge and MSW, or MSW alone. The wastes were loaded into lysimeters designed to simulate the leachate and gas generation of landfilling. The codisposal and MSW-only cells were 6 feet in diameter and 9 feet in height. The City of Cincinnati provided approximately 50 tons of MSW to the project; the research team mixed the waste by tearing open plastic garbage bags before they were placed into the lysimeters. The staff also removed certain large and unrepresentative items (including a piano and some tires) before the waste was placed into the lysimeters. A sample of the MSW used in the tests revealed plastics as 8.1% of the MSW, with paper (except telephone books) accounting for 45.4%; textiles, 11.9%; yard waste, 10.5%; ferrous metals, 6.3%; telephone books, 4.6%; and a variety of other materials, 13.2%.

The vessels were loaded with waste in July 1982 and quarterly monitoring of leachate was performed for a period of four years. Thirty-four parameters were measured to determine leachate and gas quality and quantity. SCS included nine priority pollutants, including three phthalates, in the leachate studies. (Testing for the phthalates was discontinued after three years, however, in order to conserve project funds.) The researchers also added water in differing quantities to the lysimeters.

The findings of most interest for this research are the levels of phthalates in the leachate from the MSW-only lysimeters. The study authors did not comment on these results directly, however, since the MSW-only cells were used as study controls. Table 4-10 summarizes the results for each of the three phthalates averaged over four lysimeters. The four lysimeters received two different levels of moisture during the year. The table shows that the highest annual means of the leachate levels of phthalates were 138 ppb for Bis(2-ethylhexyl)phthalate (also called di(2-ethylhexyl)phthalate), 49 ppb for dibutyl phthalate, and 162 ppb for dimethyl phthalate. The highest individual leachate tests were 424 ppb for dimethyl phthalate. (The sludge samples are also included in the table as an indication of the presence of these chemicals in the other material tested.) This test, it should be noted, does not replicate landfill conditions in that only one load of the wastes was placed in the cell. In an operating landfill, wastes would be added continuously. It does provide, however, an indication of the presence of phthalates in MSW leachate. The levels of Bis(2-ethylhexyl)phthalate are also consistent with the range reported in the collected landfill monitoring studies performed above.

In conclusion, the municipal landfill data indicate some generation of leachate from phthalates; these chemicals are commonly used in plasticizer additives. Among inorganics, the potential contribution of plastic additives cannot be determined.

4.3   INCINERATION

Municipal waste incineration is currently the subject of significant analytical efforts and an active EPA regulatory program. At least two current Federal initiatives are addressing MSW incineration:

EPA Regulation of Municipal Waste Combustor (MWC) Emissions. On November 30, 1989, EPA proposed regulations under sections 111(a), 111(b), and 111(d) of the Clean Air Act (CAA) to control emissions from new and existing municipal waste combustors (MWCs). These regulations address three classes of MWC emissions: (1) MWC organics (including dioxins/furans); (2) MWC particulates (including metals such as lead and cadmium); and (3) MWC acid gases (including hydrogen chloride [HCl] and sulfur dioxide [SO_2]). The regulations also address nitrogen oxide (NO_x) emissions from certain new MWCs. For existing MWCs, the regulations provide guidelines for the development of state plans to control MWC emissions. For new MWCs, the regulations are proposed as New Source Performance Standards (NSPS) to limit dioxin/furan, particulate, HCl, SO_2, and NO_x emissions; the proposed NSPS also impose operating standards to provide further assurance of control of dioxin/furan emissions. For both new and existing facilities, the proposed regulations also require that 25% by weight of the waste stream be separated for recovery prior to combustion. EPA is accepting public comments on these proposed regulations until March 1, 1990, and will promulgate final regulations on MWC emissions later this year. (54 FR 52209 and 52251, December 20, 1989).

Table 4-10

PHTHALATE LEVELS IN LEACHATE FROM SIMULATED LANDFILL STUDY (ppb)

Year	MSW-Only		MSW-Sludge (a)		Sludge-Only	
	Mean	Highest Value	Mean	Highest Value	Mean	Highest Value
Bis (2-ethyl hexyl) Phthalate						
First	96.5	235.5	11.2	22.9	4.4	7.5
Second	58.4	182.2	13.7	22.7	4.9	6.9
Third	138.3	238.0	84.9	206.1	130.7	191.0
Three-yr. avg.	97.7	--	36.6	--	46.7	--
Dibutyl Phthalate						
First	10.0	19.0	11.1	16.5	7.1	9.6
Second	11.9	22.9	10.3	12.2	5.3	7.0
Third	49.3	94.8	53.3	102.7	59.3	101.7
Three-yr. avg.	23.7	--	24.9	--	23.9	--
Dimethyl Phthalate						
First	142.7	340.0	14.6	54.0	513.1	1,800.0
Second	162.6	424.3	100.8	156.5	198.0	420.0
Third	0.0	0.0	0.4	1.6	1.0	3.8
Three-yr. avg.	101.8	--	38.6	--	237.4	--

(a): Wastes tested were 70% MSW and 30% municipal wastewater treatment sludge.

Source: SCS Engineers, Inc., 1988

The regulations specify plastics as one of the MSW constituents that may be targeted for materials spearation, but do not specifically address the contribution of plastics to MWC emissions. The three major classes of emissions addressed (dioxins/furans, metals, and acid gases) are, however, those which have been most often associated with concerns about plastics combustion.

FDA Environmental Impact Statement Regarding Polyvinyl Chloride. The Food and Drug Administration (FDA) has filed a Notice of Intent to prepare an Environmental Impact Statement (EIS) on the impacts of increased consumer use and disposal of polyvinyl chloride (PVC) (53 FR 47264, November 22, 1988). The EIS will consider the impacts of PVC combustion, including impacts on MWC operations as well as direct and indirect impacts on human health and the environment.

These initiatives follow upon EPA's "Municipal Waste Combustion Study," a Report to Congress published in June 1987 (U.S. EPA 1987a). The Report and its companion technical volumes provide a comprehensive overview of issues and concerns related to MSW combustion. Although not focused specifically on plastics, the Report discusses at length the primary health and environmental concerns related to plastics incineration, including polychlorinated dioxins (PCDD) and furans (PCDF) emissions, HCl emissions, and the generation and toxicity of MWC ash. The Report also discusses the efficiency, availability, and cost of control technologies to reduce MWC emissions.

Although they constitute only some 7.3% of MSW (see Table 2-16), plastics have figured heavily in the controversy surrounding MSW incineration. In particular, polyvinyl chloride (PVC) has been subject to significant attention because of its potential contribution to the formation of PCDDs and PCDFs during MSW combustion, and its potentially deleterious impact on incinerator operation (through formation of HCl gas). Plastics have also received attention because of their contribution to heavy metals in MSW and the consequent potential impacts on MWC ash toxicity.

The following discussion focuses on these concerns. It does not, however, present firm conclusions regarding the human health and environmental impacts of plastics incineration. Development of EPA's conclusions awaits completion of EPA's MWC rulemaking and FDA's EIS on PVC incineration. EPA will share the results of these analytical efforts with Congress when they are complete. As noted above, EPA's proposed regulations are scheduled for release by the end of 1989. FDA's draft EIS is expected to be released in 1990.

4.3.1 Introduction

The following paragraphs provide an introduction to the existing and projected population of municipal waste combustors (MWCs) in the United States. They also describe the combustion properties of plastics relevant to their use as an MWC fuel and their impact on MWC operation, emissions, and ash. A brief description of the combustion process and of available pollution control technologies for MWC is also provided.

4.3.1.1 Number, Capacity, and Types of Incinerators

In 1986, 6% of municipal solid waste was incinerated (Franklin Associates, 1988a). Primarily because it reduces MSW disposal requirements by 70 to 90% (by volume), incineration has become an increasingly attractive disposal option for many communities, especially those facing dwindling landfill capacity and rapidly increasing tipping fees. Table 4-11 traces the development of MSW incineration in the U.S. since 1955. There are currently approximately 160 incineration facilities (320 units) online, representing nearly 68,000 tons per day (tpd) of capacity. Of this total, nearly 48,000 tpd (78%) has come online since 1980 (Radian, 1989). EPA projects continued rapid growth for this disposal option; based on facilities under construction, under contract or contract negotiation, or formally proposed, the Agency projects nearly 200 additional facilities, representing approximately 175,000 tpd of capacity, to come online in the next few years (U.S. EPA, 1987b).

Three types of incinerators comprise virtually all of current U.S. MWC capacity (Radian 1989):

- **Mass burn (57% of current U.S. MWC capacity).** Mass burn combustors accept all MSW except items too large to go through the feed system. Unsegregated refuse is placed on a grate that moves through the combustor. Air in excess of that required for combustion is forced into the system below and above the grate.

- **Refuse-derived fuel (RDF) (29%).** RDF combustors require that waste be processed before combustion; processing typically consists of shredding and removal of most noncombustibles (e.g., glass, aluminum, and other metals). RDF may be co-fired with coal.

- **Modular combustors (10%).** These combustors are typically smaller than mass burn facilities and also accept waste without processing. One type of modular combustor is similar to the mass burn units in that excess air enters the primary combustion chamber. A second type of modular system uses starved-air primary combustion; excess air is added to the partially combusted gases in a secondary chamber to achieve complete combustion. The lower air velocities in the starved air system suspend less ash and reduce the problem of fly ash control.

Over 81% of MWC capacity represents waste-to-energy facilities equipped with heat recovery boilers. Most facilities that do not recover heat to generate steam or electricity are older incinerators -- over 95% of capacity brought online since 1980 has included heat recovery boilers (Radian, 1989).

Table 4-11

MUNICIPAL WASTE COMBUSTORS OPERATIONAL
IN THE UNITED STATES, 1988

Year of Plant Startup	Facilities	Units	Capacity (tons/day)
Not Available	19	23	1,309
pre-1955	1	2	200
1956-1960	2	8	1,960
1961-1965	4	6	1,005
1966-1970	8	21	5,008
1971-1975	23	41	7,610
1976-1980	21	34	2,897
1981-1985	43	102	27,227
1986-1989	38	83	20,614
Total	159	320	67,830

Source: Radian, 1989.

4.3.1.2 Combustion Properties of Plastics

The various types of plastics have quite different combustion properties. Categorized by combustion properties, plastics can be described as follows (see Table 4-12) (Leidner, 1981):

Polyolefins (e.g., polyethylenes, polypropylenes, polystyrene). All of these have high heats of combustion (generally over 17,000 Btu/lb) and combust primarily to carbon dioxide and water (under good incinerator operating conditions). They may contain additives (e.g., pigments or flame retardants), but generally yield relatively small amounts of ash, or corrosive or toxic gases.

Oxygen-containing plastics (e.g., polycarbonates, polyacetals, polyethers, polyesters, polyacrylates, and polymethacrylates). These typically have lower heats of combustion (11,000 - 17,000 Btu/lb) but also combust primarily to carbon dioxide and water. Possible additives (e.g., flame retardants or pigments) can produce ash, or corrosive or toxic gases.

Nitrogen-containing plastics (e.g., polyacrylonitriles, polyamides, and polyurethanes). These are similar in heating value to the oxygen-containing plastics.

Halogen-containing plastics (e.g., polyvinyl chloride and other polyvinyl halides). These plastics have low to moderate heats of combustion (e.g., less than 11,000 Btu/lb) and may not be flammable under ambient conditions. The potentially large quantities of additives (e.g., plasticizers) in these plastics, however, enhance flammability. Upon combustion, the polymers yield hydrogen chloride (HCl) or hydrogen fluoride (HF), which dissolve in water to produce corresponding hydrohalic acids that are corrosive to metal and other materials. Moreover, without sufficient flammable material and air (oxygen) to ensure a high flame temperature, these materials tend to produce soot when they burn (Tsuchiya and Williams-Leir, 1976).

Despite their relatively small contribution by weight to post-consumer waste, plastics contribute disproportionately to the Btu content of incinerated MSW. Assuming an average heat of combustion for mixed plastics of 14,000 Btu/lb, plastics contribute over half again as much to the fuel value of MSW as a comparable mass of paper or wood, and have a fuel value three times that of typical MSW (see Table 4-12). Magee (1989) has estimated that the 7.3% by weight of plastics in MSW may contribute nearly 25% to the total Btu content of the waste.

4.3.1.3 Plastics Combustion and Pollution Control

COMBUSTION -- Plastics burn in two phases: pyrolysis and combustion (see Figure 4-3). During pyrolysis, the complex plastic solids are chemically decomposed by heat into gases, the composition of which is strongly dependent on the plastic involved (Boettner et al., 1973) and on the conditions (temperature, pressure, etc.) under which pyrolysis occurs. The mixture of pyrolysis gases then enters the flame, where combustion takes place. Because one plastic can produce dozens of different pyrolysis products, a variety of volatile compounds enter the flame. (In this, plastics are no different from other organic materials -- wood, paper, and food wastes all produce a wide variety of pyrolysis products.) In contrast, a limited number of combustion products leave the flame (under good combustion conditions; see below). Regardless of the

Table 4-12

HEATING VALUES FOR PLASTICS AND OTHER MSW COMPONENTS

Material	Examples	Heating Value (Btu/lb)
Polyolefins	Polyethylene Polypropylene Polyisobutylene Polystyrene	17,870–20,150
Halogen–containing plastics	Polyvinyl chloride Polyvinylidine chloride	7,720 4,315
Oxygen–containing plastics	Polycarbonates Polyacetals Polyethers Polyesters Polyacrylates Polymethacrylates	11,470–13,410
MSW (typical)	N/A	4,500–5,500
Nitrogen–containing plastics	Polyacrylonitrile	13,860
Paper	N/A	7,590
Wood flour	N/A	8,520

Note: Heating values represent the amount of energy released during combustion of a substance, and can be used to compare the relative efficiencies of different substances as fuels during incineration. For comparison, the heating value of fuel oil is approximately 17,700 Btu/lb.

N/A – not applicable.

Source: Leidner 1981

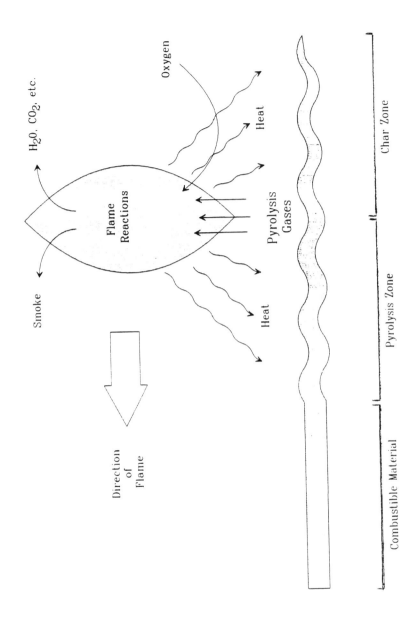

Figure 4-3. Flame Dynamics Showing Separation of Pyrolysis and Oxidation

material being burned, combustion gases are typically small, chemically stable (two- or three-atom) molecules (Dynamac, 1983b). Typical combustion gases include water (H_2O), carbon dioxide (CO_2), carbon monoxide (CO), sulfur dioxide (SO_2), and nitrogen oxide (NO).

The composition of the dominant combustion gases is determined by the ratios of elements (C/H/O/N) entering the flame and the temperature and pressure of the flame. Other elements present as plastics constituents (e.g., chlorine) or additives (e.g., lead, cadmium, tin) experience a variety of fates. Some are released primarily as gaseous emissions (e.g., chlorine and mercury), while others are entrained either in fly ash or bottom ash.

Incomplete combustion (caused by either insufficient oxygen or low flame temperature) may lead to the emission of more complex products -- typically mixtures of pyrolysis products that are not completely oxidized. The organic emissions of most concern in MSW incineration (chlorobenzenes, chlorophenols, PCDDs, PCDFs, and others) are the products of incomplete combustion. Incomplete combustion may also lead to the emission of large volumes of particulates (soot), which may disrupt the operation of particulate collection devices.

POLLUTION CONTROL -- Proven pollution control technologies are available to effect greater than 99% capture of MWC particulate emissions and greater than 90% capture of acid gas emissions (HCl, HF, and SO_2). The most effective identified combination of particulate/acid gas controls consists of a dry alkaline scrubber coupled with a fabric filter or electrostatic precipitator (U.S. EPA, 1987b). As part of its current regulatory development process for MWCs, EPA is considering a requirement that these or other technologies be installed at new MWCs; emissions controls may be imposed on existing MWCs through state guidelines developed pursuant to Section 111(d) of the Clean Air Act (U.S. EPA, 1987b).

Among the current population of MSW incinerators, particulate control devices are widespread, but very few facilities include acid gas control technologies. Table 4-13 describes the distribution of pollution control technologies among operational MSW incinerators. Over 97% of MWC capacity is equipped with particulate controls; electrostatic precipitators are the dominant technology (installed on nearly 75% of MWC capacity), followed by fabric filters (12%) and a variety of other technologies (Radian, 1989). Information available to EPA does not allow a precise estimate of the online MWC capacity equipped with acid gas controls; it appears, however, that not more than about 15% of current MWC capacity is fitted with technologies capable of effective acid gas emission control (Radian, 1989). Virtually all of the 2,157 tpd of MWC capacity with no installed pollution control technologies consists of small modular units.

Table 4-13

POLLUTION CONTROL EQUIPMENT INSTALLED
ON OPERATIONAL U.S. MUNICIPAL WASTE COMBUSTORS

Pollution Control Equipment	Capacity (tons/day) Employing Specified Pollution Control Equipment, Listed by Year of Plant Startup									
	1955	1956-1960	1961-1965	1966-1970	1971-1975	1976-1980	1981-1985	1986-1989	Not Available	TOTAL
None	0	0	90	100	284	457	773	100	353	2,157
Electrostatic precipitator (ESP)	200	960	75	4,360	6,141	1,824	25,486	9,643	50	48,739
Spray dryer/ESP	0	0	0	0	0	0	0	1,500	0	1,500
Cyclone/ESP	0	0	0	0	0	400	0	0	0	400
Fabric filter	0	0	0	0	0	0	508	506	56	1,070
Spray dryer/Fabric filter	0	0	0	500	0	0	0	6,721	0	7,221
Wet scrubber/Fabric filter	0	0	0	0	0	0	0	80	0	80
Duct sorbent injection	0	0	0	0	0	0	0	200	0	200
Wet scrubber	0	0	600	48	0	60	0	0	450	1,158
Venturi wet scrubber	0	1,000	0	0	1,175	0	0	94	0	2,269
Cyclone/Venturi wet scrubber	0	0	0	0	0	0	0	400	0	400
Cyclone	0	0	0	0	0	156	0	0	0	156
Electrified gravel bed	0	0	0	0	0	0	360	400	0	760
Wetted baffles	0	0	240	0	0	0	0	0	0	240
Not Available	0	0	0	0	10	0	100	970	400	1,480
TOTAL	200	1,960	1,005	5,008	7,610	2,897	27,227	20,614	1,309	67,830

Source: Radian, 1989.

4.3.2 Incinerator Management Issues

A number of issues related to incinerator management and operations have been associated with the combustion of plastics as a component of MSW. These include:

- Excessive flame temperature
- Formation of incomplete combustion products
- Formation of slag
- Formation of corrosive gases

Of these issues, the formation of corrosive gases is of greatest concern. PCDDs, PCDFs, and other potentially harmful emissions constituents or their precursors are typically products of incomplete combustion. Because they are of concern as a potential environmental release rather than as an incinerator management/operation issue, they are addressed in Section 4.3.3, below.

4.3.2.1 Excessive Flame Temperature

Excessive flame temperature can damage incinerator construction materials and lead to increased emissions of some pollutants (e.g., carbon monoxide). Excessive flame temperature may result from the combustion of high-Btu fuel in the presence of sufficient oxygen. Limited anecdotal evidence has suggested that plastics occasionally contribute to excessive MWC flame temperature (Wirka, 1989), but EPA's literature review and solicitation of industry opinion for this report have not suggested that this problem is serious or widespread. On balance, the positive contribution of plastics to MSW fuel value outweighs concerns related to the possible occurrence of excessive flame temperature. However, as the percentage of plastics in the waste stream increases, this concern may need to be re-examined.

4.3.2.2 Products of Incomplete Combustion (PICs)

Low flame temperature and/or insufficient oxygen can lead to emission of carbon monoxide, pyrolysis gases, and/or soot. Incineration of plastics raises this management issue for two reasons: 1) combustion of some plastics (e.g., halogen-containing plastics or plastics containing flame retardants) may reduce flame temperature; or 2) large concentrations of some high-Btu plastics may overwhelm the local air supply in the combustion chamber, resulting in the formation of pockets of volatile PICs that may be emitted from the incinerator if insufficient secondary air is available to complete combustion. Such occasional incidences of PIC emissions have been termed "transient puffs."

There is little reason for concern that plastics in MSW may cause operating conditions leading to the formation of incomplete combustion products. The mixtures of plastics introduced into MSW incinerators are unlikely to have the effect of reducing flame temperature. Concerns regarding transient puffs of PICs related specifically to plastics appear to be largely unsupported. Given existing and projected concentrations of plastics in MSW, it is unlikely that plastics combustion could exhaust local air supplies and result in significant PIC emissions. Proper incinerator operation (e.g., mixing the incinerator feed, maintaining adequate primary and secondary air) is far more important to controlling PIC emissions than is the presence of plastics or any other single MSW constituent. If conditions conducive to incomplete MSW combustion do arise for any reason, combustion products associated with plastics feed pose management and operation problems no different than products associated with other MSW constituents (wood, paper, food wastes, etc.)

4.3.2.3 Formation of Slag

Slags form when substances melt under incinerator operating conditions and travel as liquids to relatively cool zones of the incinerator, where they resolidify. These substances become an operational concern if (for example) they clog air inlets or interfere with the operation of grates or stoking devices. Incinerator feed material can contribute to slag formation 1) if the material itself is prone to form a slag, or 2) if the material contributes to the development of operating conditions conducive to slag formation (low temperature, low oxygen concentration) from other feed constituents.

EPA has not seen any evidence to suggest that plastics contribute to slag formation by either of these pathways. The fact that plastics are a high-Btu incinerator fuel implies both that they are extremely unlikely themselves to form a slag under almost any incinerator operating conditions, and that they are unlikely to contribute to the development of the low-temperature conditions conducive to slag formation from other constituents of MSW.

4.3.2.4 Formation of Corrosive Gases

The introduction of rising quantities of plastics into MSW incinerators has led to concern regarding the generation of corrosive gases such as hydrogen chloride, and organic acids (e.g., acetic acid). As an incinerator management and operational issue, hydrogen chloride gas generated upon introduction of polyvinyl chloride (PVC) and related halogenated plastics into incinerator feed provokes the greatest concern (FDA, 1988a; Seelinger, 1984); this concern is related primarily to corrosion of incinerator and boiler internal surfaces and to its impact on incinerator reliability and lifespan. PVC and related chlorinated polymers are plastics especially implicated in this concern.

EPA and FDA are both currently investigating the impact of chlorinated plastics on MWC operation; pending the outcome of these initiatives, it would be premature for EPA to present definitive conclusions at this time. The following paragraphs describe some of the preliminary results of these and other analyses, highlighting the major sources of debate and the most significant questions awaiting resolution.

The controversy surrounding the impact of hydrogen chloride gas on incinerator operations is defined by two opposing lines of analysis. On the one hand, PVC is a minor constituent of MSW. PVC accounts for approximately 15% of U.S. plastics production (Table 2-2). But because many applications of PVC are in long-lived construction applications (many of which do not enter the MSW stream), it has been estimated that PVC contributes only approximately 9-11% to total MSW plastics discards in the U.S. (Alter, 1986). These estimates place bounds of approximately 0.6-1.1% on the contribution of PVC to MSW. Because PVC is only one of many potential sources of chlorine in MSW (other significant sources include paper, food wastes, lawn and garden wastes), these estimates lead some investigators to conclude that PVC cannot be a significant contributor to total HCl in MWC emissions (e.g., Magee, 1989).

On the other hand, PVC may be one of the major sources of chlorine in MSW. Alter et al. (1974) reported that all plastics accounted for approximately 36% -- and were the largest single source -- of total chlorine in MSW samples in Wilmington, Delaware. Other significant chlorine sources in these samples were paper (23% of total chlorine), food waste (17%), and rubber and leather (14%) (Table 4-14). Churney et al. (1985) analyzed MSW in Baltimore County, Maryland, and Brooklyn, New York. In Baltimore County, all plastics contributed approximately 30% to total chlorine content; the major chlorine source in these MSW samples was paper (Table 4-14). In Brooklyn all plastics contributed 51% to total MSW chlorine; paper was also the other major chlorine source in these samples, contributing 25% to total MSW chlorine (Table 4-14). Total chlorine in the Brooklyn samples was also nearly double that in the Baltimore County samples. Because PVC is by far the major chlorine-containing plastic, some investigators have concluded from these statistics that PVC is potentially the largest single source of HCl emissions from incinerators, and contributes significantly to the potential for corrosion under at least some operating conditions.

It is clear that there is a correlation between the PVC content of incinerator feed and uncontrolled HCl emissions. Kaiser and Carotti (1971) added varying amounts of PVC to incinerator feed and identified an apparently linear relationship between the mass of PVC added and HCl emissions (Table 4-15). Their work has been corroborated in a series of test runs at a Pittsfield, Massachusetts, incinerator (MRI, 1987).

The presence of hydrogen chloride gas does not cause significant corrosion under all operating conditions. At low operating temperatures, HCl may condense and form hydrochloric acid, which will attack metal surfaces. Under some conditions at higher temperatures, a series of reactions may occur between chlorine, steel, and oxygen to result in the formation of iron oxide (rust). Many modern incinerators are constructed with corrosion-resistant materials (refractories, ceramic-coated metals, reinforced plastics); and most include operating controls sufficient to ensure that conditions conducive to HCl corrosion occur infrequently, if ever. Nonetheless, because a significant proportion of U.S. incinerator capacity has neither of these safeguards against corrosion, the contribution of PVC to incinerator HCl formation is a potentially significant concern.

4.3.3 Environmental Releases

Plastics combustion products and residues may be released either to gaseous emissions or to ash (including fly ash and bottom ash). Historically, three environmental releases from incinerated plastics have caused the greatest concern:

- Hydrogen chloride (HCl) emissions

- Dioxin (i.e., polychlorinated dibenzodioxin (PCDD) and dibenzofuran (PCDF)) emissions

- Heavy metals in ash

TABLE 4-14

CONTRIBUTION OF MSW CONSTITUENTS TO TOTAL MSW CHLORINE

MSW Constituent	Contribution to Total MSW Chlorine		
	Baltimore County Maryland (1)	Brooklyn New York (1)	New Castle County Delaware (2)
Paper	55%	25%	23%
Plastics	30%	51%	36%
Organics, Total	1%	6%	
Wood			1%
Garden Waste			4%
Food Waste			17%
Textiles	4%	2%	6%
"Fines"	9%	15%	
Rubber and Leather	ND	ND	14%
	----	----	----
TOTAL	100%	100%	100%
Total Chlorine in MSW (% by Weight)	0.46%	0.89%	ND

Sources:

(1) Churney et al., 1985

(2) Alter et al., 1974. Note: Alter et al. analyzed only the organic fraction of MSW. Therefore, these percentages overstate the contribution of each listed constituent to total MSW chlorine, since some chlorine is found in inorganic MSW constituents.

Table 4–15

EFFECT OF INCREASING THE PVC CONTENT
OF MUNICIPAL SOLID WASTE
ON MEASURED CHLORINE LEVELS

Percentage of Polyvinyl Chloride Added	Chlorine Content of Flue Gas(a)
None	455
2 percent	1,990
4 percent	3,030

(a) ppm by volume of dry gas corrected to 12% carbon dioxide.

Source: Kaiser and Carotti, 1971.

MWC emissions have caused particular concern because MWC facilities are frequently located in or near heavily populated areas with potentially poor dispersion characteristics for emitted pollutants. Concerns relate both to health impacts (e.g., from HCl and dioxins) and to the potential for corrosion of exposed surfaces (e.g., from HCl).

Concerns over hydrogen chloride (HCl) and dioxin emissions relate exclusively to the combustion of PVC (and related chlorinated polymers), and not to other plastics present in MSW; the source of concern is the chlorine content of these polymers and its potential contribution to HCl and dioxin formation. As stated above, both EPA and FDA are in the process of completing analyses of the contribution of PVCs to MWC emissions and of their impacts on human health and the environment. This section outlines the most significant areas of uncertainty regarding the impacts of PVCs on MWC emissions, and summarizes evidence tending to reinforce or to rebut concerns about these impacts. But pending the completion of the EPA and FDA analyses, this discussion does not present Agency conclusions about health or environmental impacts related to the presence of PVCs in MSW incinerator feed.

Heavy metals are present in some plastics as additives -- generally colorants or heat stabilizers. The metals of greatest concern are lead and cadmium. Again, significant controversy surrounds not just the contribution of plastics to the heavy metal content of MWC ash, but also the more-encompassing questions related to the impacts of toxic MWC ash constituents from all sources. This section focuses on the contribution of plastics to the total concentrations of lead and cadmium in MWC ash, but does not address the larger issues related to the overall toxicity of MWC ash.

Additional issues occasionally related to plastics combustion are emissions of phosphorous and sulfur compounds, other products of incomplete combustion, and aerosols generated by flame retardants and other plastics additives. These issues are addressed briefly in this section, but in general they are considered to be much less significant than concerns related to PVC combustion and to the contribution of plastics to heavy metals in MWC ash.

4.3.3.1 Emissions from MSW Incinerators

Five classes of emissions are addressed in the following paragraphs: hydrogen chloride, dioxins and furans, sulfur and phosphorus compounds, products of incomplete combustion, and aerosols. Of these, hydrogen chloride and dioxin emissions related to the combustion of PVC are considered the most consequential.

HYDROGEN CHLORIDE - Polyvinyl chloride (PVC) and other chlorine-containing plastics yield hydrogen chloride (HCl) gas when combusted. Chlorine emissions from MSW incinerators have been correlated with the amount of PVC in MSW feed (see Table 4-15). Section 4.3.2.4 (above) focused on the impacts of HCl on MWC management and operation, but HCl emissions to MWC exhaust gases are also a significant issue. EPA has estimated current MWC HCl emissions to be approximately 24,000 metric tons per year, and emissions from projected MWC facilities to be an additional 97,000 metric tons per year (assuming no acid gas controls are installed) (U.S. EPA, 1987b). Potential concerns relate both to the impact of emitted HCl on exposed materials (e.g., corrosion of metals and other exposed surfaces) and to health impacts on human and animal populations.

Section 4.3.2.4 reviewed a variety of evidence relating the generation of HCl to the concentration of PVCs in MSW, and cited apparently contradictory conclusions reached by a number of researchers regarding the contribution of PVCs to HCl emissions. Ongoing analyses by both EPA and FDA will formalize these Agencies' conclusions regarding the significance of this contribution and potential means to address any identified problems.

DIOXINS AND FURANS - Because of their toxicity, dioxins (PCDDs) and furans (PCDFs), in either emissions or ash, have been among the greatest causes for public concern regarding MSW combustion. PVCs and other chlorinated polymers are the plastics implicated in MWC dioxin and furan generation. [Note: Most of the commodity plastics (e.g., PS, PET, PP, HDPE, and LDPE) do not contain chlorine and are not implicated in dioxin or furan emissions.]

A voluminous technical literature exists examining the possible mechanisms of dioxin/furan formation during combustion and the possible role of PVCs in dioxin/furan formation. The following discussion summarizes the evidence (frequently contradictory) developed in this literature and the disparate conclusions reached by a number of researchers. Pending completion of ongoing EPA and FDA analyses of this issue, however, the Agency cannot present definitive conclusions regarding the contribution of PVCs to MWC dioxin/furan emissions.

The chemistry of dioxin/furan formation during MSW incineration is unclear. Four theories have been developed to explain the presence of dioxins and furans in incinerator emissions:

- Dioxins may be present in MSW constituents and may not be destroyed in the incinerator. At least one study has cited evidence that dioxins may be present in MSW incinerator feed in concentrations equal to or greater than those observed in stack emissions, although the feed could not account for stack emissions of furans (Magee, 1989).

- Dioxins and furans may be formed from chlorinated organic precursors in the incinerator. A variety of potential precursors may be present in MSW, including PCBs, PCPs, and chlorinated benzenes.

- Dioxins and furans may be formed from organic compounds and a chlorine donor in the incinerator. A wide variety of materials in MSW may yield the postulated organic substrate (including petroleum products, wood and paper, and food wastes), while the chlorine could be derived either from an organic donor (e.g., PVC) or an inorganic chloride salt.

- Dioxins and furans may be formed from organic compounds and a chlorine donor as a result of catalyzed reactions on fly ash in incinerator exhaust. Again, a wide variety of materials in MSW may provide either the organic substrate or the chlorine donor.

As a chlorine donor, chlorinated plastics may contribute to the third and/or fourth of these mechanisms. Virtually any organic material (e.g., paper, wood, food waste) may provide the postulated organic precursors. There is no indication that PVCs or any other plastics contribute disproportionately to postulated organic substrates.

All but the first of these mechanisms relate the formation of dioxins and furans to incomplete combustion of organic compounds. Dioxin/furan formation by these mechanisms demands the presence of complex aromatic organic substrates -- compounds that may be released during pyrolysis of plastics, wood, paper, food wastes, leather, or of almost any other organic MSW constituent. But these compounds are amenable to complete oxidation. With proper operating conditions, therefore, including the presence of sufficient oxygen and a high flame temperature, the potential for dioxin and furan formation can be very much reduced by destruction of the required precursors. One of the primary goals in the design of current MSW incinerators is to ensure that both combustor design and operational control are such that complete combustion is facilitated and complex organic compounds are completely destroyed during incineration.

Evidence Refuting a Relationship between PVC and Dioxin/Furan Formation -- Experiments conducted at a modular, excess air incinerator in Pittsfield, Massachusetts, addressed the relationship between PVC feed concentration and PCDD/PCDF emissions. Under varying operating temperatures and feed compositions, the study failed to establish a statistically significant correlation between the amount of PVC in the incinerator feed and the levels of PCDDs or PCDFs at any of a number of measurement locations (incinerator exhaust up- and downstream of the boiler and in the stack). The study did identify a negative correlation between PCDD/PCDF concentration and incinerator temperatures, and a positive correlation between PCDD/PCDF concentration and carbon monoxide levels; these results tend to confirm the influence of operating conditions on dioxin and furan formation (MRI, 1987).

Magee (1989) reviewed a number of studies on the impact of PVC on incinerator emissions and concluded that the weight of evidence refutes any hypothesized correlation between PVC and dioxin/furan emissions. For example, Benfenati and Gizzi (1983, cited in Magee, 1989) attempted to correlate PCDD/PCDF emissions with HCl, SO_2, NO_x and CO emissions from a refuse incinerator using multiple regression techniques on data gathered over a nine-month period. No significant correlation was found. Ballschmiter (1983, cited in Magee, 1989) studied PCDD/PCDF emissions from six incinerators in Germany over one year of operation. His conclusions were as follows: "We have tried to correlate this wide range of dioxin content in fly ash with other measurable parameters... our particular concern was focused on HCl emissions, but there is no simple correlation with PCDD formation. The results even suggest no correlation at all."

A similar conclusion was reached by Visalli (1987, cited in Magee, 1989), who reviewed three incinerator test programs (including the Pittsfield study mentioned above). Visalli commented that because chlorine availability from all MSW sources is thousands of times greater than that required to account for measured PCDD/PCDF concentrations, it is extremely unlikely that dioxin and furan formation can be correlated with any single chlorine source. This conclusion was corroborated by Karasek et al. (1983, cited in Magee, 1989), who added sufficient PVC to triple the concentration found in unamended MSW but identified no increase in dioxin or furan emissions.

Rankin (Rutgers, 1986) has also summarized a number of studies and reviews that examined the relationship between chlorinated plastics and PCDD/PCDF emissions; and he found no evidence that such a relationship exists. Rankin also pointed out that even if all chlorinated plastics were removed from mixed MSW, the concentration of chlorine available from other sources is many times greater than that required to account for all the dioxins and furans present in municipal waste combustor emissions.

Evidence Supporting a Relationship between PVC and Dioxin/Furan Formation -- Contradicting this evidence, a number of studies suggest that PVCs do play a role in the formation of dioxins and furans during MSW incineration -- and that this role is potentially significant enough to warrant regulatory concern. A number of bench scale and laboratory studies (e.g., Markland et al., 1986; Liberti and Brocco, 1982) have reported the formation of PCDDs/PCDFs both when chlorinated plastics are pyrolized alone and when PVC is added as a chlorine donor to combustion mixtures consisting of pure vegetable extracts.

In an analysis conducted in 1988, FDA (1988b) made the following statement regarding the Pittsfield, Massachusetts, incineration study cited above:

> Although no statistically significant effect was found between the amount of PVC in waste and emissions of PCDD, [and] PCDF ...., the Pittsfield data suggest the possibility that a relationship exists. When the data are normalized for waste feed and airflow rates, mean concentrations of PCDD and PCDF usually increased when PVC was added to the feed. The minimal replication as well as substantial variability suggest that the statistical power of the tests to determine the effect of PVC spiking of feed was low.

FDA (1988) was also uncertain of the extent to which the Pittsfield study results could be extrapolated to the variety of existing incinerator conditions. EPA agrees with FDA's analysis of the Pittsfield study results. At least one reviewer of the Pittsfield study (Clarke, 1988) has suggested that the apparent relationship between PVC concentration and furan emissions was more pronounced than the apparent relationship with dioxins.

Summary -- Given these conflicting experimental results and the very different interpretations sometimes imposed on a single set of experimental data, it is hardly surprising that the contribution of PVCs to dioxin/furan formation remains a very controversial issue. A number of judgments can be made on the basis of existing evidence, however:

- Incinerator operating conditions are more important to PCDD/PCDF formation than the presence or absence of any single MSW constituent. However, it appears unlikely that operating conditions can be controlled adequately to ensure that dioxins are never formed during normal incinerator operations (Linak et al., 1987).

- PVCs can serve as a chlorine donor for PCDD/PCDF formation.

- There are multiple sources of chlorine in MSW, which in sum provide chlorine concentrations many times those sufficient to account for observed PCDD/PCDF emissions.

- PVC may be one of the major sources of chlorine in MSW.

The analytical efforts currently underway at both EPA and FDA should result in a compilation of the best evidence available to address the importance of PVCs to MWC dioxin and furan emissions, the further development of these Agencies' positions regarding the contribution of PVCs to dioxin and furan emissions, and appropriate strategies to address any identified problems.

SULFUR AND PHOSPHORUS EMISSIONS - No major commercial plastic polymers contain high percentages of sulfur or phosphorus, and EPA has not identified any significant concerns related to the contribution of plastics to MWC sulfur or phosphorus emissions. Phosphorus may be present in organophosphate flame retardant additives used in polyurethanes in furniture and bedding applications, and in organophosphate plasticizers employed in a variety of plastics (Radian, 1987; Dynamac, 1983a). In the experiments of Kaiser and Carotti (1971), the presence of 4% plastics in the MSW stream had no significant effect on sulfur (SO_2) emissions; only polyurethane plastic had any effect on the phosphate (PO_4^{3-}) emissions. These results suggest combustion of a chloro-organic phosphate flame retardant, which may have been present in the polyurethane (Dynamac, 1983a).

PRODUCTS OF INCOMPLETE COMBUSTION - Products of incomplete combustion may be emitted if MWC combustion conditions are inadequate to allow the complete oxidation of MSW pyrolysis products. The most common causes of incomplete combustion are inadequate oxygen and low flame temperature. Products of incomplete combustion from all sources in MSW include soot, carbon monoxide, hydrogen cyanide, organonitriles, olefins, and chlorinated aliphatics and aromatics (Dynamac, 1983b). Many of these compounds are toxic, and even the nontoxic compounds can contribute to smog formation.

Any discussion of the contribution of plastics to incomplete combustion products must be placed in the context of the entire MSW stream. Pyrolysis of virtually any organic material results in the formation of a wide variety of compounds, toxic and nontoxic. EPA's research has not generated any evidence to suggest that plastics pyrolysis products are any more (or less) toxic than the pyrolysis products of other MSW constituents.

Virtually all pyrolysis products can be oxidized under proper incinerator operating conditions (adequate oxygen combined with a flame temperature in the incinerator's designed operating range). Given this fact, MSW constituents that tend to promote the maintenance of proper operating conditions are unlikely to contribute to the formation of incomplete combustion products; conversely, MSW constituents that tend to quench the incinerator flame may tend to be responsible for the formation of such products. Against this standard, virtually all plastics in MSW appear to promote the maintenance of conditions conducive to complete combustion, and so to reduce the possibility that pyrolysis products will be emitted. The possible exceptions to this conclusion pertain to plastics containing significant concentrations of flame retardant additives, but EPA has seen no evidence suggesting that these are a significant concern.

AEROSOLS FROM PLASTICS ADDITIVES - Noncombustible plastics additives may be a source of a variety of aerosol pollutants from MSW combustion, including species of bromine, phosphorus, antimony (all used for flame retardants), and others.

No direct evidence links plastics to observed ambient concentrations of these aerosols, and indirect evidence is inconclusive. Gordon (1980) examined the elemental composition of urban aerosols and suggested that contributions come from various source materials (e.g., soil, coal, limestone, oil, motor vehicle exhaust, sea salt, and MSW). Gordon's findings suggest that much of the airborne particulate content of antimony, cadmium, and zinc can be attributed to MSW. Plastics, in turn, may be the source of a significant proportion of antimony and cadmium in MSW.

EPA has estimated current MWC cadmium emissions to be approximately 10.4 metric tons per year (U.S. EPA, 1987b). Data developed by Franklin Associates (1988b) suggest that plastics may contribute over 30% of all cadmium in MSW and as much as 88% of all cadmium in the combustible fraction of MSW. Although to EPA's knowledge no concerns have been raised regarding MWC cadmium air emissions nor of the contribution of plastics to such air emissions, the contribution of MSW plastics to ambient cadmium concentrations may merit further research.

U.S. EPA (1987b) has estimated that 341 metric tons per year of lead are emitted from municipal waste combustors. However, plastics contribute only some 1.7% to all lead in MSW (Franklin Associates, 1988b), and EPA is not aware of any empirical or theoretical evidence linking MSW plastics combustion to potential concerns regarding MWC lead air emissions. Therefore, EPA does not consider MWC lead air emissions associated with plastics combustion to be of concern.

4.3.3.2 Plastics Contribution to Incinerator Ash

MWC ash is the subject of significant controversy. Public and Congressional concern has focused on the toxicity of MWC ash, primarily on fly ash. Ash, by definition, includes the noncombustible, or refractory, fraction of MSW that is not susceptible to pyrolysis and combustion during incinerator operations. Incinerator ash consists of two fractions:

- **Bottom ash** consists of relatively large particles removed from the grate or bed of the incinerator.

- **Fly ash** includes very fine particles entrained in incinerator exhaust gases -- the "particulates" captured by air pollution control devices. Because of its high surface area:volume ratio, fly ash typically holds more leachable compounds on particle surfaces than bottom ash and is more susceptible to leaching. Fly ash may also provide sites for the catalysis of reactions in flue gases; for example, one proposed mechanism for dioxin formation postulates a catalyzed reaction between hydrocarbons and a chlorine donor on the surface of fly ash particles.

Plastics may have two impacts on MWC ash generation: 1) they may affect the volume of ash generated, and 2) they may affect the toxicity of either fly or bottom ash.

VOLUME OF ASH GENERATED - EPA has estimated that between 3.2 and 8.1 million tons of MWC ash were generated in 1988, representing between 20 and 50% of the mass of incinerated MSW. Bottom ash constitutes 90 to 95% of total MSW ash; fly ash constitutes 5 to 10% of all ash generated (U.S. EPA, 1988b).

Virtually all of the carbon, hydrogen, nitrogen, and halogens in plastics combust to gaseous compounds and are emitted with stack gases during MSW combustion. Noncombustible plastics additives may produce refractory residues that contribute to fly and bottom ash generation.

Based on the concentration of additives in plastics (see Section 2), EPA has no evidence that plastics contribute disproportionately to the volume of MSW ash generated. The concentration of refractory materials is somewhat less in plastics than in MSW as a whole, suggesting that their relative volumetric contribution to incinerator ash generation is less than their contribution to the raw MSW waste stream.

INCINERATOR ASH TOXICITY - Toxic metals are the constituents of concern in plastics ash. These metals, used as additives in a variety of plastics products, include antimony, lead, cadmium, zinc, chromium, tin, and molybdenum. Of these substances, lead and cadmium have generated the most debate in relation to their contribution to MWC ash toxicity and are the focus of the following discussion.

As plastics additives, lead and cadmium are dispersed in a combustible medium. As such, they tend to be driven from the solid plastic during pyrolysis and to become entrained in the combustion gases and exhaust stream; ultimately, a substantial proportion presumably contribute to incinerator fly ash. Because most other lead and cadmium in MSW is contained in noncombustible items (see discussion below), the relative contribution of plastics to lead and cadmium in fly ash is probably greater than their contribution to lead and cadmium concentrations in unprocessed MSW.

Franklin Associates (1988b) has generated estimates of the contribution of plastics to total MSW discards of lead and cadmium (see Table 4-6). Franklin Associates estimates that 98% of lead discards are in noncombustible items. Of the 2.4% of lead contained in combustibles, plastics contribute an estimated 71%; therefore, plastics contribute about 1.7% of total lead discards. For cadmium, the situation is markedly different. Franklin Associates estimates that 36% of discarded cadmium is dispersed in combustible items and that 88% of this subtotal is represented by plastics. Thus, plastics account for nearly 32% of all cadmium discards, and for the bulk of discards in combustible items.

Using a different methodology, Considine (1989) has generated similar estimates of the contribution of plastics items to total discards of cadmium. For lead, Considine's estimates suggest a somewhat greater proportion of lead in the combustible fraction (15%), and a greater (5%) contribution of plastics to lead in MSW.

4.4   LITTER

Improper disposal practices are a subject of concern. This section examines the characteristics of general and plastic litter and compares types of litter for their impact on the solid waste stream.

4.4.1 Background

The content of litter is described in data collected from agencies with responsibilities for public roads or lands and from beach cleanup activities. For example, the Michigan Department of Transportation has conducted collection surveys of litter along state roads and highways (Michigan DOT, 1986). Table 4-16 indicates the types of litter generated in a variety of places. e.g., near roads and in parks. Plastic articles range from 13.4 to 21.1% of the total in the various areas. (These statistics on the relative share of plastic waste are not comparable to other analyses of solid waste, such as municipal garbage, because they are generated by a count of articles rather than by weight or volume measurements.)

For marine and beach wastes, information was presented in Sections 2 and 3 on the results of large beach and harbor cleanup and survey efforts. For one of the beach cleanup efforts. plastic wastes (again measured by a count of items) represented nearly two-thirds of the debris collected. This quantity is consistent with the expectation that plastic materials are most likely to be transported to the beach and thus will be highly represented in beach cleanup.

The Michigan research also can be used, although with some caution, to indicate the change in the composition of litter over time. Table 4-17 presents summaries of the litter collection results during 30-day tallying periods conducted for several years between 1968 and 1986. The share of plastics in the items collected grew from 10.0% in 1968 to 21.2% in 1986. Bottles and cans have both declined as a percentage of total waste. It should be noted, however, that Michigan passed a "bottle bill" in 1978, a change that would influence the relative shares of plastic wastes and other materials in litter. Thus, the change in the relative share of plastic items is indicative in unknown proportions to their increased share of numerous end use markets and the effect of the bottle bill legislation.

The growth in plastic litter may be particularly influenced by developments in lifestyle that increase the use of plastics in certain activities or situations which are prone to generation of litter. An indication of the relationship of plastic litter to the activities in which it is generated can be provided by the Michigan data on the plastic items accumulated in a 30-day collection period in 1986. Table 4-18 shows the distribution of plastic articles. It is interesting to note that fast-food containers and drink cups represented nearly 40% of the items. A common perception that fast-food packaging is responsible for much of the glut of solid wastes -- a perception noted by Rathje (et al., 1988) -- could originate partly from the observations of littered fast-food wastes. The largest category of plastic wastes, however, could be grouped only as miscellaneous items.

Litter is generated, however, from sources other than the casual fast-food patron. Keep America Beautiful (KAB), a national nonprofit public education organization that endeavors to improve community waste handling practices, has examined litter and other solid waste

Table 4-16

COMPOSITION OF LITTER AT
VARIOUS MICHIGAN STUDY SITES
(1986)

Litter Type	Percent of Items					
	Highway	County Roads	City	State Parks	Roadside Parks	Rest Area
Cans	4.3	6.5	5.7	8.7	2.6	1.4
Glass	2.8	2.6	6.6	9.9	3.6	0.4
Plastic	21.1	13.4	14.9	23.0	15.6	15.3
Paper	51.4	73.1	66.6	53.5	78.2	81.5
Miscellaneous	20.4	4.4	6.2	4.9	0.0	1.4
TOTAL	100.0	100.0	100.0	100.0	100.0	100.0

Source: Michigan DOT, 1986.

Table 4-17

NUMBERS AND TYPES OF ITEMS ACCUMULATED PER MILE(a)
ALONG MICHIGAN STATE HIGHWAYS IN A 30–DAY PERIOD
(1968–1986)

Type of Item	1968		1977		1978		1979		1980		1986	
	No.	%	No.	%	No.	%	No.	%	No.	%	No.	%
CANS – beer, soft drink & food	74	9.9	162	15.1	180	13.8	34	4.4	21	3.0	35	4.3
BOTTLES – beer, soft drink, food & liquor	49	6.5	47	4.4	48	3.7	11	1.4	8	1.1	23	2.8
PLASTIC – packages & containers	75	10.0	159	14.8	156	12.0	122	15.9	130	18.6	172	21.2
PAPER – newspaper, packages & containers	392	52.3	506	47.1	795	61.1	519	67.8	412	59.0	418	51.4
MISCELLANEOUS – incl. auto parts	159	21.2	201	18.7	123	9.4	80	10.4	127	18.2	165	20.3
TOTAL	749	100.0	1,075	100.0	1,302	100.0	766	100.0	698	100.0	813	100.0

(a) Extrapolations based upon monitoring of 36 permanent study sites.

Source: Michigan DOT, 1986.

Table 4-18

NUMBERS AND TYPES OF PLASTIC ITEMS ACCUMULATED PER MILE(a)
ALONG MICHIGAN ROADSIDES IN A 30-DAY PERIOD
(1986)

Type of Item	Number	Percentage of Total
Fast food containers	34.7	20.2
Fast food drink	30.5	17.8
Other ready-to-consume drinks(b)	6.3	3.7
Non-returnable soft drink	1.9	1.1
Returnable soft drink	1.3	0.8
Returnable beer	0.3	0.2
Non-returnable beer	0.3	0.2
Ready-to-consume liquor drink	0.2	0.1
Wine cooler	0.0	0.0
Other plastic items	96.2	56.0
TOTAL	171.7	100.0

(a) Extrapolations based upon monitoring of 36 permanent study sites.
(b) e.g. juice containers.

Source: Michigan DOT, 1986.

problems. KAB officials have stated that substantial litter is generated by wastes falling or blowing from uncovered truck beds. In many communities, KAB supports local ordinances requiring that truck beds be covered (Tobin, 1989).

4.4.2 Analysis of Relative Impacts of Plastic and Other Litter

As a step in analyzing the litter problem, these wastes can be usefully categorized as follows:

- Litter discarded in areas, such as urban areas, where there is likely to be litter collection

- Litter discarded in areas where there is little or no policing for solid waste collection

- Litter that accumulates in beach and other shore areas

These categories allow generalizations about issues that determine the impact and significance of litter. This discussion differentiates among impacts generated by plastic litter and by other types of litter.

For the first category of waste, the litter is assumed to be routinely collected and thus to add to the general municipal solid wastes. The impact of such plastic litter is therefore considered in the context of regular collection efforts. For litter classified in the first category:

- The discarded objects represent an aesthetic loss to the community; the loss may be marginally greater for plastic wastes due to qualities of this waste (such as bright, unnatural coloring and tendency to be blown around by wind). Glass litter, however, may represent a greater nuisance value due to breakage.

- Collection places a burden on community services for waste pickup. No incremental impact, however, from plastics waste (as opposed to paper or glass) could be defined.

- The persistence of plastic waste is not significant because waste collection is assumed to occur well before degradation occurs for any waste except food wastes.

For the second category of litter, the lack of waste collection efforts influences the impacts of the plastic litter. For litter classified in the second category:

- The discarded wastes also represent an aesthetic loss to the community; this loss could be marginally greater for plastic wastes due to qualities of this waste (such as bright, unnatural coloring and a tendency to be blown by wind). Glass litter, however, may represent a greater nuisance due to breakage.

- The slow degradation of the plastic waste increases the aesthetic loss because the litter is assumed to remain uncollected for long periods. The incremental loss due to plastics litter remains modest, however, relative to other materials, which also require some time to degrade.

Beach and marine litter represents a separate category because it is the result of both land activities and wastes from commercial and recreational use of the seas. Plastic wastes from the latter selectively survive ocean currents and wash up on beaches. In this category:

■ Plastics generate a disproportionate share of the litter problem due to the likelihood that plastics litter from marine activities will be washed up on the beach.

■ The beach cleanup efforts of some communities can be taxed by the constant volume of plastic wastes reaching the beach. Because of the uneven patterns of waste deposition on beaches, some communities face unusually large beach clean-up requirements. The Texas Gulf Coast areas, for example, receive large amounts of ocean debris.

■ Plastic litter on beaches represents an aesthetic loss to the community, particularly because of the high value placed on the beach resource, as is evident in high property values and in the popularity of beaches as resort areas.

The high value of beach resources, and of clean beaches in particular, has been investigated a number of times. Researchers have utilized several techniques to develop measures of the value of clean beaches to area residents and to tourists. (For more information on the valuation of beach areas, see Silberman and Klock, 1988; Boyle and Bishop, 1984; Bell and Leeworthy, 1986.)

REFERENCES

Alter, H. 1986. Disposal and Reuse of Plastics. In: Encyclopedia of Polymer Science and Engineering. John Wiley and Sons, Inc. New York, NY.

Alter, H., G. Ingle, and E.R. Kaiser. 1974. Chemical analyses of the organic portions of household refuse; the effect of certain elements on incineration and resource recovery. Solid Wastes Management 64(12):706-712. Dec 1974.

Ballschmiter, K. et. al. 1983. Occurrence and absence of polychlordibenzofurans and polychlordibenzodioxins in fly ash from municipal incinerators. Chemosphere 12(4/5):585-594. As cited in Magee, 1989.

Barcelona, M.J. 1989. Telephone communication between M.J. Barcelona of Illinois State Water Survey and Eastern Research Group. Mar 17.

Barcelona, M.J. et al. 1985. Practical Guide for Ground-Water Sampling. Robert S. Kerr Environmental Research Laboratory, U.S. Environmental Protection Agency.

Bell, F.W. and V.R. Leeworthy. 1986. An Economic Analysis of the Importance of Saltwater Beaches in Florida. Florida Sea Grant Report No. 82. Feb 1986.

Benfenati, E. et. al. 1983. Polychlorinated dibenzo-p-dioxins and polychlorinated dibenzofurans in emissions from an urban incinerator. Correlation between concentration of micropollutants and combustion conditions. Chemosphere 12(9/10):1151-1157. As cited in Magee, 1989.

Boettner, E.A., G.L. Ball, and B. Weiss. 1973. Combustion Products from the Incineration of Plastics. National Technical Information Services. Springfield, VA. EPA 670/2-73-049. NTIS PB 222-001.

Boyle, K.J. and R.C. Bishop. 1984. A Comparison of Contingent Valuation Techniques. University of Wisconsin-Madison. Agricultural Economics Working Paper No. 22. Jul 1984.

Center for Plastics Recycling Research. 1986. Environmental Impacts of Plastics Disposal in Municipal Solid Wastes. Technical Report #12. Rutgers University. Piscataway, NJ.

Churney, K.L. et al. 1985. The Chlorine Content of Municipal Solid Waste from Baltimore County, MD, and Brooklyn, NY. National Bureau of Standards. Gaithersburg, MD. NBSIR 85-3213.

Clarke, M.J. 1988. Improving Environmental Performance of MSW Incinerators. Paper presented at Industrial Gas Cleaning Institute Forum (Washington DC, Nov 1988).

Considine, W.J. 1989. The Contribution of the Plastic Compmonent of Municipal Solid Waste to the Heavy Metal Content of Municipal Solid Waste and Municipal Waste Combustor Ash. Report prepared for the Society of the Plastics Industry, Washington, DC. Apr 1989.

Curran C.M. and M.B. Tomson. 1983. Leaching of trace organics into water from five common plastics. Ground Water Monitoring Review. Summer 1983. pp. 68-71.

Dunlap, W.J. et al. 1976. Organic Pollutants Contributed to Groundwater by a Landfill. Gas and Leachate from Landfills. U.S. EPA Solid and Hazardous Waste Research Division. Cincinnati, OH.

Dynamac. 1983a. An Overview of the Exposure Potential of Commercial Flame Retardants. EPA Contract 68-01-6239. Dynamac Corporation. Rockville, MD.

Dynamac. 1983b. An Overview of Synthetic Materials Producing Toxic Fumes During Fires. EPA Contract 68-01-6239. Dynamac Corporation, Rockville, MD.

FDA. 1988a. U.S. Food and Drug Administration. Vinyl Chloride and Other Chlorinated Polymers; Intent to Prepare an Environmental Impact Statement. 53 FR 47264. Federal Register. Nov 22, 1988.

FDA. 1988b. U.S. Food and Drug Administration. Documentation supplied in support of a letter from Richard J. Ronk, U.S. Food and Drug Administration Center for Food Safety and Applied Nutrition, to Richard Sanderson, U.S. EPA Office of Federal Activities, requesting EPA assistance in the environmental review of a proposed rule on vinyl chloride polymers. Feb 2, 1988.

Forester, W. 1988. Solid waste: There's a lot more coming. EPA Journal 14(4):11-12. May. U.S. Environmental Protection Agency. Washington, DC.

Franklin Associates. 1988a. Characterization of Municipal Solid Waste in the United States, 1960 to 2000 (update 1988). Prepared for U. S. Environmental Protection Agency. Contract No. 68-01-7310. Franklin Associates, Ltd. Prairie Village, KS.

Franklin Associates. 1988b. Characterization of Products Containing Lead and Cadmium in Municipal Solid Waste in the United States, 1970 to 2000. Prepared for U.S. Environmental Protection Agency. Franklin Associates, Ltd. Prairie Village, KS.

Gordon, G.E. 1980. Receptor Models. Environ. Sci. Technol. 14(7):792-800.

Kaiser, F.R. and A.A. Carotti. 1971. Municipal Incineration of Refuse with 2% and 4% Additions of Four Plastics: Polyethylene, Polystyrene, Polyurethane, Polyvinyl Chloride. Society of the Plastics Industries. New York, NY.

Karasek, F.W. et al. 1983. Gas chromatographic-mass spectrometric study on the formation of polychlorinated dibenzo-p-dioxins and polyvinyl chloride in a municipal incinerator. J. Chrom. 270:227. As cited in Magee, 1989.

Kinman, R. et al. 1985. Evaluation and Disposal of Waste Within 19 Test Lysimeters at Center Hill. U.S. EPA Solid and Hazardous Waste Research Facility. Cincinnati, OH.

Leidner, J. 1981. Plastics Waste, Recovery of Economic Value. Marcel Dekker, Inc., New York, NY. 317 p.

Liberti, A. and Brocco. 1982. Formation of Polychlorinated Dibenzo-dioxins and Polychlorinated Dibenzofurans in Urban Incineration Emissions. In: Chlorinated Dioxins and Related Compounds: Impact on the Environment. Pergamon Press. New York, NY.

Lu, J.C.S., B. Eichenberger, and R.J. Stearns. 1985. Leachate From Municipal Landfills. Production and Management. Noyes Publications. Park Ridge, NJ.

Magee, R.S. 1989. Plastics in Municipal Solid Waste Incineration: A Literature Study. Prepared by the Hazardous Substance Management Research Center of the New Jersey Institute of Technology for the Society of the Plastics Industry. Washington, DC. 54 p.

Markland, S. et al. 1986. Determination of PCDDs and PCDFs in Incineration Samples and Pyrolytic Products. In: C. Rappe, G. Choudhary, and L.H. Keith, (eds). Chlorinated Dioxins and Dibenzofurans in Perspective. Lewis Publishers. Chelsea, MI.

Michigan DOT. 1986. Michigan Dept. of Transportation. Michigan Litter Composition Study. Maintenance Division.

Modern Plastics. 1988. Solid waste becomes "crisis." Modern Plastics 65:25-26. Jan 1988.

MRI. 1987. Midwest Research Institute. Results of the Combustion and Emissions Research Project at the Vicon Incinerator Facility in Pittsfield, MA. Midwest Research Institute. Kansas City, MO. pp S-1 to S-4.

Murthy, A. et al. 1989. Biochemical Transformations Within Municipal Solid Waste Landfills. Proceedings of the Sixth National RCRA/Superfund Conference. April 12-14, 1989. New Orleans, LA. Hazardous Materials Control Research Institute.

NSWMA. 1988. National Solid Wastes Management Association. Landfill Capacity in the U.S.: How Much Do We Really Have? Washington, DC.

Radian Corporation. 1987. Chemical Additives for the Plastics Industry: Properties, Applications, Toxicologies. Noyes Data Corporation, Park Ridge, NJ.

Radian Corporation. 1989. Database of Existing Municipal Waste Combustion Studies. Database maintained for the U.S. Environmental Protection Agency. Supplied by Ruth Mead, Radian Corporation, Research Triangle Park, NC.

Rauch Associates, Inc. 1987. The Rauch Guide to the U.S. Plastics Industry.

Rathje, W.L. et al. 1988. Source Reduction and Landfill Myths. Le Project du Garbage. Dept of Anthropology, University of Arizona, Tucson, AZ. Paper presented at Forum of the Association of State and Territorial Solid Waste Management Officials on Integrated Municipal Waste Management, July 17-20, 1988.

Rathje, W.L. 1989. Telephone conversation between Eastern Research ·Group and Dr. W.L. Rathje of the University of Arizona. Mar 6.

Repa, E. 1989. Telephone conversation between Eastern Research Group and Edward Repa of the National Solid Wastes Management Association. Mar 3.

Sax, I.N. and R.J. Lewis, Sr. 1987. Hawley's Condensed Chemical Dictionary. 11th Ed. Van Nostrand Reinhold Co. New York, NY.

Schlegel, J. 1989. Telephone conversation between Eastern Research Group and Jack Schlegel of International Plastics Consultants Corporation. May 3.

SCS Engineers. 1988. Pilot Scale Evaluation of Sludge Landfilling: Four Years of Operation. NTIS PB88-208434. May.

Seelinger, R.W. 1984. Comments submitted by Richard Seelinger, Ogden Martin Systems, Inc., to Dr. Buzz Hoffman, U.S. Food and Drug Adminstration, Environmental Impact Section, in response to proposed expansion of the use of polyvinyl chloride in consumer product applications. July 2, 1984.

Silberman, J. and M. Klock. 1988. The recreation benefits of beach renourishment. Ocean and Shoreline Management 11:73-90.

SPI. 1985. Society of the Plastics Industry. Position Paper for Exempting Compounded Plastics from the California Hazardous Waste Regulations. Society of the Plastics Industry, Inc. Washington, DC.

Sykes, A.L. et al. 1986. Sorption of organics by monitoring well construction materials. Ground Water Monitoring Reports. Fall:44-47.

Tobin, K. 1989. Telephone communication between Eastern Research Group, Inc. and Kit Tobin. Manager of Network Services, Keep America Beautiful, Stamford, CT. Apr 5.

Tsuchiya, Y. and G. Williams-Leir. 1976. Equilibrium composition of fire atmospheres in smoke and products of combustion. In: Hilado C.J. (ed). Fire and Flammability Series Vol. 15, part II. Technomic Publishing Co. Westport, CT. pp. 381-392.

U.S. EPA. 1987a. U.S. Environmental Protection Agency. Municipal Waste Combustion Study. Report to Congress. EPA/530-SW-87-021a. U.S. Environmental Protection Agency, Washington, DC.

U.S. EPA. 1987b. U.S. Environmental Protection Agency. Assessment of Municipal Waste Combustor Emissions Under the Clean Air Act. Advance Notice of Proposed Rulemaking. 52 FR 25399. July 7, 1987.

U.S. EPA. 1988a. U.S. Environmental Protection Agency. Report to Congress: Solid Waste Disposal in the United States. EPA/530-SW-88-011B. Washington, DC.

U.S. EPA. 1988b. U.S. Environmental Protection Agency. Fact Sheet: Municipal Waste Combustion Ash. U.S. Environmental Protection Agency, Washington, DC.

U.S. EPA. 1989. U.S. Environmental Protection Agency. The Solid Waste Dilemma: An Agenda for Action. EPA/530-SW-89-019, U.S. Environmental Protection Agency Office of Solid Waste, Washington, DC.

Visalli, J.R. 1987. A Comparison of Some Results from the Combustion-Emission Test Programs at the Pittsfield, Prince Edward Island and Peekskill Municipal Solid Waste Incinerators. Annual Meeting of the Air Pollution Control Association (New York, June 1987). As cited in Magee, 1989.

Webster, I. 1989. Telephone communication between Eastern Research Group, Inc. and Dr. Ian Webster, Unocal, Inc., Environmental Affairs Division. Los Angeles, CA. May 3.

Wilson, D.C. et al. 1982. Leaching of cadmium from pigmented plastics in a landfill site. Environ. Sci. Technol. 16(9): 560-566.

Wirka, J. 1989. Comments submitted by Jeanne Wirka, Environmental Action Foundation, in response to a draft version of EPA's Report to Congress, Methods to Manage and Control Plastics Waste. April 20, 1989.

5. Options to Reduce the Impacts of Post-Consumer Plastics Wastes

This section explores the options available for improving the management of plastic wastes. The principal topics covered include source reduction, plastics recycling, and use of degradable plastics. The section examines the potential role for and the major environmental and other implications of each of the options. An additional discussion covers the potential methods to improve controls over the releases of plastics to the marine environment.

5.1 SUMMARY OF KEY FINDINGS

5.1.1 Source Reduction

- Source reduction, which includes activities that reduce the amount or toxicity of waste generated, may be achieved in many ways, including:

  -- modifying the design of a product or package to decrease the amount of materials used,

  -- substituting away from toxic constituents, and

  -- using economies of scale with product concentrates or larger size containers.

- Some source reduction efforts, particularly those involving material substitution, generate changes in resource costs, manufacturing processes, product use and utility, and waste disposal. Such source reduction opportunities should be systematically evaluated to assess potential impacts.

- Opportunities for volume reduction should be sought in all components of MSW. Plastics are thought to be a potential candidate for consideration.

- Toxicity reduction through the decreased use of lead- and cadmium-based additives in plastics is possible, but substitution away from these additives must be done carefully, with consideration of a wide range of factors.

- EPA identified four studies of the energy and environmental effects of source reduction possibilities involving material substitution. While none of the studies covered the entire range of factors of interest, the studies are indicative of the type of research needed. The studies indicated that plastics did not generate exceptional environmental releases relative to possible alternative materials, such as paper and glass.

- Source reduction efforts aimed at plastic wastes (e.g., bans on polystyrene foam) have been initiated by Federal, State and local governments and by industry. EPA is not aware of a full systematic analysis of the potential benefits or impacts of these efforts.

208

5.1.2 Recycling

■ Current estimates indicate that approximately 1% of the post-consumer plastic waste stream is recycled.

■ Some recycling technologies employ inputs of relatively homogeneous recycled resins to yield products that compete with those produced from virgin plastics resins. They offer the greatest potential to reduce long-term requirements for plastics disposal. There are no foreseeable limitations on markets for products of these technologies; their deployment is currently constrained by limited supplies of clean, homogeneous recycled resins.

■ Other recycling technologies use inputs of mixed, potentially contaminated plastics to yield products that compete not with virgin resin products, but with commodities like lumber and concrete. Unless the products of this recycling are recycled themselves, this process will not ultimately reduce requirements for plastics disposal. Markets exist for the products of these technologies, but continued growth of the mixed resin recycling industry may depend on the identification of additional markets, technological developments to increase product quality, and reduction of costs to increase cost-competitiveness in identified markets.

■ Recycling processes that are often termed "tertiary" employ a wide variety of inputs, ranging from mixed plastics and nonplastics to very pure resins, to yield products consisting of hydrocarbon fuels and possible chemical feedstocks. Only the latter outputs result in effective plastics recycling, but their production demands the use of nearly pure resins, which are in limited supply, as inputs to the tertiary process.

■ Curbside collection of plastics (and other recyclable components of MSW) provides the vehicle for capturing the greatest variety and amount of plastic waste. But this strategy is not universally applicable, imposes relatively high costs for collection, and may result in collection of a mixed plastics waste stream that may not be amenable to the processing alternatives that produce the highest quality products.

■ Container deposit legislation, originally adopted as litter control legislation, will not divert a significant amount of plastic waste. Soft drink containers, the usual target of deposit legislation, represent only 3 percent of all plastics in the municipal solid waste stream. Deposit legislation, however, typically captures a large percentage of targeted items and yields well-characterized plastics.

■ No significant deleterious environmental impacts are known to be associated with plastics recycling.

■ Because recycling of post-consumer plastic wastes is fairly new, its long-term viability and its ability to reduce plastic disposal requirements are, at this time, unknown.

5.1.3 Degradable Plastics

- Various mechanisms are technically viable for manufacturing degradable plastics, but photodegradation and biodegradation are the principal mechanisms that have been explored.

- A variety of technologies have been developed to enhance photodegradation or biodegradation of certain plastic materials although thus far, very little data are available regarding degradation byproducts and residues and their effect on the environment.

- Degradable plastics will generally sell at a price premium to nondegradable plastics and may generate additional costs in processing and distribution.

- According to current limited data, photodegradation (i.e., loss of structure and strength) of small plastic items takes less than a year, but biodegradation (i.e., degradation of the filler material such as cornstarch) of plastic products requires several years.

- The limited data available suggest that photodegradation rates are somewhat reduced in marine environments.

- Uncertainty surrounds the effect different degradable technologies, when applied commercially, will have on the post-consumer recycling process.

- Most commercial application of enhanced degradable technologies for plastics has been encouraged by the legislative initiatives in this area; inherent product cost or product quality considerations are generally unfavorable to the use of degradable plastics.

5.1.4 Additional Efforts to Mitigate Impacts of Plastic Waste

- Among the options available to EPA for controlling sewer, stormwater, and nonpoint source discharges of plastic wastes to the marine environment are increased enforcement and/or regulatory development under the Clean Water Act.

- Implementation of the MARPOL Annex V prohibitions on deliberate disposal of plastic wastes from all vessels should help reduce volumes of plastic waste disposed of from vessels, although the absence of controls on fishing gear losses and uncertain regulatory compliance levels among U.S. and foreign vessels make the degree of improvement uncertain.

- A variety of measures are needed to reduce losses of fishing nets, traps, and other gear to the marine environment, because these losses are not regulated under MARPOL Annex V. Several methods are being considered by NOAA.

- Incineration and landfilling, which will still be needed for disposal of plastic and other wastes, are coming under increased regulatory control under their respective programs.

5.2   INTRODUCTION TO THE EXAMINATION OF PLASTIC WASTE MANAGEMENT STRATEGIES

This section examines several strategies for reducing or mitigating impacts of plastic waste disposal. The strategies are geared to resolving the specific issues identified in Sections 3 and 4.

Table 5-1 summarizes the principal waste management issues identified in the earlier sections. One important marine problem, entanglement, is mainly associated with various fishing-related wastes. Other implicated products are uncut strapping bands, plastic sheeting, and beverage container ring carrier devices. Ingestion is also a concern; plastic pellets (unprocessed resins), plastic bags and sheeting, and polystyrene spherules (crumbled polystyrene foam) are the items most commonly consumed by marine life. A broad spectrum of plastic wastes contribute to solid waste management issues on land.

Table 5-2 lists the strategies that are analyzed in this section and describes their potential influences on the various plastic waste management issues. Source reduction and recycling methods, by reducing the amount of gross or net discards of waste, can help mitigate the downstream effects of plastic waste disposal (assuming that the methods are applied to those resins or products that are posing problems in the environment or discouraging alternative waste management strategies). In some applications, the use of proven degradable plastics could reduce litter and entanglement problems. However, the indiscriminate use of degradable plastics may impede other waste management strategies or pose additional environmental concerns.

Options for reducing the plastic wastes introduced into the marine environment from urban runoff, combined stormwater overflows, separate sewer systems, and vessels are also described in this section. Section 5.6.4 outlines potential steps for reducing the release of plastic pellets into the marine environment.

Finally, the waste management strategies of incineration and landfilling are also discussed. These strategies are included here because proper incineration and landfilling must play a role in integrated solid waste management.

These waste management strategies are closely related to those presented in the recent publication of the EPA Municipal Solid Waste Task Force, entitled "The Solid Waste Dilemma - An Agenda for Action" (U.S. EPA, 1988). As discussed in Section 4, this document outlines an integrated waste management system to better manage municipal solid waste. The document describes the major waste management techniques in order of overall desirability, as follows:

- Source reduction

- Recycling

- Incineration with energy recovery, and landfilling

Table 5-1

PLASTIC WASTE MANAGEMENT ISSUES

Media	Potential or Actual Solid Waste Management Issue	Plastic Products Implicated
Marine	Entanglement of marine life	Fishing nets & lines Crab & lobster traps Uncut strapping Beverage container ring carrier devices
Marine	Ingestion by marine life	Plastic pellets Polystyrene beads Plastic bags and sheeting Wastes in combined sewer and stormwater runoff
Marine (and beach)	Aesthetic losses due to litter	Plastic wastes in combined sewer and street runoff; plastic waste dumped from vessels and by beach-goers
Land	Leachate generated from plastic in MSW landfills and from MWC ash	Polyvinyl chloride (with additives) Plastic products with colorants or other metal-based additives
Land	Consumption of landfill capacity	All plastics in MSW
Land	Aesthetic losses due to litter	Disposable plastic products, especially fast food packaging
Air	Incremental emissions from incineration of plastic waste	Halogenated polymers and some plastic additives

Note: Nonplastic wastes also contribute to some of the issues cited.

Table 5–2

RELATIONSHIP OF WASTE MANAGEMENT
STRATEGIES TO PLASTIC WASTE
MANAGEMENT ISSUES

Potential Strategies	Intended Effect on Plastic Pollution	Specific Plastic Waste Issue Addressed
Source reduction	Reduces gross discards and toxicity of certain additives in plastic wastes	All problems related to proper and improper disposal
Recycling	Reduces net discards of plastics	All problems except possible releases of pellets in manuf. and transportation of raw pellets
Degradable plastics	Reduces long–term impacts of improperly discarded plastics	Marine, beach, and other litter
Control of urban runoff and sewers	Reduces release of plastic floatable wastes generated from land sources	Marine and beach litter; Ingestion
Implementation of MARPOL Annex V regulations	Prohibits overboard disposal of plastic wastes by vessels	Entanglement; Ingestion; Marine and beach litter
Control of emissions from incineration with energy recovery	Reduces emissions	Incremental emissions from plastic wastes
Control of leachate from landfills	Prevents contamination of groundwater	Leachate from additives

Source: Eastern Research Group estimates.

As noted, this section examines each of these strategies.

Public education programs can support all of the methods discussed in this chapter. Programs that provide information on the proper disposal for wastes and highlight concerns that arise from improper disposal (e.g., littering) may be extremely effective.

5.3   SOURCE REDUCTION

5.3.1   Definitions and Scope of the Analysis

Source reduction refers to actions that decrease the amount or toxicity of materials entering the municipal solid waste stream. By reducing waste quantity, source reduction efforts influence the rate of generation of gross discards of waste. (As will be shown, recycling affects the rate of generation of net discards of waste.) To the extent that a smaller volume of materials is used in the manufacturing of products, the technique reduces the downstream disposal issues or difficulties. By reducing waste toxicity, source reduction efforts directly reduce or eliminate disposal concerns.

Source reduction encompasses certain activities by manufacturers that are designed to reduce the amount or toxicity of solid waste generated. This report is oriented toward post-consumer solid waste; thus, the study examines changes in the manufacturing processes (including the design and production of packaging) that result in reductions in the amount or toxicity of solid waste generated after the useful life of products. For post-consumer wastes, source reduction also encompasses activities at the consumer level. For example, reuse of a product by a consumer reduces the amount of waste generated and, therefore, is considered source reduction.

Using this definition, source reduction activities are considered separate from recycling activities. However, the two options may sometimes overlap and sometimes be at odds. For example, reducing the toxicity of an item in the waste stream may improve its recyclability. On the other hand, attempts to improve the recyclability of a product may increase the amount of waste generated. Because of this interaction, some policy makers define source reduction to include designing for recyclability. Although EPA has not adopted this broader definition for source reduction, EPA recommends that the impact on recyclability be evaluated when a source reduction activity is considered.

Some source reduction efforts, particularly those involving material substitution, require a careful, systematic analysis. Changes may be generated in areas such as energy and natural resource use, process-waste management, and consumer safety or utility by such source reduction activities. To ensure that environmental impacts are not merely shifted or actually increased by a source reduction activity, an analysis of these and other affected areas must be completed. This type of analysis is described in Section 5.3.4.

The sections below present various aspects of MSW source reduction. First, the discussion covers opportunities for reducing the amount of waste produced (Section 5.3.2) and for reducing waste toxicity (Section 5.3.3). Both of these sections focus on plastic waste source reduction efforts. The following section presents the factors that need to be evaluated when analyzing some source reduction efforts (Section 5.3.4). The methodology is not specific to plastic waste but is appropriate for any MSW source reduction efforts that involve material substitution. Finally, the last section addresses current plastic-specific source reduction initiatives (Section 5.3.5).

5.3.2 Opportunities for Volume Reduction of Gross Discards of Waste

In order to determine where source reduction efforts should be focused, the make-up of the waste stream must be examined. Section 2 presents estimates of the distribution of various materials in the MSW stream (see Table 2-16). Paper and paperboard (35.6% by weight) represents a large portion of MSW and is therefore considered an excellent candidate for a source reduction effort. By product category, packaging (all materials) accounts for 30% by weight of the waste stream. Thus, packaging is a target for source reduction. All components of the waste stream need to be evaluated for possible source reduction opportunities.

As shown in Section 2, plastics account for 7.3% by weight of the MSW stream. By volume, however, plastics are a more significant component of the waste stream (see Section 4.2.1.1 for a discussion of attempts to measure plastic waste volumes). In addition, the plastic waste component is expected to increase to 9.2% by weight by the year 2000, with a corresponding increase in volume. Thus, plastic waste presents potential waste reduction opportunities. As plastics increase in the waste stream, other components of the waste stream (e.g., metals) may decrease. The impact of plastic source reduction efforts on other waste stream components must be evaluated.

The first step in this investigation of opportunities for source reduction of plastic wastes is to select those categories of plastic products that appear most amenable to volume reduction. Not all categories of plastic materials are included in municipal solid waste. Further, some categories of plastic are so highly engineered for special purposes that modification of the product characteristics or production techniques may sharply reduce the product value. Table 5-3 rates the major categories of plastic products as candidates for volume reduction efforts. The criteria used (see column heads) provide a means of distinguishing among product categories and a method of focusing the subsequent analysis. The criteria and the rationale for their use are:

- **The share of the product market** - While volume reduction can be justified in any product category for which it is effective, the larger the volume of the product category, the greater the potential benefits of volume reduction efforts. Growth trends, such as the strong growth in the packaging category, are also considered.

- **The predominance of disposable products in the product category** - Disposable products (e.g., those items with lifetimes of less than one year) are added most directly into the solid waste stream, and thus they may be considered good candidates for volume reduction.

- **The significance of consumer preference attributes relative to technical performance attributes** - Products engineered to high technical performance criteria may be less

Table 5-3

POTENTIAL ROLE OF
VOLUME REDUCTION IN PLASTIC MARKETS

Market Category	% of U.S. Plastics Market	Share of Disposable Items	Ratio of Consumer/ Performance Elements	Potential for Volume Reduction
PACKAGING	33.5	High	Varied, but often high	High
BUILDING AND CONSTRUCTION	24.8	Low	Low	Low
CONSUMER AND INST. PRODUCTS	11.1	Moderate to high	Varied, but often high	Fairly high
ELECTRICAL AND ELECTRONIC	6.1	Low	Low	Low
FURNITURE AND FURNISHINGS	4.9	Low	Low	Low
TRANSPORTATION	4.5	Low	Low	Low
ADHESIVES, INKS, AND COATINGS	4.8	High	Varied	Moderate
ALL OTHER	11.0	Low	Low	Low
TOTAL	100.0	--	--	--

Source: Market shares from The Society of the Plastics Industry, 1988a. Other data estimated by
         Eastern Research Group.

adaptable to modification for the purposes of volume reduction. In contrast, products designed primarily for consumer preferences may allow some latitude for modification of product design or substitution of materials without a fundamental loss in product performance.

These criteria are applied here only at an aggregate level; specific candidates for volume reduction can emerge only from a comprehensive, product-specific assessment, in which each of these criteria receives careful analysis. For example, tradeoffs between consumer preference and technical performance attributes (the third criterion) may be extremely complex; color -- not usually considered a "functional" aspect of a product -- may be related to its safe use and therefore may not be a trivial attribute. Additional criteria not specified here may also apply. For example, available waste management methods may be important to a source reduction decision; volume reduction may be less important for products that are recycled than for those that cannot be recycled. The analysis presented here also does not examine the possibility that source reduction among other materials could lead to increases in some uses of plastics; for this reason, increased use of plastic may sometimes be a practical component of more broadly focused source reduction activities. Such a situation may develop where plastics are lighter, smaller, and/or less toxic than other materials. With these caveats, the three criteria are adequate to differentiate the best potential product areas for volume reduction, which are packaging and consumer products.

Source reduction of waste volumes can be accomplished by a variety of methods (see Table 5-4). The options include substitution away from plastics in manufacturing, reduction in the quantity of plastic used for given applications, use of economies of scale in packaging, and consumer reuse of plastic products. Attempts to apply any of the options to unsuitable applications (e.g., increasing the useful lifetime of rapidly obsolescent products) are likely to be ineffective and potentially harmful. Thus, careful consideration is needed.

One source reduction option, the substitution of other materials for plastics, has received widespread attention recently. A number of observers have suggested that substitution is beneficial and appropriate for a wide variety of plastic products because many such products, particularly plastics packaging, were composed of "traditional" (nonplastic) materials in years past. However, substitution of other materials must be carefully analyzed. Some information comparing plastic and nonplastic materials is presented in Section 5.3.4. This type of information can be used to support a complete, comprehensive analysis of material substitution efforts.

5.3.3 Opportunities for Toxicity Reduction

As a complement to efforts to reduce waste volumes, EPA also seeks to reduce the toxicity of wastes requiring disposal. This source reduction option is intended to decrease the risks posed by disposal of toxic constituents. An analysis of toxicity reduction options must include consideration of the toxicity of any proposed substitutes.

EPA's initial efforts in MSW toxicity reduction has been focused on lead and cadmium. The toxicity of these constituents, which have been found in MSW landfill leachate samples and incinerator ash samples, is well known; thus, they pose disposal concerns. In Section 4 the

Table 5–4

METHODS OF REDUCING THE VOLUMES OF
PLASTIC MATERIALS CONSUMED

Method	Possible Product Application
Substitute away from plastics packaging	Some plastic packaging
Modify designs to increase useful lifetime	Products of some inherent durability
Modify designs to decrease quantity of resin used	Some plastic packaging
Modify designs by using fewer environmentally damaging resins	Some plastic packaging or consumer items
Utilize economies of scale w/larger packages	Products with substantial shelf life
Utilize economies of scale w/product concentrates	Water–based solutions
Combine products into a single container	Products used in combination
Consumer reuse of plastic items	Containers of nontoxic materials

Note: Application of any method to a specific product would require a systematic analysis as described in Section 5.3.4.

Source: Categorization developed by Eastern Research Group.

sources of lead and cadmium in the waste stream were discussed. Plastics account for approximately 2% of total lead discards and 36% of discarded cadmium (see Table 4-6). Most of the lead and cadmium found in plastics is in heat stabilizers (used in polyvinyl chloride) and colorants.

EPA is currently investigating substitutes for lead- and cadmium-based plastic additives; Appendix C presents some preliminary findings of this study. EPA's study is examining the factors that determine the potential for successful substitution of less toxic additives and the most prominent substitution candidates for the lead and cadmium additives used in plastics. The practicality of substitutes depends on the nature of the demand for the additive, the relative cost of alternative, less toxic additives, and the performance characteristics of the alternatives. Several additional factors may also be relevant in particular manufacturing situations, such as whether a substitute additive would require changes in manufacturing techniques. In some markets, manufacturers have successfully moved away from use of heavy metal additives in favor of other competitively priced materials. In other areas, however, such as the use of cadmium additives as colorants, the available alternatives have poorer performance characteristics. One should note that none of the potential substitutes identified in Appendix C has been fully analyzed. In particular, the toxicity of these potential substitutes has not yet been evaluated. EPA will address this and other outstanding issues as the Agency continues its analysis of substitutes for lead- and cadmium-based plastic additives.

Other constituents of concern may also be candidates for toxicity reduction efforts. For example, Section 4 noted a number of other additives that could pose some environmental concerns, including phthalates used as plasticizers in PVC and flame retardants composed of antimony oxide. The potential for reducing the use of these additives in plastics is a complex issue that will require thorough investigation.

5.3.4 Systematic Analysis of Source Reduction Efforts

As mentioned above, an analysis of source reduction efforts is extremely important. Critical trade-offs may be overlooked without a comprehensive analysis. For example, some source reduction efforts could generate increases in use of other scarce resources, increases in production costs, or declines in the safety or utility of the product to consumers. These changes, which can be caused by material substitution, could be easily overlooked without a careful examination of several factors.

Source reduction efforts must always be considered in the context of the entire MSW stream. The goal of source reduction is to reduce the amount or toxicity of the entire waste stream, not just of one component. It is relatively easy to reduce the amount of one component in the waste stream by substituting other materials. Such actions, however, may not reduce the size of the total waste stream; in fact, they may increase it. Conversely, while the substitution of lighter materials may result in a decrease in the total mass of MSW, it may also reduce the recyclability of the waste stream.

The factors that must be considered in any analysis are listed in Table 5-5. This table enumerates the range and variety of effects generated by source reduction actions. These

Table 5–5

TOPICS FOR CONSIDERATION IN ANALYSIS OF
SOURCE REDUCTION EFFORTS

Stage of Product Lifecycle	Topics for Consideration
Production of resins and manufacture of products	Natural resource extraction Raw material use Energy use Production process waste streams (quantity and toxicity) Management of process waste streams Labor costs (including social costs of worker displacement) Requirements for importing of production inputs
Distribution	Energy use in transport Labor costs
Use	Consumer utility Consumer safety Cost to consumer
Disposal	Volume and weight in landfilling Toxicity in incineration or landfilling Compatibility with recycling practices or other waste management strategies

include changes in manufacturing processes, changes in the utility of a product to consumers, and changes or effects from product reuse or disposal. A comprehensive assessment must consider the range of environmental releases generated by raw materials exploitation, manufacturing, and transport. Air, water, and solid waste profiles of the processes involved, as well as the solid waste management requirements for the discarded products must be considered. Consideration of these factors generates a "lifecycle analysis" for different products and materials.

EXAMPLES OF SOURCE REDUCTION STUDIES -- EPA identified four studies that examine effects of source reduction options directed primarily at the reduction of the volume of plastic materials. None of the studies provides a comprehensive study of all variables, or a complete lifecycle analysis. For example, none of the studies examined consumer-related issues (e.g., effects on consumer product safety or utility). All the studies focus on only one category of source reduction technique, i.e., direct substitution of other materials for plastics. The first study was prepared by Midwest Research Institute (MRI, 1974) for the Society of the Plastics Industry. Initiated in 1972 and published in 1974, this work is now somewhat dated. The second study was sponsored by the National Association for Plastic Container Recovery (NAPCOR) and focuses on soft drink containers (Franklin Associates, 1989). The third and fourth studies analyze packaging practices in West Germany, with one prepared by an industry association and the other by a government environmental agency.

In the first study, MRI compared environmental information related to the production of 1) seven varieties of plastic products, and 2) seven products made of alternative materials, including glass, paper, aluminum, and steel (see Table 5-6). The study compared results for the categories of raw materials used, energy consumed, process water used, process solid waste generated, atmospheric emissions produced, waterborne wastes, and post-consumer wastes.

There are several important limitations to this study. First, the MRI study did not consider any raw materials, such as additives, which aggregated to less than 5% of weight of the finished product. Second, the relative toxicity of any of the wastes produced was not considered. Thus, the cost and risks of disposing of the various waste streams was not compared. No details were available on any assumptions concerning the compression of post-consumer solid wastes. In addition, the authors noted that no credit was given to post-consumer wastes for energy recovery for that portion of the solid waste stream (9%) that is incinerated.

The study concluded that using plastic products was more favorable for conservation of raw materials and reduction in the amount of environmental emissions produced than using the competing nonplastic products in six of the seven categories. In the remaining category (production of a nine-ounce vending cup from either high-impact polystyrene or paper), the competing products were roughly equal in resource utilization and quantity of environmental releases. The MRI study appears to reflect the underlying economies of manufacture that have led to the steady growth of plastic materials in consumer and industrial product areas. MRI's estimates for each category are summarized in Table 5-6. The authors of the MRI study note that their results were affected by the relative weight of the products compared. The plastic products were lighter than the competing products in every case but one, where both containers were of equal weight. The glass container (for a half-gallon bottle) weighed almost nine times

Table 5-6

SUMMARY OF TOTAL RESOURCE AND ENVIRONMENTAL
IMPACTS FOR PLASTICS AND ALTERNATIVE PRODUCTS

Container	Material	Raw Materials (a) (pounds)	Energy (Million Btu)	Water (Thousand gallons)	Solid Wastes (cubic feet)	Atmospheric Emissions (pounds)	Waterborne Wastes (pounds)	Post-Consumer Solid Wastes (cubic feet)
Half-gallon bottle	PVC	200,426	12,177	2,007	965	57,363	8,914	5,317
	Glass	3,919,809	25,739	6,981	17,279	126,755	14,337	15,452
Gallon milk container	HDPE	8,712	7,515	726	306	27,385	4,081	3,175
	Paper	190,375	7,204	6,755	918	34,054	16,527	4,762
Gallon produce bag	LDPE	384	540	44	21	1,983	248	194
	Paper	22,542	612	532	81	3,506	1,371	536
8-Ounce dairy tub	ABS	1,631	1,928	491	75	6,892	1,135	706
	Aluminum	32,183	5,813	1,032	2,026	24,764	18,095	239
9-Ounce vending cup	HIPS	577	550	215	13	1,689	418	226
	Paper	8,315	324	304	38	1,515	740	226
Gallon oblong container	HDPE	25,925	16,093	1,824	918	60,437	7,973	5,952
	Steel	1,140,789	20,328	21,126	40,066	80,596	196,923	1,570
Meat trays	PS Foam	303	879	118	37	3,691	327	266
	Pulp	35,559	847	339	130	3,509	1,759	806

Note: (a) Crude oil and natural gas raw materials are included in energy category.

Source: Midwest Research Institute, 1974. Estimates based on production of 1 million containers of each type.

as much as the plastic container. Table 5-7 presents the relative weights of the plastic and nonplastic containers.

In summary, the MRI study presents an investigation of the major effects of source reduction via substitution of materials. Its older publication date and incomplete treatment of waste toxicity and product additives limit its usefulness as a guide to the environmental effects of material substitution. The study represents, however, a good example of the kind of research needed to analyze some source reduction strategies.

The second study, funded by NAPCOR, followed a similar methodology, although it only examined soft drink beverage containers. This effort was designed to assess the energy consumption and the environmental releases associated with producing nine different types of these containers. Table 5-8 shows the containers selected for study -- four sizes of plastic containers of polyethylene terephthalate, a 12-ounce aluminum can, and four types of refillable (one type) and non-refillable (three types) glass containers. One refinement in this study was the factoring of the rate of recycling or reuse into the estimates produced. This research examined energy and environmental impacts from raw material extraction through processing, manufacturing, use, and final disposal. Like the previous study, however, it did not consider the relative toxicity of the environmental releases.

Table 5-8 summarizes the comparisons of energy consumption and environmental releases for the nine containers using three assumptions about recycling rates (no recycling, recycling at current rates, and 100% recycling or reuse). Overall estimates are also given for plastic, aluminum and glass that represent averages weighted by the market average of each type of container made from that material (the market shares are given in Table 5-9). In general, the polyethylene terephthalate containers generated lower energy and environmental impacts. The refillable glass container, however, produced lower impacts in several of the measurements; but the savings in energy and environmental impacts (per gallon of beverage delivered) were partly offset by the greater weight of the container. In terms of the solid waste volumes generated, the polyethylene terephthalate containers generated lower volumes of solid waste when virgin raw materials were used, but slightly larger waste volumes than aluminum cans if 100% recycling was assumed.

The NAPCOR study represents another interesting example of the type of analysis needed to determine the value of source reduction possibilities. An even more complete analysis, however, is still needed to address all the possible aspects of such an analysis.

The third effort to investigate the consequences of replacing plastics with alternate materials was conducted by a West German trade association, the Society for Research Into the Packaging Market (1987). This group examined the gross implications of the complete replacement of plastic packaging with alternative materials. The study is based on reviews of aggregate data and industry averages on the amount of energy use per unit of production, the value (cost) of materials, and the weight of averages for broad container categories. The authors concluded that replacement of plastics with other materials (glass, paper, steel, aluminum, and others) would generate the following changes: 1) packaging waste would increase by 256% by volume and 413% by weight, 2) energy consumption for packaging production would increase by 201%, and 3) the cost of packaging would increase by 211%. The authors

Table 5-7

RELATIVE WEIGHTS OF
PLASTIC AND NONPLASTIC
PRODUCTS IN MRI STUDY

Type of Container	Material	Container Weight (grams)	Ratio of Nonplastic to Plastic Items
Half-gallon	PVC	134.0	8.86
bottle	Glass	1188.0	
Gallon milk	HDPE	80.0	1.50
container	Paper	120.0	
Gallon	LDPE	4.9	2.78
produce bag	Paper	13.5	
8-Ounce	ABS	17.8	1.03
dairy tub	Aluminum	18.4	
9-Ounce	Polystyrene	5.7	1.00
vending cup	Paper	5.7	
Gallon oblong	HDPE	450.0	2.37
container	Steel	356.0	
Meat	PS Foam	6.7	3.03
trays	Pulp	20.3	

Source: Midwest Research Institute, 1974.

Table 5-8

SUMMARY OF ENERGY AND ENVIRONMENTAL IMPACTS FOR CONTAINERS
AND FOR MATERIAL CATEGORIES IN NAPCOR RESEARCH

Container/ Assumption About Level of Recycling	Total Energy Consumed (million Btu)	Atmospheric Emissions (pounds)	Waterborne Wastes (pounds)	Solid Wastes (pounds)	Solid Wastes (cu. ft.)
Containers Manufactured From Virgin Raw Materials					
Polyethylene Terephthalate	21.2	62.0	10.8	513.1	31.1
16–oz PET	33.9	98.7	16.6	939.7	56.2
1–liter PET	27.3	78.9	13.6	687.9	42.9
2–liter PET	20.1	59.0	10.3	478.9	29.0
3–liter PET	19.7	57.4	10.4	463.8	28.1
Aluminum	50.0	137.0	44.1	1,938.0	40.4
12–oz aluminum	50.0	137.0	44.1	1,938.0	40.4
Glass	49.1	217.4	21.1	7,000.0	142.8
10–oz nonrefillable	42.0	189.6	20.7	5,725.7	117.4
16–oz nonrefillable	35.1	157.0	16.9	4,721.2	96.9
16–oz refillable	61.7	271.5	24.8	9,066.3	184.4
1–liter nonrefillable	37.0	172.1	17.5	5,354.6	110.1
Containers Manufactured at Current Recycling Rates					
Polyethylene Terephthalate	NA	NA	NA	NA	NA
16–oz PET	31.6	92.3	15.9	814.6	46.1
1–liter PET	25.5	74.1	13.1	592.1	35.1
2–liter PET	18.9	55.8	10.0	415.1	23.9
3–liter PET	18.6	54.2	10.1	403.3	23.0
Aluminum	32.9	91.7	26.9	1,068.1	21.5
12–oz aluminum	32.9	91.7	26.9	1,068.1	21.5
Glass	NA	NA	NA	NA	NA
10–oz nonrefillable	41.7	183.8	20.4	5,273.2	109.2
16–oz nonrefillable	34.8	152.0	16.6	4,347.6	90.2
16–oz refillable	15.4	53.8	8.2	1,505.5	29.7
1–liter nonrefillable	36.7	165.2	17.2	4,915.7	100.9

(cont.)

Table 5–8 (cont.)

SUMMARY OF ENERGY AND ENVIRONMENTAL IMPACTS FOR CONTAINERS AND FOR MATERIAL CATEGORIES IN NAPCOR RESEARCH

Container/ Assumption About Level of Recycling	Total Energy Consumed (million Btu)	Atmospheric Emissions (pounds)	Waterborne Wastes (pounds)	Solid Wastes (pounds)	Solid Wastes (cu. ft.)
Containers Manufactured from 100 Percent Recycled or Reused Materials					
Polyethylene Terephthalate	14.6	44.8	9.2	189.5	4.0
16–oz PET	22.3	66.9	13.4	363.6	8.5
1–liter PET	18.1	54.6	11.3	232.5	4.9
2–liter PET	14.0	43.0	8.8	176.6	3.7
3–liter PET	13.9	42.5	9.1	173.3	3.6
Aluminum	15.9	46.3	9.7	198.2	3.2
12–oz aluminum	15.9	46.3	9.7	198.2	3.2
Glass	20.9	73.5	10.6	762.3	12.6
10–oz nonrefillable	38.7	130.0	17.0	1,198.4	19.4
16–oz nonrefillable	32.4	107.7	13.8	985.2	16.2
16–oz refillable	11.6	37.9	6.4	521.3	8.8
1–liter nonrefillable	33.8	102.3	14.0	965.6	13.9

Source: Franklin Associates, 1989.
NA = Not available.

Table 5-9

SOFT DRINK CONTAINERS COMPARED
IN NAPCOR RESEARCH

Material/ Container	No. of Containers Required to Deliver 1,000 Gallons of Beverage	Market Shares, By Material (%)	Estimated Current Recycling Rate (%)
Polyethylene Terephthalate		100.0	20
16–oz bottle	8,000	5.7	–
1–liter bottle	3,785	4.3	–
2–liter bottle	1,893	83.2	–
3–liter bottle	1,262	6.8	–
Aluminum			
12–oz aluminum can	10,667	100.0	50
Glass		100.0	10
10–oz nonrefillable bottle	12,800	6.1	–
16–oz nonrefillable bottle	8,000	41.5	–
16–oz refillable bottle	8,000	50.8	8 trips
1–liter nonrefillable bottle	3,785	1.6	–

Source: Franklin Associates, Ltd. 1989

did not explore the differences in manufacturing processes between plastic and alternative materials in detail equivalent to that of the MRI study.

The fourth study was prepared by the West German Federal Office of the Environment and focuses only on alternatives for the manufacture of shopping bags (1988). The authors developed energy and environmental impact comparisons for low-density polyethylene shopping bags, unbleached kraft paper bags, and bags made from either polyamide fibers or from jute fibers. The first two categories were assumed to be single-use bags, while the bags of polyamide or jute fibers were assumed to be reusable 100 or 50 times, respectively.

Table 5-10 presents the estimates of energy use and of environmental releases from production of an equivalent number of each bag. The polyethylene bag required less energy consumption for production and also produced lower amounts of most of the air and water pollutants than either of the other categories shown. The authors also qualitatively assessed the solid waste disposal requirements and found no significant difference in disposal requirements between the single-use bags. The authors concluded that there was no ecological basis for requiring a switch from single-use polyethylene to paper bags. They also concluded that switching toward the plastic bags would not produce a "significantly lower burden" to the environment because of the significance of the solid waste burden created by either single-use bag. They suggested instead that reusable bags were the preferred alternative and would result in net energy and environmental benefits. If the values shown in Table 5-10 are converted to a per-use basis, the lowest values would be for reusable bags. These bags would also produce substantially less solid waste.

5.3.5 Current Initiatives for Source Reduction

Some momentum toward source reduction has been generated by various regulatory requirements at the state, local, Federal, and international levels. These restrictions have included complete bans on certain plastics and selective bans on nondegradable plastics. Industry also has considerable incentives to reduce packaging costs and this objective often coincides with that of source reduction.

5.3.5.1 State and Local Initiatives

A variety of laws have been directed at plastics packaging that limit or prohibit the use of plastics in packaging or other consumer goods. Most of these laws have apparently not been based on a lifecycle analysis of the plastic and substitute materials. The laws are also not focused specifically on articles found in the investigation of the marine or other effects from plastic disposal noted in Sections 3 and 4. They are instead the result of more general concerns about plastic wastes.

Minneapolis and St. Paul recently passed local ordinances that prohibit food establishments from using food packaging that is not "environmentally acceptable." Environmentally acceptable packaging is defined to include that which is degradable (not including degradable plastics),

Table 5-10

COMPARISON OF ENVIRONMENTAL IMPACTS
FROM PRODUCTION OF 50,000 BAGS OF
COMPETING MATERIALS

	Bag Material		
Energy/Pollution Parameter	Poly – ethylene (a)	Unbleached Kraft Paper	Paper Combinations(c,d)
Energy, GJ			
Production Processes	29	67	69
Contained in Material	<u>38</u>	<u>29</u>	<u>29</u>
Total Energy Consumption	67	96	98
Air Polluting Emissions, kg			
Sulfur dioxide	9.9	19.4	28.1
Nitrogen oxides	6.8	10.2	10.8
Organic materials	3.8	1.2	1.5
Carbon monoxide	1	3	6.4
Dust	0.5	3.2	3.8
Waste Water Burdens, kg			
Chemical oxygen demand	0.5	16.4	107.8
Biological oxygen demand (e)	0.02	9.2	43.1
Organic materials,			
except phenols	0.003	NA	NA
Phenols	0.0001	NA	NA
Chloro–organic compounds	NA	NA	5

GJ= Gigajoules

Notes:
(a) 0.4 square meters of PE film and thickness of 50 microns (18g)
(b) 0.4 square meters of paper with surface weight of 90 grams per square meter (36 grams)
(c) This material consisted of 60% white kraft paper, 25% brown kraft paper, 15% white sulphite paper.
(d) The energy consumption for the process includes 29 GJ that is obtained from burning residual materials (waste liquor, etc.); this and the materials portion derive from the wood raw material.
(e) BOD within 5 days

Source: West German Federal Office of the Environment, Berlin, 1988.

returnable, or recyclable. A package is considered recyclable only if it is part of a municipally sponsored program within the Twin Cities (City of Minneapolis, 1989).

A Suffolk County, NY, ordinance prohibits the use or sale of polystyrene or polyvinyl chloride food containers at retail food establishments. Non-biodegradable food packaging is also prohibited at retail food establishments (EAF, 1988). The consequences of these changes on overall source reduction are uncertain. The New York State Supreme Court recently required the county to conduct a thorough environmental impact study before implementing this ban (Plastics Recyc. Update, June 1989).

State actions have included bans on plastic cans (Minnesota and Connecticut; Wirka, 1988) and certain packaging made of foamed polystyrene (Florida, Maine; EAF, 1988). Many states have considered some form of source reduction legislation. Connecticut recently (June 1989) passed an extensive source reduction bill.

Federal legislation has not directly called for source reduction, though some measures encourage this approach. Congress passed the Marine Plastic Pollution Research and Control Act (1987), which amends the existing Act to Prevent Pollution from Ships. The amendments implement Annex V of the international Marine Pollution treaty (MARPOL), which prohibits all deliberate disposal of plastics from vessels and offshore oil and gas platforms. The Coast Guard issued interim final regulations for this program on April 28, 1989. While the Coast Guard regulations are restrictions on disposal practices, one method of compliance is source reduction -- i.e., restricting or eliminating the presence and use of plastics onboard vessels. In anticipation of regulatory promulgation, at least one major U.S. shipping line, Lykes Bros., has experimented with the elimination of plastic containers for all food stores on its vessels (Castro. 1988). Several bills have recently been proposed at the Federal level that would provide incentives (e.g., packaging taxes) for source reduction.

EPA in its "Agenda for Action" has strongly encouraged source reduction activities. Current EPA efforts in the source reduction area are described in Section 6 of this report.

Regulatory measures that encourage or directly force substitution away from plastics have been more widely employed in Europe. Italy has banned the use of nonbiodegradable packaging (Claus, 1987). Several European countries have adopted packaging control laws that authorize direct restrictions on packaging that creates problems for recycling, reuse, or eventual disposal. Denmark, Netherlands, and West Germany all have fairly broad authority to restrict packaging methods (Wirka, 1988).

5.3.5.2 Industry Initiatives

Industry has made many efforts to reduce the amount of plastics and to modify the types of plastic used in products and packaging. The principal thrust of these efforts has been to reduce production costs. Industry also is pursuing source reduction efforts as the result of regulation-forced changes in markets and presumably enlightened self-interest. The items below provide a sampling of efforts taken by manufacturers to reduce the volumes of waste materials:

- Procter & Gamble has marketed a fabric softener in Europe that is sold in a reusable container. Consumers buy replacement concentrate pouches and mix the concentrates with water in the original container. P&G plans to test market a variety of products using similar packaging approaches in the U.S. market (Rattray, 1989).

- General Electric has developed a refillable polycarbonate plastic bottle that can be used for dairy products, juice, and water (Wirka, 1989).

- Polaroid has reduced the amount of disposable materials in their film packs (Popkin, 1989).

- Digital Equipment Corporation (DEC) has eliminated the use of styrofoam "free flow" packing materials at its two "DEC Direct" operations, which supply computer accessories to Digital customers. Previously all orders were shipped in the same size box with foamed polystyrene used as filler. This procedure resulted in the use of 200,000 cubic feet of free flow per year. The company now ships products in appropriately sized boxes using mechanically crumpled paper as filler (O'Sullivan, 1989).

- DEC has also succeeded in substituting die-cut fiber board inserts for styrofoam in the packaging of its computer "mouse" (O'Sullivan, 1989).

It is not known to what extent these or other firms conducted analyses of the effects of their source reduction efforts.

5.4 RECYCLING

This section examines the impact of recycling methods as a possible strategy for amelioration of plastic waste issues. Recycling is a method of reducing the quantity of net discards of municipal solid waste by recapturing selected items for additional productive uses. Although these benefits have not been quantified, plastics recycling also offers the potential to generate demonstrable savings in fossil fuel consumption, both because recycled plastics can displace virgin resins produced from refined fossil fuels, and because the energy required to yield recycled plastics resins may be less than that consumed in the production of resins from virgin feedstocks. Recycling is one of EPA's preferred solid waste management strategies, as described in the publication "The Solid Waste Dilemma: An Agenda for Action" published by the Agency's Office of Solid Waste.

The Congressional mandate for this Report to Congress specifies that the potential for recycling to reduce plastic pollution is to be addressed. Included in the sections below are analyses of the current types of recycling systems, the array of technical and operational difficulties evident in the wider use of recycling for plastic products, and means to enhance the growth of recycling methods.

The analysis also distinguishes recycling of plastics from recycling technologies as they are applied to other solid waste streams, such as glass or aluminum. As will be shown, recycled plastics represent a mixed batch for recycling due to the variety of resins in the waste stream.

This contrasts with the relatively homogeneous recycled materials that can be derived in glass or aluminum recycling. The mixed nature of most post-consumer plastics has significant influence on the methods adopted in plastics recycling programs.

5.4.1 Scope of the Analysis

This section focuses on the recycling of post-consumer plastic solid waste. It does not address the recycling of plastic materials by industry during manufacturing and processing operations. Such industry recycling of unprocessed resins is extensive and considerably reduces the manufacturing losses of plastic resins. Nevertheless, the focus of this study is plastics generated from post-consumer solid waste.

Not all plastics in MSW are amenable to recycling. For example, trash bags by definition are intended to facilitate MSW disposal, and so are unavailable for recycling. Many or most plastics films used in food contact applications may be inappropriate for recycling because currently practicable collection alternatives require consumers to store plastics before collections, yet valid concerns regarding odors and potential health risks from food-contaminated wastes may make storage of such items impractical.

5.4.2 Status and Outlook of Plastics Recycling Alternatives

Recycling plastics from MSW encompasses four phases of activity:

- Collection. As with all other recyclable materials, plastics must be segregated from other MSW constituents and collected for transfer to processors.

- Separation. Plastics segregated from MSW include a variety of resins. It is not necessary to separate plastics by resin type to allow their recycling, but separation by resin allows the production of the highest-quality recycled products.

- Processing/Manufacturing. A number of processes are used to manufacture recycled plastic products. They are generally grouped into three categories:

  Primary processes are defined as industrial recycling of manufacturing and processing scrap. Typically, such scrap is blended with virgin resins and re-introduced into plastics production processes. Primary plastics recycling is not addressed in this Report to Congress.

  Secondary processes encompass a continuum of processing alternatives. One end of this continuum is defined by processes that consume clean, homogeneous resins that can be used to manufacture products interchangeable with those produced from virgin plastic resins. At the other end of this continuum are processes that consume mixed recycled plastics in the manufacture of products that do not replace or compete with virgin plastic products, but replace structural materials such as wood and concrete in product applications.

Tertiary processes involve the chemical or thermal degradation of recycled plastics into chemical constituents that serve as fuels or chemical feedstocks. Tertiary processes may use either homogeneous or mixed plastics as inputs.

■ Marketing. Homogeneous recycled resins may be processed into products that compete in markets with virgin plastics. With currently available technologies, most mixed recycled plastics are processed into generally lower value products that compete in markets with materials such as lumber and concrete.

These four phases of recycling activity are closely related. For example, the extent of separation among plastic resins achieved during collection largely determines the types of processing available and the products that can be manufactured from the recycled resins. Marketing considerations, in turn, determine the marketability and value of these products, and drive the economic calculations by which the viability of the entire recycling chain is evaluated.

The following paragraphs provide an introduction and background to the detailed discussion of the four recycling phases that follows. A number of characteristics affect all phases of plastics recycling and tend to differentiate the technical, economic, and policy considerations relevant to plastics from those that affect the recycling of other MSW constituents:

QUALITY OF THE RECYCLED RESINS -- Only homogeneous resin streams can be recycled into products that compete with virgin resins. All plastics recycling processes result in some degradation of the physical and chemical characteristics of the plastic resin(s). For this reason, recycled plastics may not be suitable to replace virgin resins in many applications with exacting product specifications (particularly in food-contact applications). However, with good separation into clean, homogeneous resins, recycled plastics may be used to make a broad range of products that would otherwise be fabricated from virgin resins, or may be incorporated into mixes with virgin resins in a variety of product applications. With current recycling technologies for mixed plastics, however, recycled resins are incorporated into products with less demanding physical characteristics, for which market competition comes not from virgin plastics but from other commodities like lumber or cement. This fact has implications on estimates of the long-term benefits of mixed plastics recycling, which are addressed below.

LONG-TERM IMPACTS OF PLASTICS RECYCLING -- Some concern surrounds the long-term impacts of mixed plastics recycling processes. Whereas processes using homogeneous resins displace consumption (and disposal) of virgin plastics, mixed plastics recycled products do not displace the use, nor ultimately the disposal, of virgin plastics. Instead, they compete with, and displace consumption, use, and disposal of other commodities like lumber or cement. Ultimately, the mixed plastic recycled products must themselves be disposed of. The benefits of mixed plastics recyling may therefore be most appropriately measured in terms of the long-term deferment, rather than the elimination, of plastics disposal requirements (Curlee, 1986; IEc, 1988).

A number of technical and policy considerations frame the potential role and impact of mixed plastics recycling:

■ **Mixed plastics recycling does not reduce demand for virgin plastics.** Because its products do not compete with products manufactured from virgin plastics, mixed plastics recycling does not reduce the demand for or the consumption of virgin plastics.

■ **Recyclability of mixed plastics recycled products.** It is difficult to determine if the products of mixed plastics recycling will themselves be recyclable. For the following reasons, it appears that they may not be recyclable:

-- The unknown composition and the physical characteristics of mixed plastics recycled products may prevent their recycling. Mixed plastics recycling processes generally result in the marked degradation of the physical characteristics of their constituent resins. As a result, it appears that mixed plastics recycled products may not be acceptable as inputs to further recycling efforts.

-- A collection infrastructure for mixed plastics recycled products has not been established. Many or most of the current products of mixed plastics recycling are not targeted for consumer applications, but for commercial or industrial use. In these applications, it is unlikely that the recycled products will be captured for further recycling.

If the products are not recyclable, mixed plastics recycling will not reduce the ultimate requirement for plastics disposal, but will delay that requirement for the lifetime of the recycled product. When that product is disposed of, all of the plastic content of the product enters the waste stream.

■ **Mixed plastics recycling reduces total waste disposal requirements.** Even if it has no long-term impact on plastics disposal requirements, mixed plastics recycling does reduce total long-term waste disposal. For example, if one cubic yard of recycled post-consumer plastics displaces consumption of one cubic yard of lumber in a product application (e.g., for fencing), the total disposal requirement at the end of the plastics lifecycle is one cubic yard; if plastics recycling is not implemented, however, total disposal requirements are two cubic yards (one cubic yard of post-consumer plastics plus one cubic yard of lumber from the fencing application). (A related topic potentially deserving further investigation is the relative environmental impact of mixed recycled plastics disposal compared to disposal of displaced nonplastic products; for example, the potential environmental impacts of plastic "lumber" disposal appear to be qualitatively different from those that may be associated with disposal of pressure-treated wood.)

For these reasons, measuring the benefits of mixed plastics recycling is complex. If mixed plastic recycled products cannot themselves be recycled, then the benefits of mixed plastics recycling must be measured in terms of deferring, rather than eliminating, long-term plastics disposal requirements. However, this delay in itself may be a substantial benefit; for example, it puts recycled plastics to productive use for a number of years, during which recycling technologies may be expected to improve, and so to allow the further recycling of the initial recycled product. And even if mixed recycled plastics products cannot ultimately themselves be recycled, and so have no long-term impact on plastics disposal, their use does reduce total disposal requirements for all wastes.

This situation is in marked contrast to recycling scenarios for glass and metals from MSW. For these MSW constituents, the recycled raw material is indistinguishable from the virgin raw material, and the benefits of recycling can be measured directly in terms both of reducing the demand for the raw materials used in the recycled product and of reducing short- and long-term disposal requirements.

VARIETY OF PLASTICS WASTES -- Plastics in MSW are a very heterogeneous collection of materials. "Plastics" encompasses an extremely broad range of materials. Plastic products in MSW include not only items made from a single resin, but an increasing number of items that include a blend of resins. The blending of resins in individual items may involve the simple physical joining of two or more resins (e.g., PET drink containers with HDPE base cups) or the chemical bonding of different resins in a single plastic film. Further, the nature of the additives incorporated to yield specific plastic product qualities is diverse.

Mixed resin products and the presence of a variety of additives may significantly affect recycling options. For example, many mixed resin products are amenable only to mixed plastics processing technologies, while the presence of some additives may complicate the use of some or all recycling technologies for some plastic items.

DIFFICULTY OF SORTING PLASTICS RESINS -- It is technically difficult to separate relatively pure resins from mixed plastics collected for recycling. Commercially demonstrated separation technologies are almost exclusively limited to processes that separate PET and HDPE. A number of promising technologies to effect separation of mixed plastics are under active development, including infrared analysis, laser scanning, gravity separation, and incorporation of chemical markers into different resins. Successful development and implementation of one or more of these technologies may allow reliable separation of mixed plastics into homogeneous resins.

LOW DENSITY OF POST-CONSUMER PLASTICS WASTES -- Plastics have a high ratio of volume/weight compared to other recyclable constituents of MSW. This fact adversely affects the practicality of plastics collection in municipal MSW recycling programs and the economics of transporting recycled plastics to processors. The problem may be addressed by shredding or crushing at the point of collection, but these alternatives can reduce the practicality of separation into homogeneous resins.

LIMITED HISTORY OF PLASTICS RECYCLING -- Nearly all of the collection, separation, and processing alternatives outlined below have been successfully implemented in at least a few locations across the country. For many of these alternatives, however, only limited data exist from which to extrapolate costs, participation rates, technological or institutional barriers, and other factors that will determine their long-term viability. For this reason, much of the following discussion of the outlook for each alternative is qualitative, and is based on the experience and opinion of participants in ongoing recycling efforts.

This analysis also makes no assumptions about the imposition of any of these alternatives as Federal policy. The outlook for each alternative is discussed presuming the absence of any Federal law or regulation concerning plastics recycling.

5.4.2.1   Collecting Plastics for Recycling

Plastics may be segregated from MSW either before or after MSW collection. That is, consumers may be required to segregate plastics from MSW, or collection agencies may attempt to segregate plastics (and other recyclables) after MSW has been collected from consumers. Technologies exist to segregate metals and glass from MSW after collection, but currently available technologies are much less effective at capturing plastics. For this reason, nearly all discussions of plastics collection alternatives have focused on possibilities of capturing plastic recyclables before they enter the municipal solid waste stream. The following discussion reflects this focus.

Five alternative strategies have been implemented to segregate plastics, as well as all other commodities, from MSW for recycling. These are:

- Curbside pickup

- Drop-off recycling centers

- Voluntary container buy-back systems

- Reverse vending machines

- Container deposit legislation

These strategies are explained and compared in the following discussion.

This discussion does not directly address shipboard collection of plastic wastes. Vessels, however, may become a reliable source of mixed plastics for recycling as MARPOL Annex V regulations are implemented by U.S. and international fleets. Under MARPOL Annex V, ships are prohibited from disposing of plastics overboard. Because one of the most cost-effective means of compliance with these regulations is to store plastics for onshore disposal, and because ports are being required to provide collection facilities for these plastics, there should be a steady supply of plastics from port collection facilities.

Table 5-11 summarizes the major advantages and disadvantages of each of the five major collection strategies for recyclable plastics in MSW. Please note that all of these strategies and many of the advantages and disadvantages apply to other components of the waste stream as well as to plastics. No significant technical barriers exist to implementation of any of the collection alternatives discussed below. The principal obstacles are economic or institutional. As policy alternatives, some also capture only a small percentage of recyclable plastics, and so tend to have only a minor impact on plastics waste disposal requirements. Arrayed against these hurdles are the benefits each strategy offers in reductions in plastics disposal requirements and production of high-value recycled products.

Table 5-11

FEATURES, ADVANTAGES,
AND DISADVANTAGES OF
ALTERNATIVE COLLECTION
STRATEGIES

Features	Advantages	Disadvantages
Curbside collection – Pickup of recyclables as part of MSW collection	Consumer convenience; no travel to recycling center required	Possible net cost to municipalities
	High participation rates in many implementation scenarios	Not feasible in localities with no centralized MSW collection
	Facilitates collection of a variety of recyclables other than plastics	Requires in-home storage of recyclables by consumers
	Facilitates collection of a wide variety of plastic products	Inconvenient for consumers if requires separation of plastics from other recyclables
	Documented record of successful implementation	May result in collection of mixed plastics wastes not amenable to high-value recycling applications
	Potentially greatest reduction in MSW disposal requirements	Implementation difficult in areas with many multi-family dwellings
Drop-off recycling center – Recyclables collected at centralized, municipal, or privately operated facility	Low cost to implementing municipalities	Inconvenient for consumers who must both store and transport recyclables
	Small manpower requirements	Relatively low participation rates
	Facilitates collection of a wide variety of plastic products	Not amenable to implementation as mandatory programs (difficult to enforce)
	Facilitates collection of a variety of recyclables other than plastics	
	May allow separation of recyclable plastic by resin, allowing processing into high-quality recycled products	
Voluntary container buy-back program – Consumers voluntarily return designated recyclables to recycling centers operated by private parties or government agencies, receive payment for recycled articles	Little cost to government agencies if implemented by private parties	Relatively low participation rates
	Payment provides incentive to consumers	Generally focused on only a small percentage of recyclable plastic articles (high-value, single resin items)

(cont.)

Table 5-11 (cont.)

FEATURES, ADVANTAGES,
AND DISADVANTAGES OF
ALTERNATIVE COLLECTION
STRATEGIES

Features	Advantages	Disadvantages
Voluntary container buy-back (cont.)	Allows collection of relatively pure resins amenable to processing into high-quality recycled products	Inconvenient for consumer who must both store and transport recyclables Cost to implementing agency; payment sufficient to induce consumer participation may exceed value of recycled plastics
Reverse vending machines – Consumers deposit recyclables in machine, receive case or other payment; machine typically grinds and stores plastics for pickup	Potentially no cost to government agencies Payment provides incentive to consumers Allows collection of relatively pure resins amenable to processing into high-quality recycled products Available as implementation option for other recycling strategies (e.g., drop-off centers, container deposit legislation) Machine shredding reduces space requirements for recyclables	Inconvenient for consumer who must both store and transport recyclables Captures only a small percentage of recyclable plastic articles Payment sufficient to induce consumer participation may exceed value of recycled plastics
Container deposit legislation – Consumers pay deposit at time of purchase; deposit is redeemed when recyclable article is returned to collection center (retail outlet or other designated facility)	Very high rates of return may be obtained for designated articles Allows collection of relatively pure resins amenable to processing into high-quality recycled products Typically includes collection of additional high-value recyclable containers (glass, aluminum) Documented record of successful implementation	Potentially significant costs on collection "middleman" (e.g., distributors, retailers) Captures only a small percentage of recyclable plastic articles Inconvenient for consumer who must both store and transport recyclables May have negative impact on curbside or drop-off recycling programs

Source: Compiled by Eastern Research Group.

CURBSIDE PICKUP -- Curbside pickup involves the separation of recyclables from other MSW by consumers and the pickup of recyclables as part of a municipality's solid waste collection activities. Use of this recycling option has been growing rapidly in recent years, and approximately 600 communities (ranging from small rural towns to cities like Seattle, Washington, and San Jose, California) have implemented curbside recycling programs to date. At this time, most of these programs do not include plastics, however. Implementation of a curbside collection program involves choices regarding the following factors:

- Mandatory or voluntary participation

- Frequency of collection (weekly, bi-weekly, monthly, etc.)

- Timing of collection (same or different day as MSW collection)

- Degree of recyclable separation required; alternatives include:

  - All recyclables placed in one container
  - Paper separated from all other recyclables
  - Paper, metals, glass, and plastics separated into individual containers

Only limited analyses have been completed of factors that tend to promote the success of curbside collection programs; in general, participation rates in curbside collection programs are greatest when programs are mandatory and when the programs are designed to maximize convenience to consumers in sorting and storing recyclables. Success factors related to consumer convenience include:

- High frequency of collection (removes the need for long-term storage of recyclables)

- Collection on the same day as collection of other MSW

- Minimal requirement for separation of recyclables -- three or four categories appears to be a practicable maximum, from the standpoint both of consumers and of municipal collection teams

- Provision to consumers by the municipality of containers for recyclable storage and curbside set-out

Appendix B summarizes program characteristics of 22 successful curbside recycling programs across the country. Of the available collection alternatives, this strategy tends to divert the largest proportion of MSW from disposal (including glass, metal, and newspapers in addition to plastics). Thus, use of curbside pickup may be expected to increase among states and individual municipalities, especially those in densely populated areas of the country where landfilling costs are currently greatest and landfill capacity is most rapidly dwindling.

Curbside pickup programs face a number of institutional and economic barriers: It is probably not feasible in communities that do not currently provide municipal waste collection (or curbside pickup by private haulers) and/or have low population densities. Private parties may implement curbside pickup, but there are apparently not enough profits in recycling operations to support such "middlemen" unless a means is found to allow them to participate in the savings realized from reduced landfilling costs (Brewer, 1988a).

Collection programs may also face significant challenges in urban environments. Most of the curbside collection programs currently operating are in suburban or rural settings with few multi-family dwellings. Unique difficulties are imposed by the presence of a large number of multi-family dwellings and by the congestion of the urban environment. These must be addressed and overcome by program planners if curbside collection is to capture a significant proportion of urban MSW. Among the difficulties faced by urban collection programs are:

- Lack of storage space -- Many urban residences are small, and very few have garages or other unused space for recyclables storage.

- Use of dumpsters -- Many multi-family residences use one or more large containers for MSW collection. Implementation of a recycling program implies using additional containers for recyclables collection, for which little space may be available in urban settings.

- Difficulty of access -- Narrow streets and alleyways may impede vehicle access to collected recyclables, and may make collection a very slow process, adding significantly to program costs.

Program cost may also slow the growth of curbside collection programs in some areas. IEc (1988) reported data from a number of communities in which the net cost of recycling programs (after recyclable sales and savings in disposal fees) ranged from $40 to over $100 per ton of material collected. On the other hand, six out of eight programs reviewed during preparation of this Report were either breaking even or showing an economic benefit associated with their recycling programs; revenue-to-cost ratios ranged as high as 1.8, or $81 per ton of material collected. Section 5.4.3 presents a detailed review of available information on the costs associated with curbside recycling.

In most practicable implementation scenarios, curbside programs collect a mixture of plastics wastes. In many current programs, mixed plastics are also commingled with recyclable nonplastics. For this reason, implementation of curbside programs either demands that efficient plastics separation strategies be implemented to allow the capture of homogeneous resin streams, or implies that only mixed plastics technologies will be available as processing options for the collected mixed plastics.

DROP-OFF RECYCLING CENTERS -- Drop-off centers require the consumer to transport recyclables to a central location. Their primary advantage over curbside recycling is their relatively low cost to the implementing community. They may also be the only practicable collection alternative in communities that do not provide for MSW collection but that require consumers to bring their wastes to a central collection facility. Drop-off centers face the disadvantage that participation rates are generally much lower than for curbside programs.

The primary variables defining implementation strategies for drop-off centers include:

- Degree of recyclables separation required

- Number and location of recycling centers in the community

- Hours of operation

As with curbside programs, participation rates tend to increase when implementation strategies are designed to minimize any inconvenience associated with recycling.

Drop-off recycling centers are likely to continue to be implemented among states and municipalities that are hesitant to face the costs and institutional requirements of curbside recycling or in which curbside recycling is infeasible. Past experience with drop-off centers suggests, however, that after initially high participation rates, consumer use diminishes significantly unless the sponsoring agency implements continuing public relations efforts. And with low voluntary participation, drop-off centers may not divert a large proportion of MSW from disposal.

VOLUNTARY CONTAINER BUY-BACK PROGRAMS -- In a voluntary buy-back system, consumers bring designated recyclable items to a central facility where they receive a cash payment on a per item basis. These systems differ from container deposit systems in that the designated items are purchased without a deposit. Buy-back programs may be implemented by private organizations (e.g., beverage industry groups) or by government authorities.

These programs are not likely to divert significant quantities of MSW plastics to recycling programs, although they can be successful at the local level. Like drop-off centers, these systems face the disadvantage that they require consumers to store recyclables and bring them to a central recycling location. Buy-back systems also may be impeded by the need to balance payments made to consumers with the economic value of the recycled products. Payments to consumers sufficient to induce high participation rates are likely to impose serious financial burdens on the sponsor of the program. The economics of these programs remain poor, however, because the sponsor does not participate in savings attributable to reduced landfill requirements.

REVERSE VENDING MACHINES -- Reverse vending machines are not an independent policy option for plastics recycling, but an implementation option available to support drop-off recycling centers, voluntary buy-back programs, or container deposit legislation. A single machine accepts a specific class (or a few classes) of container and returns cash, a reduced-price coupon for a subsequent consumer purchase, or a receipt redeemable for cash or merchandise. Most machines incorporate a compactor or shredder to minimize internal storage requirements for the recycled material. The primary advantage of reverse vending machines is that they require no human involvement at the point of recycling; they can therefore be widely distributed (e.g., at supermarkets and other retail outlets) and so can greatly increase the convenience of consumer participation in non-curbside recycling programs.

These machines are particularly attractive as a collection option in support of container deposit legislation because they reduce the cost, space, and manpower requirements associated with collection of recyclables by retailers or other collection centers. Reverse vending machines are currently being deployed in at least three "bottle bill" states (Connecticut, New York, and Massachusetts) and have been legally recognized as recycling centers under California's recycling program (Brewer, 1988a).

Reverse vending machines also allow discrimination between resin types. Feasible technologies exist that can allow machines to differentiate among resins, either to limit the plastics accepted or to sort plastics for processing. Current use of reverse vending machines has been largely limited to PET soft drink containers, but the technology may be applied to other plastic containers (e.g., milk and laundry detergent bottles).

CONTAINER DEPOSIT LEGISLATION -- Deposit legislation is now viewed as an option to divert plastic and other recyclable containers from the MSW stream, although it was originally implemented as a means to reduce roadside litter. Container deposit legislation (the "bottle bill") has been enacted in nine states (see Table 5-12). Deposits apply to soft drink, beer, and some bottled water containers, and several states also include deposits on a few other beverage containers. None of the current state laws recovers milk jugs, juice or most other beverage containers, or containers for non-beverage liquids (e.g., bleach, cleansers). Nor do any state laws apply to plastic/cardboard containers (e.g., milk cartons). It has been estimated that the PET containers targeted by most deposit legislation represent only 3% of the plastic waste stream, or only 0.2% of the entire municipal solid waste stream (IEc, 1989).

California also has legislation that provides an incentive for consumers to recycle beverage bottles, although not a deposit system. In California, consumers are given a refund (equal to the redemption value) for every container they return. The beverage industry pays the redemption value.

Container deposit legislation has proven to be very effective at capturing targeted items. Table 5-12 presents data on compliance rates in several "bottle bill" states and California; state authorities estimate compliance rates ranging from 50 to over 90%. Not all containers captured by deposit legislation are recycled, however. For example, New York estimates that only 57% of collected PET containers are recycled; the balance are disposed of as part of the MSW stream. Iowa and Massachusetts report that even smaller percentages of collected plastic containers are recycled. In contrast, virtually all glass and aluminum containers collected in these states are apparently recycled (IEc, 1989).

Deposits are typically 5 cents per container (except in Michigan, where the deposit is 10 cents). State programs may differ in the number of classes of containers covered, the organizational structure enacted to facilitate the return of containers to processors, and the flow of payments to distributors and retailers. There has been significant retail and beverage industry resistance to deposit legislation, however, because of the allegedly high costs to "middlemen" for providing the required collection, storage, and (sometimes) transportation facilities for collected recyclable

Table 5-12
ESTIMATED CONSUMER RETURN RATES OF PET
BEVERAGE CONTAINERS RESULTING FROM BOTTLE
BILL LEGISLATION
(Connecticut and Delaware not included)

State	Year Passed	Primary Collection Method	% Recovery of All Containers	% Recovery Targeted Plastics	Deposit Minimum (cents)	Program Features
California	1986	Redemption Centers	>53	5	1 (b)	More than 2000 "convenience zone" collection centers. Wine coolers will be added in 1990.
Iowa	1979	Retailers	91-95	--	5	Includes wine coolers and other alcoholic beverage containers.
Maine	1978	Redemption Centers	56	50	5	Includes wine coolers.
Massachusetts	1983	Retailers	87-99	60-90	5	
Michigan	1978	Retailers	92-93	90	10	Includes wine coolers. Proposed legislation will expand the variety of recycled materials.
New York	1983	Retailers	74	70	5	
Oregon	1971	Retailers	95	80-90	5	Very high public acceptance of recycling for this well-established program
Vermont	1973	Redemption Centers	80-90	65-70	5	Experienced a much lower return rate on larger containers. Proposed bill to expand to include alcoholic beverage containers.

(a) These figures are estimates. Many states with bottle bills have no established reporting system or requirements.
(b) The California return incentive increases proportionately depending upon the total amount of scrap collected in the state. Also added is an amount equalling the current scrap value of the container.

Sources: Bree, 1989; Calif., 1988; Maine DECD, 1988; Mass DEQE, 1988; Gehr, 1989; Koser, 1989; MacDonald, 1989; Phillips, 1989; Schmitz, 1989; Wineholt, 1989

containers. The costs of container deposit legislation are discussed in more detail in Section 5.4.3.4 (below).

Although deposit legislation captures a high percentage of targeted containers, these containers represent only a small fraction of all plastics in MSW. A few states (e.g., Michigan) are considering extension of deposit legislation to a broader spectrum of plastic products (i.e., not only beverage containers), but nationally little momentum is apparent toward such policies. Deposit legislation allows collection of homogeneous resin streams because it targets specific categories of containers. "Bottle bill" states are currently the primary suppliers for plastic recycling processors.

Some potential exists for conflict between deposit legislation and curbside collection programs (and drop-off recycling centers). Deposit legislation is generally targeted at easily characterized containers that economically are among the most valuable plastic items in MSW. To the extent that it succeeds in capturing a large proportion of these items, deposit legislation may tend to reduce both the quantity and the economic value of plastics available for curbside collection. This, in turn, may have a negative impact on the costs and benefits of curbside plastics recycling, and may influence some communities to exclude plastics from their recycling programs.

SUMMARY: COLLECTION ALTERNATIVES -- Curbside collection offers to divert the most significant quantities of MSW from disposal. Thus, use of this collection alternative is likely to expand, especially in states and/or municipalities facing high landfill costs and capacity constraints. One disadvantage of curbside collection is that it can yield mixed plastics (if many are collected) that are difficult to sort by resin type with currently available technologies. Curbside collection programs also face significant hurdles to implementation, both in urban areas with large numbers of multi-family residences, and in rural areas with no centralized MSW collection services. Container deposit legislation is very successful at capturing a large proportion of targeted plastic beverage containers, yielding homogeneous recycled resins amenable to high-value processing applications. But deposit legislation typically affects only a very small proportion of MSW plastics. Especially if broadened to include additional categories of recyclable plastic items, deposit legislation may tend to adversely impact the viability or success of curbside recycling programs. Drop-off recycling centers and voluntary buy-back programs are likely to remain minor contributors nationally to plastics recycling. Drop-off centers, however, may be a successful recycling option in rural areas. Reverse vending machines are likely to become much more prevalent as an implementation option in support of drop-off centers, buy-back programs, and/or container deposit legislation.

5.4.2.2   Separation of Plastics by Resin Types

Recycled plastics may be processed either as homogeneous resins or as mixtures of resins. Mixed resin processes currently yield products that only rarely displace virgin resins. The following discussion presents a number of alternatives currently or potentially available to facilitate the separation of collected recycled plastics into homogeneous resin types. The greatest long-term diversion of plastics from the waste stream promises to be realized if separation techniques are available that make homogeneous resins available to recycled plastics processors. The following discussion reflects both widespread interest and active efforts to refine such techniques.

The primary alternatives available to allow separation of homogeneous resins from collected recyclable plastics include:

- Separation after compaction or shredding

- Container labeling and automated separation

- Manual segregation by resin at the point of collection

- Collection focused on specific resin or container types

- Standardization of resin contents of recyclable products

The advantages and disadvantages of these alternative separation strategies (see Table 5-13) are discussed in the following paragraphs.

SEPARATION AFTER COMPACTION OR SHREDDING -- The most cost-effective means to collect a large volume of plastics for recycling and delivery to processors is simply to segregate mixed plastics from MSW and shred or compact them at the point of collection.

Separation of mixed shredded resins into homogeneous streams is technically difficult, however. For well-characterized mixtures of two known resins (e.g., PET and HDPE from beverage bottles) density separation may be possible; this technology is currently employed to segregate shredded PET/HDPE bottles into their constituent resins for recycling. But the wide variety of resins present in commingled plastics wastes, and the very similar densities of many of these resins, effectively preclude the use of density separation techniques for assorted mixed plastics, and no other technologies currently available or under development appear capable of achieving reliable separation for shredded plastic wastes.

Separation of crushed containers may be feasible, however. The following section, describing technologies available and under development to automatically separate intact or crushed plastic containers, describes a number of existing or promising technologies that may facilitate the segregation of homogeneous resin streams from mixed, crushed MSW plastics.

CONTAINER LABELING AND AUTOMATED SEPARATION -- The Society of the Plastics Industry (SPI) has instituted a voluntary labeling system for recyclable plastic containers (Figure 5-1); the molded label contains a code specifying the primary resin incorporated into the product. These codes have been voluntarily adopted by much of the plastics processing industry and are currently beginning to appear on containers distributed in consumer markets (IEc, 1988). Fifteen states have made use of the SPI codes mandatory on rigid plastic containers distributed in the state (SPI, 1989). Several other states are considering such actions.

No insurmountable technical barriers apparently stand in the way of the development of automated scanning and sorting systems that read an encoded label and divert products to

TABLE 5-13

ADVANTAGES AND DISADVANTAGES OF ALTERNATIVE STRATEGIES TO ALLOW
SEPARATION OF RESIN TYPES FROM MIXED RECYCLABLE PLASTICS

Strategy	Advantages	Disadvantages
Separation after compaction or shredding	Convenience to consumers; does not require consumers to separate wastes	Currently not possible to effect separation into homogeneous resins after shredding
	Minimizes sorting, storage, and transportation requirements for collecting agencies	Shredding yields mixed plastics not amenable to processing into products displacing virgin resins
	Allows collection strategies capturing large volume and variety of MSW plastics	
Container labeling and automated separation	Convenience to consumers; does not require consumers to separate wastes	Technology not currently in place
	Promises to allow separation into homogeneous streams	May imply requirement for centralized storage and separation facility, with associated costs
	Promises to allow collection strategies capturing large volume and variety of MSW plastics	Possible requirement to transport collected recyclables to centralized storage and separation facility
	Minimizes manpower requirements required for sorting	
Manual separation by consumer or collection agency	Simple technology	Potentially prohibitive manpower requirements
	Convenience to consumers if collecting agency performs separation	May imply large storage and transportation requirement for collecting agency
	Allows collection strategies capturing large volume and variety of MSW plastics	Inconvenience to consumers if they are required to perform separation
Collection focused on specific resin or container types	Facilitates collection of homogeneous resin streams	Inconvenience to consumers if they are required to store and transport recyclables to central collection point
	Allows recycling efforts to focus on high-value, high-volume recyclable products	

(cont.)

TABLE 5-13 (cont.)

ADVANTAGES AND DISADVANTAGES OF ALTERNATIVE STRATEGIES TO ALLOW
SEPARATION OF RESIN TYPES FROM MIXED RECYCLABLE PLASTICS

Strategy	Advantages	Disadvantages
Collection focused on specific resin or container types (cont.)	Convenience to consumers, who are required to collect only a subset of plastics wastes	Captures only a small portion of potentially recyclable plastics
	Relatively low cost to recycling agencies	
	Consistent with collection strategies offering financial incentives to recycle	
Standardization of resin use for certain product applications	Facilitates collection of homogeneous resin streams	May imply significant governmental intervention in private markets
		May be difficult to enlist voluntary industry cooperation
		May be applicable to only a small percentage of recyclable products

Source: Compiled by Eastern Research Group.

Figure 5-1

PROPOSED CODING SYSTEM FOR PLASTICS RESIN

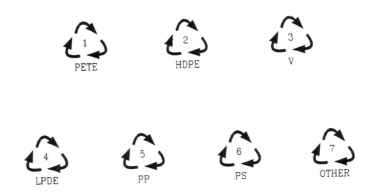

1. Polyethylene terephthalate
2. High-density polyethylene
3. Vinyl
4. Low-density polyethylene

5. Polypropylene
6. Polystyrene
7. Other, including multilayer

Source: SPI, 1988.

separate shredding and storage lines, although this technology has not yet been implemented (Medeiros, 1989).

Other technologies potentially available to effect separation of mixed plastics need not rely on an encoded label. The Center for Plastics Recycling Research is investigating an infrared sorting system that may be applicable to crushed containers (Dittmann, 1989). Density separation techniques are currently employed to separate some resins (e.g., PET and HDPE recovered from beverage containers). These techniques might find wider application, but their ability to effectively sort the wide variety of resins and resin/additive combinations found in mixed MSW plastics is questionable. For example, PET and PVC are of similar density and thus difficult to separate. By industry agreement or regulatory requirement, chemical markers might also be incorporated in commodity resins in consumer applications; these markers could facilitate separation by spectrographic or other means.

There is significant industry interest in these technologies, and a number of implementation alternatives are under active development. These technologies face foreseeable barriers, however, primarily economic and institutional.

Economic barriers include: 1) the potential cost of such systems; and 2) costs imposed on municipalities or other recycling agencies to transport uncrushed (with some technologies), unshredded containers to the sorting facility. An institutional barrier is also associated with these economic considerations, in that the expense of the systems may make them feasible only if implemented in regional (e.g., county-wide) processing centers, which in turn may require a coordinated infrastructure among governments in a region.

This option is most compatible with curbside collection programs (and drop-off recycling centers) because these programs promise to provide large volumes of mixed plastics wastes. Automated separation is also compatible with container deposit legislation; this is especially true if deposit legislation is extended to a broad range of recyclable containers.

Development of this alternative may also be determined, to some extent, by the growth of markets for the products of homogeneous plastics recycling processes. If these markets continue to develop, processors may demand greater quantities of homogeneous recycled resins. Such demand may drive the development and implementation of automated plastics separation.

SEGREGATION BY RESIN AT THE POINT OF COLLECTION -- If a uniform labeling convention is in place, plastics may be segregated manually by resin as they are collected for recycling. In a curbside collection program, separation may be performed by consumers before setting materials out for recycling, or by the MSW collection agency either at curbside or at a central processing facility. In a centralized collection scheme, separation may also be required of the consumer, or may be performed during or after the transfer of recyclables from consumer to the collection center.

While this technique is technologically simple, it is labor intensive. The inconvenience to consumers of scanning and separating products by resins suggests that participation in this separation scheme would be low. If collecting agencies also must sort the wastes, significant labor costs will be imposed; costs will also be imposed at the point of collection for the storage of recyclables, and potentially for the transport of sorted recyclables to processing facilities (although shredding or compaction at the point of collection may allow this expense to be avoided).

Nevertheless, a number of communities perform manual sorting of recyclables. Typically, their collection and separation efforts have focused on only one or a few classes of plastic articles (e.g., HDPE milk jugs, PET/HDPE beverage containers). Some of these communities have worked in conjunction with human services agencies to employ handicapped citizens for sorting tasks. These citizens provide a low-cost work force for the recycling program, and benefits are also measured by the provision of meaningful work for this segment of the population.

In conclusion, manual sorting is not the most efficient of sorting alternatives, but it offers benefits that will undoubtedly encourage its use by a number of community recycling programs.

COLLECTION FOCUSED ON SPECIFIC RESIN OR CONTAINER TYPES -- A number of municipal recycling programs, as well as most "bottle bill" plastics recycling efforts, focus on a limited subset of all recyclable plastic containers. For example, some communities (e.g., Naperville, Illinois (Massachusetts DEQE, 1988) have focused on HDPE milk jugs in their recycling efforts, while most container deposit legislation affects primarily PET/HDPE beverage containers.

Such focused recycling efforts by definition yield an easily characterized, homogeneous stream of recyclables. Compared to other separation alternatives, they also offer the advantages of consumer convenience and relatively low cost to recycling agencies. But they result in the collection of only a small subset of potential recyclables, and so offer limited benefits in terms of total reduction of the volume of plastics in MSW requiring disposal.

Nonetheless, based on the purchasing activity of recycled plastic processors, this strategy has proven very effective in capturing the homogeneous resin streams required for plastics recycling technologies dependent on homogeneous input streams. By definition, use of this strategy will continue to expand with any expansion of bottle deposit legislation, use of reverse vending machines, or voluntary buy-back programs. If states begin to expand the scope of deposit legislation, however, such legislation may result in the collection of more mixed plastics waste streams. In this case, deployment of alternative separation strategies may be required if these states are to continue to be sources of homogeneous resin streams.

STANDARDIZATION OF RESIN CONTENTS OF RECYCLABLE PRODUCTS -- One of the most intractable problems in mixed plastics recycling is the great variety of resins in MSW. In the face of this diversity, it may be desirable to apply uniform standards for resin content across at least some classes of plastic containers to facilitate their separation into a homogeneous stream of recyclable plastic. This option is not really a separation strategy in itself, but facilitates the coding and separation of a potentially wide selection of plastics products. This strategy has been used in West Germany and the Netherlands, where the Coca Cola company has worked with a bottle producer and government agencies to develop a single-resin beverage container to support recycling programs (NOAA, 1988).

Barriers to growth are significant for this option. It affects the business decisions of potentially thousands of producers and marketers. Resin and additive contents are often dictated by specific product needs (e.g., for vapor impermeability, transparency or translucence, chemical resistance to specific compounds), and so may be impractical for government authorities to review or assess. Nonetheless, for a limited range of items with common characteristics (e.g., beverage containers, milk jugs, detergent bottles), standardization may spread through voluntary industry agreements (based on the perceived public relations value of marketing in recyclable containers), which might be encouraged by government involvement.

SUMMARY: OUTLOOK FOR SEPARATION OF PLASTICS INTO RESIN TYPES -- No technologies are currently widely employed to effect the separation of resins from mixed plastics wastes. The most effective means currently employed to yield homogeneous recycled resin streams is to focus collection efforts on one or a few products containing a correspondingly small number of resins. Two additional strategies may facilitate the collection of homogeneous resin streams: 1) development of standard container labeling and automated sorting equipment, and 2) voluntary use of standardized resin contents in some classes of plastic products. Significant industry efforts are underway to develop automated sorting technologies. Within a few years these may allow mixed recycled plastics to be sorted efficiently and cost effectively.

5.4.2.3 Processing and Manufacturing of Recycled Plastics

Depending on the nature and homogeneity of resins available from collected (and possibly sorted) recycled plastics, a number of processes are available to produce recycled plastic products. Discussions of many of these processes are available in a number of sources (e.g., Plastics Recycling Foundation, 1988; Mass. DEQE, 1988; IEc, 1988; Brewer, 1988a; Curlee, 1986); the following discussion provides an overview of the principal distinguishing characteristics of these processes, including their inputs and the nature and quality of products they yield.

Processing technologies available for post-consumer plastic wastes may be grouped into two categories:

- Secondary Processes -- include a variety of technologies distinguished by the nature of required inputs and by the characteristics of their products. They are commonly differentiated by the nature of resins input to the process:

  -- Secondary Processes/Homogeneous Resins -- yield products that compete with the products of virgin resins.

  -- Secondary Processes/Mixed Resins -- yield massive or thick-walled products that may replace lumber, concrete, or ceramics.

- Tertiary Processes -- use either pure or mixed resins to yield monomers or oligomers used as fuels (mixed plastics inputs) or as chemical feedstocks (pure resin inputs).

As noted in Section 5.4.2, "Primary" recycling processes refer to industrial reprocessing of manufacturing and processing scrap; these processes do not affect post-consumer wastes and are not addressed in this Report. Some analysts also define a fourth (or "quaternary") category of recycling processing technology (e.g., Curlee, 1986). Quaternary processes are defined as the pyrolysis and combustion of plastics in an energy-recovery incinerator; as such, they are not recycling processes as defined in this Report.

Processing technologies are defined primarily by the purity of their required input streams and the quality of their products. As has been noted, homogeneous inputs are required for technologies that can use recycled plastics in blends with virgin resins or that can produce products competitive with products manufactured from virgin resins. As input quality falls, output products tend not to displace consumption of virgin plastics, but to compete in markets with lower-value commodities such as lumber and concrete. The products of tertiary recycling processes (monomers and oligomers resulting from the nearly complete breakdown of plastics resins) do not compete with plastics strictly defined, but with the raw input materials to plastics (and other chemical) production processes.

SECONDARY PROCESSING TECHNOLOGIES/HOMOGENEOUS RESIN INPUTS -- These processing technologies are generally the same as or similar to those used to process virgin plastic resins, and demand inputs of high resin quality and homogeneity.

Secondary recycling processes for homogeneous resins typically heat recycled plastics (or a blend of recycled and virgin resins) into their melt range and use any of a number of production processes (e.g., injection molding, extrusion) to yield a final product. To date, such processes have been employed primarily with homogeneous resin streams from recycled PET/HDPE beverage containers and HDPE milk jugs. Table 5-14 presents a number of the products currently produced from recycled PET and HDPE, with estimates of the size of current and projected markets for these products.

These processing technologies are the same as or very similar to those employed with virgin resins; as such, they are "mature," cost-effective, and well characterized. They are capable of processing inputs of recycled resins into high-value products and are currently supply-limited.

SECONDARY PROCESSING TECHNOLOGIES/MIXED RESIN INPUTS -- Secondary processing technologies using mixed resin inputs yield products with relatively non-demanding physical and chemical characteristics. Typically, mixed resins are heated (generally by pressure and friction) above the melt points of the dominant resins in the blend and extruded or molded into desired product shapes. Plastics that do not melt in the blend (and other contaminants) are encapsulated and serve as filler in the final product; other materials (fillers, colorants, stabilizers, flame retardants, etc.) may be added during the blending process to yield desired product qualities.

Some of the products of mixed resin secondary processes include (Brewer 1987):

- Plastic "lumber" (suitable for boat docks, fence posts, animal pens, landscaping applications, etc.)

Table 5-14

ESTIMATED MARKETS FOR RECYCLED PLASTIC RESINS
(Millions of pounds)

Polyethylene Terephthalate Product Applications	Market Size		High-Density Polyethylene Product Applications	Market Size	
	1987	1992		1987	1992
Fiber	90	180	Bottles (nonfood)	--	115
Injection molding	25	160	Drums	--	25
Extrusion	25	130	Pails	20	65
Non-food grade containers	--	30	Toys	--	15
Structural foam molding	--	30	Pipe	30	80
Paints, polyols, other chemical uses	10	20	Sheet	--	25
Stampable sheet	--	30	Crates, cases, pallets	--	105
Other	--	10	All other	4	130
Total – polyethylene terephthalate	150	590	Total – high-density polyethylene	54	560

Source: Center for Plastics Recycling Research, 1987.

- Car stops and railroad ties

- Pallets

- Gratings and man-hole covers

- Cable reels

Mixed resin secondary processes are currently available and have been deployed by a number of firms in the United States. European countries (especially Germany) and Japan have been leaders in developing and implementing these technologies. They continue to face a number of technical and economic barriers, however. Technically, these technologies face the challenge of producing higher-quality, higher-value products from mixed plastics inputs. Their current range of products competes with low-value commodities in relatively limited markets; both market diversity and product value must increase if these technologies are to fulfill their promise to absorb a large proportion of recyclable mixed plastics. Economically, the costs of these processing technologies must be reduced to allow their products to compete effectively in established markets; the long lifespans and maintenance-free qualities of their products may not be sufficient to overcome consumer resistance to high initial purchase prices.

TERTIARY PROCESSING TECHNOLOGIES -- Tertiary processes recover basic chemicals and fuels from waste plastics. By far the most common tertiary process is pyrolysis, in which wastes are heated in the absence of oxygen, driving off volatile components of the inputs (plastics monomers and oligomers and other products) and leaving a "char" consisting mainly of carbon and ash. The mix of products and their potential uses are determined both by the nature of the input stream and by pyrolysis conditions; they can include combustible gases useful as chemical feedstocks and gases and liquids that can be used as fuels (Curlee, 1986).

Tertiary processes can be employed with a wide variety of inputs, including mixed organic wastes (e.g., all combustible fractions of MSW), mixed plastics wastes, or homogeneous plastic resin streams. Control over outputs is greatest when inputs are well characterized and consist of only one or a few known constituents. Only if these conditions are met do tertiary processes yield products of sufficient quality and purity to be used as chemical feedstocks; as input quality declines, tertiary products are generally useful only as fuels (and the distinction between tertiary "recycling" and simple incineration tends to be obliterated).The primary advantage of tertiary processes is their ability to be used with mixed plastics or with mixed plastic/nonplastic wastes. If used with such wastes, however, tertiary processes tend to become a disposal rather than a recycling alternative. Because tertiary processing technologies can also be employed with homogeneous plastics waste streams to yield high-value chemical products, they may also compete with homogeneous resin secondary processing technologies as an option to recycle sorted and well-categorized plastics resins separated from MSW.

SUMMARY: PROCESSING AND MANUFACTURING OF RECYCLED PLASTICS --
Secondary processing technologies using homogeneous resins as inputs generally yield products that displace virgin plastic resins in product applications. They therefore result in the most significant reduction of plastics use and the long-term reduction of disposal requirements. But

their required inputs of homogeneous, well-characterized resins may be difficult or impossible to obtain with curbside collection or drop-off recycling centers (unless these collection options are focused on a limited set of plastics wastes, e.g., PET/HDPE containers or plastic milk jugs). Mixed resin secondary technologies, on the other hand, can operate with mixed plastics wastes. But they produce relatively low value products that do not compete with virgin plastics but with lumber, concrete, ceramics, and metals. Because they do not reduce virgin plastics use, and because they must themselves eventually be disposed of, the products of mixed resin secondary processes delay, but may not ultimately reduce, plastics disposal requirements (see Section 5.4.2). Tertiary recycling processes may use either mixed or pure resin streams as inputs. If mixed plastics are inputs, tertiary processes generally yield hydrocarbon fuels (and can best be classed as a disposal option); if homogeneous resins are inputs, however, tertiary processes may yield well-characterized products that can be used as feedstocks in the production of plastics or other chemical products.

5.4.2.4 Marketing of Recycled Plastics Products

The presence of adequate markets for recycled plastics products will be a critical determinant of the potential for recycling to divert a significant proportion of plastics from MSW disposal. Available information indicates that substantial markets exist for the products of secondary processes employing homogeneous resin inputs and for some tertiary processing technologies, and that market opportunities should not limit the growth of these technologies in the foreseeable future. The products of mixed resin secondary processes, however, may face significant marketing challenges; these processes may need to overcome cost and product quality hurdles to be assured of adequate long-term markets.

A number of other changes may help to improve current and potential markets for recycled plastics. These include changes in consumer preferences (e.g., through marketing efforts stressing the environmental benefits of recycled plastics), the development of cooperative marketing associations, increased government procurement of recycled products, and increased industrial and government research and development in all phases of plastics recycling.

The following paragraphs summarize published estimates of current and potential markets for recycled plastic products. Published quantitative estimates have focused on markets for recycled PET and HDPE products, because these resins have been those most widely targeted under currently implemented collection strategies, and on the products of mixed resin secondary recycling processes.

MARKETS FOR UNPROCESSED RECYCLED PLASTICS -- In addition to U.S. and foreign markets for the finished products of recycling processes, foreign markets may exist for unprocessed or partially processed recycled plastics. In a Massachusetts study, less developed countries were singled out as a large potential market for recycled resins (Mass. DEQE, 1988), and some recycling programs have specifically targeted foreign processors to accept recycled resins. No quantitative estimates of these markets exist, however, and some evidence suggests that these markets may be very volatile, and so may not be reliable as a market for large volumes of recycled resins (Smith, 1989).

MARKETS FOR PRODUCTS OF SECONDARY TECHNOLOGIES USING HOMOGENEOUS RESINS -- Table 5-14 provides estimates of the 1987 and 1992 markets for recycled PET products. The 1987 market of 150 million pound/year is predicted to grow by 400%, to 600 million pounds/year, by 1992 (Center for Plastics Recycling Research, 1987). The latter estimate represents slightly more than 30% of U.S. 1988 production of PET (SPI, 1988a) and would represent approximately 50% of all PET soft drink bottles.

Table 5-14 also provides estimates for products of recycled HDPE. Markets for these products are projected to grow ten-fold between 1987 and 1992, to a total of 560 million pounds/year (Center for Plastics Recycling Research, 1987). This figure represents approximately 7% of 1988 U.S. production (SPI, 1988a).

Current and potential market areas are also forecast for recycled polyvinyl chloride resins. The specific applications for this material include various building and construction applications (drainage, sewer, and irrigation pipe, pipe fittings, vinyl floor tile, fencing) and industrial uses (truck bed liners, cushioned laboratory mats). Because PVC recycling is not currently as well developed as that for PET and HDPE, reliable quantitative estimates of market size for recycled PVC products have not been generated to date.

Polystyrene is another single resin that has been the focus of recent recycling efforts. Specific applications include insulation, toys, and desk supplies. Since post-consumer PS recycling is currently very limited, reliable quantitative estimates of market size have not yet been developed.

MARKETS FOR PRODUCTS OF SECONDARY TECHNOLOGIES USING MIXED RESINS -- Table 5-15 provides qualitative estimates of the potential markets for a number of products of mixed resin processing technologies. This table reflects the fact that the greatest potential markets for these products are currently dominated by lumber, concrete, and other similar commodities. Although plastic "lumber" produced from mixed recycled plastics may be sawed, shaped, and painted, its overall potential to replace wood (or metal) in many applications is limited by its relatively poor structural properties and its relatively high price (2-3 times that of pressure heated lumber) (Bennett, 1988; IEc, 1988; Mass. DEQE, 1988).

Economic barriers currently impede further market penetration for many mixed resin recycled products. For example, plastic "lumber" may have an initial sales price 50 to 300% higher than comparable wood items (Maczko, 1989); although lifecycle savings attributable to the nonbiodegradability of the plastic item may reduce or reverse this cost differential over the product lifetime, long-term savings may be insufficient to overcome resistance to the high purchase price for many consumers.

These barriers may be reduced as mixed resin processing technologies mature. A number of American research institutions (including the Plastics Recycling Foundation and the Center for Plastics Recycling Research), as well as a number of foreign firms and government agencies, are conducting active R&D programs to increase the applicability, reduce costs, and increase product quality for mixed resin recycled plastics products. This high level of interest and commitment to additional research promises to significantly expand the market opportunities for these products in coming years.

TABLE 5-15

CURRENT AND POTENTIAL MARKETS
FOR MIXED RESIN RECYCLED PRODUCTS

Market	Key Consideration	Market Outlook
Boat docks	Continuous exposure to harsh, wet environment Plastic products currently used, accepted	Strong regional potential
Auto curb stops	Plastic currently used, cost effective Coloring throughout saves maintenance costs compared to concrete alternatives Lighter weight than concrete alternatives	Limited data available

Breakwaters	Wet environment ideal for plastic	Tight construction
		  regulations
		Regional markets
		  only

Park benches	Continued exposure to inclement weather	Strong potential
Mushroom trays	Moist conditions require plastic	Limited market data Potential food contact concerns
Horse stalls	Top and bottom rails subject to deterioration; ideal for plastic	Strong regional potential
Picnic tables	Continued exposure to inclement weather Outdoor environment ideal for plastic	Small market Limited market data

Playground
  equipment
	Outdoor environment ideal for plastic	Tight construction
		  specifications

Railroad ties	Harsh outdoor environment suitable for
	  plastics
		Potentially large
		  market
		Tight construction
		  specifications
		Depends on results of
		  ongoing long-term
		  strength tests

Source: IEc, 1988; adapted from Mass. DEQE, 1988.

MARKETS FOR PRODUCTS OF TERTIARY PROCESSING TECHNOLOGIES -- Markets for tertiary recycled plastics products vary with the process inputs. Products of tertiary processing of mixed plastics wastes represent a generally complex mixture of hydrocarbons; it is infeasible to refine such mixtures into pure product streams economically competitive with those obtained from processing petroleum or natural gas, and so these products are generally useful only as fuels. If recycling outputs are put to no other use, "tertiary processing" is no more than a synonym for incineration.

The products of tertiary processing of homogeneous, well-characterized input streams, on the other hand, can be controlled and may be economically competitive with the products of refining processes. For example, tertiary processing of PET may produce chemical feedstocks of equal quality to and at lower prices than those obtained from raw refining processes (Stroika, 1988).To date, tertiary processes that convert homogeneous resin streams into high- quality chemical feedstocks have been deployed in only a small number of installations in the United States. Although limited evidence indicates that these plants have been economically viable, little research has been conducted into the potential long-term market for these tertiary recycled products (Stroika, 1988).

SUMMARY: MARKETS FOR RECYCLED PLASTICS PRODUCTS -- Substantial markets appear to exist for the products of secondary recycling processes employing homogeneous resin inputs. In the opinion of many industry participants, the primary limitation on the development of these technologies is not current or potential market size, but assurance of a steady supply of inputs (Brewer, 1988c). Developments in homogeneous resin processing technologies suggest that they will continue to be refined to yield products that are directly competitive with those produced from virgin resins. These recycled products should be cost-competitive in appropriate markets.

Current mixed resin secondary processing technologies yield products that are competitive with relatively low cost commodities. Their long-term marketing outlook may depend on production costs as the technologies mature, and on technological developments that allow the production of higher-quality, higher-value products.

For tertiary processes operating with homogeneous inputs (which yield chemical feedstocks as products), the primary marketing consideration is the cost of the recycled outputs vis a vis the cost of feedstocks refined from fossil fuels -- these latter costs are determined jointly by fossil fuel prices, the capacity of feedstock refineries, and national and international demand for plastics. Economic scenarios combining increasing energy prices and increasing demand for plastics promise the greatest long-term markets for feedstock-producing tertiary technologies.

5.4.2.5   Summary: Integration of Plastics Recycling Alternatives

One of the most notable characteristics of plastics recycling is the variety of alternatives available to implement each of the four phases of the recycling process. But none of these phases exists in isolation; the phases, and the choices among available alternatives for each phase, are intricately interrelated. For example, implementation of mixed plastics collection strategies implies that only mixed resin secondary recycling processes will be available for the recycled resins; however, the potentially larger markets and higher value for homogeneous resin products may simultaneously spur the development of effective resin separation technologies, ultimately allowing high-volume collection (i.e., collection of mixed plastics) to be coupled with homogeneous resin recycling processes.

Table 5-16 matches the primary collection options for recyclable plastics against options to separate plastics by resin and points out the major relationships between them. The most important consideration reflected in Table 5-16 is volume of plastics collected with each collection/separation combination and the homogeneity of the resulting resin stream. Combinations of collection/separation alternatives that tend to link capture of high volumes of plastics with output of homogeneous resin streams are the most valuable in terms of opening the largest markets for recycled plastics products and providing the greatest long-term diversion of plastics from MSW disposal requirements.

Among collection options, curbside collection promises to divert the greatest proportion of MSW plastics from disposal. But unless efficient separation alternatives are employed, curbside collection may yield mixed plastics amenable only to mixed resin secondary processing. A number of promising separation alternatives are the focus of active research and development. Although no insurmountable technical barriers to implementation of one or more of these separation options are apparent, none is currently available.

Drop-off recycling centers have similar sorting requirements. But because of historically low participation rates, drop-off centers do not promise to divert a significant proportion of plastics from disposal (unless implementation is accompanied by effective, long-term public education and outreach programs).

Curbside collection (or drop-off centers) might be targeted at only a limited subset of MSW plastics, but such targeting reduces the volume of plastics collected.

Container deposit legislation as currently enacted (i.e., targeted at only a few classes of plastic containers) has the advantage that it generally yields resins pure enough to feed recycling processes demanding homogeneous resin inputs. But deposit legislation captures only a small proportion of MSW plastics.

There has been much discussion and some state action (e.g., in Michigan) to expand the range of items collected under container deposit legislation. To do so would obviously increase the proportion of MSW plastics collected under deposit programs, but this option has drawbacks. First, it will probably tend to reduce the volume of resins available to secondary processors requiring homogeneous inputs (unless effective separation technologies are implemented). Second, it may interfere with the success of curbside collection programs, primarily because of its negative impact on costs and benefits of curbside collection. Nonetheless, this collection option may be appropriate in states where demographics militate against the widespread use of curbside collection.

Combinations of deposit legislation and other collection alternatives may prove to be effective recycling policy options. For example, deposit legislation expanded to selected additional containers (of known, standardized resin content) might effectively capture a large proportion of

Table 5-16

RELATIONSHIPS AMONG RECYCLING, COLLECTION, AND SEPARATION ALTERNATIVES

Collection Alternative	Separation Alternative					
	Separation after Compaction or Shredding	No Separation	Automated Separation	Separation at Point of Collection	Collection Focused on One or a Few Resins	Other Considerations
Curbside collection; Drop-off centers; Voluntary buy-back	This separation option requires technological development to improve applications for this collection alternative; separation after shredding may not be feasible with comingled plastics	May divert the greatest volume of recyclables from MSW disposal, but results in plastics stream amenable only to mixed plastics processing	May yield resins pure enough to use homogeneous plastic processing technologies; separation technology currently unproven	This separation option probably not widely feasible for these collection alternatives; some communities employ handicapped citizens to separate recyclables	Significant reduction in potential to collect and recycle large proportion of MSW plastics	Curbside recycling offers best potential to reduce net discards; Enactment of container deposit legislation may lower participation in these collection alternatives. Unless efficient separation is implemented, collected plastics amenable only to mixed plastics processing
Container deposit legislation targeted at only a few items (i.e., most current deposit laws)	This separation option depends on technology development to apply to this collection alternative	Collected plastics may be amenable only to mixed plastics processing; however, if only one resin is targeted (e.g., PET drink containers), materials may be used for homogeneous plastic recycling processes	Should yield resins pure enough to use homogeneous plastic processing technologies; separation technology currently unproven	May be feasible with this collection option, because only a few resin and/or container types will be collected	Part of the definition of this collection alternative, yields resins amenable to homogeneous plastics processing	Results in relatively small diversion of plastics from MSW disposal; may be effectively combined with curbside collection, drop-off centers, should yield relatively pure resins amenable to homogeneous plastics processing
Container deposit legislation targeted at a wider variety of items	This separation option probably infeasible with this collection alternative	Collected plastics may be amenable only to mixed plastics processing	May be appropriate for this collection option, especially if collection focuses on containers of a relatively few resin types; technology currently unproven	This separation option probably not feasible with this collection alternative	May be part of the definition of this collection alternative; may be most effective with industry agreement to standardize resin contents of targeted items; should yield resins amenable to homogeneous plastics processing	Deposit legislation targeted at many items may tend to have a negative impact on the success of curbside collection, drop-off, and voluntary buy-back collection alternatives; unless efficient separation is implemented, plastics collected probably amenable only to mixed plastics processing

Source: Compiled by Eastern Research Group.

MSW plastics amenable to homogeneous resin processing technologies. Although curbside collection would then capture only the remaining mixed plastics, there would be no requirement to sort these wastes, and they could be fed directly into mixed plastic processing technologies. The net result might be the optimization of both the total diversion of plastics from disposal and the yield of resins amenable to homogeneous resin processing technologies. Another potentially attractive collection/sorting/processing alternative couples curbside (or other) collection alternatives that yield a mixed stream of recycled plastics with limited separation by resin types. Such limited separation might effectively skim the highest-volume or highest-value resins from mixed recycled plastics, making these resins available to processors relying on homogeneous inputs. The remaining plastics would be fed to processors employing mixed plastics processing technologies. This strategy would capture high volumes of MSW plastics and simultaneously facilitate the market expansion of homogeneous and mixed plastic recycling processes.

This discussion (and the information presented in Table 5-16) represents only a very preliminary analysis of the interaction between recycling collection, separation, and processing options. What it makes clear is that plastics recycling must be viewed and analyzed as a system of integrated components, in which decisions affecting each phase of recycling have implications on all other phases -- and on the success of a proposed recycling program as a whole. Recycling of other components is also affected by the choices made for plastics recycling. An integrated system is required.

5.4.2.6 Current Government and Industry Plastics Recycling Initiatives

Plastics recycling is a dynamic field. All four phases of recycling -- collection, sorting, processing, and marketing -- are the subject of active interest, regulatory intervention, and research and development efforts, sponsored by national governments (especially in Western Europe), state and local governments, and private industry. New developments in all phases of plastics recycling are reported almost monthly. The very rapid recent progress both in technological innovation and in governmental support for plastics recycling augurs well for the continuing success of this waste management alternative.

The following paragraphs highlight a number of recent developments in plastics recycling, compiled by Brewer (1989) and the Council for Solid Waste Solutions (1989):

RESIN SEPARATION TECHNOLOGIES --

- An industrial scale polyolefin separation plant is on-line in Coburg, West Germany. The firm responsible has announced a joint venture to site ten such separation plants in the United States.

- Sorema, an Italian firm, has sold approximately a dozen polyolefin separation plants worldwide that handle post-consumer plastics.

- Research underway at Rensselaer Polytechnic Institute is focused on a sorting system capable of isolating distinct resins from mixed plastics.

- Dupont and Waste Management, Inc. (WMI), recently announced a joint venture to collect, separate, and process recycled plastics. The joint program will be implemented in communities across the country serviced by WMI's waste collection operations.

- Wellman, Inc., America's largest processor of recycled PET, recently inaugurated a major research program to develop sorting technologies for mixed resins.

MIXED PLASTIC SECONDARY RECYCLING TECHNOLOGIES --

- Sixteen plants using the mixed plastic "ET/1" technology (producing plastic "lumber" and similar structural products) are on-line worldwide; three of these are in the United States. An additional sixteen plants have been ordered.

- Recycled Plastics, Inc., operates a mixed plastics processing plant in Iowa Falls, Iowa, and recently announced plans to site a second plant in Chicago.

- The first American plant using the mixed plastic "Recycloplast" technology went on-line in Georgia earlier this year. Siting for two additional plants, in Pennsysvania and New Jersey, is underway.

MARKETS FOR RECYCLED PLASTICS --

- Procter & Gamble is test marketing Spic & Span Pine cleaner in bottles made from recycled PET. Procter & Gamble has also announced plans to used recycled HDPE in the middle layer of bottles for its Tide, Cheer, and Downey laundry products.

- Colgate-Palmolive uses recycled PET for Palmolive Liquid dish detergent bottles.

- A New Jersey manufacturer uses recycled PET for egg cartons marketed in New York and New Jersey.

- Johnson Controls, a large PET resin producer, has guaranteed markets for a statewide network of PET buybacks in the state of Washington.

- The city of Chicago has awarded a purchase contract for recycled plastic landscape timbers and playground equipment for city-maintained playgrounds.

- The State of Illinois has entered into an agreement with DuPont to test a variety of highway construction products (e.g., roadway dividers, traffic re-routers) made from recycled PET and HDPE.

5.4.3 Costs of Plastics Recycling

Plastics are a very recent addition to recycling programs. As a result, recycling agencies and processors have little operating experience on which to base estimates of costs for any of the phases of the recycling process -- collection, separation, processing, and marketing.

This analysis covers the costs associated with collection and separation of plastics for recycling. The primary focus is on the costs of curbside collection; less detailed discussions address the costs of drop-off recycling centers and of container deposit legislation. The economics of curbside collection are emphasized because:

1. Collection costs (and revenues) will accrue primarily to local government agencies. The costs (and revenues) of processing and marketing, on the other hand, will accrue to the private sector.   Policy concern, therefore, appears most appropriately directed at the collection phase of the recycling process.

2. Of available collection strategies, curbside collection promises to divert the greatest proportion of plastics (and other recyclables) from MSW disposal requirements; this is a critical consideration especially in the densely populated regions of the country that face the highest costs for landfill or other MSW disposal alternatives.

3. The more widespread implementation of curbside collection promises to provide a reliable source of resins that will foster the continued expansion of the recycled plastics processing industry. Processing industry participants have identified the lack of such supplies as the greatest current barrier to expansion of their industry (see Section 5.4.2.4).

Most of the information presented here reflects the costs and revenues generated by the collection of mixed recyclables -- newspaper, glass, and metals in addition to plastics. Very few data exist on the cost of independent plastics collection programs, nor on the incremental costs associated with adding plastics to existing recycling efforts for other materials. A few hypothetical calculations of the incremental costs of adding plastics to curbside collection programs have been completed; they are presented below.

5.4.3.1 Costs of Curbside Collection Programs

Approximately 600 curbside collection programs have been established in the United States (Glass Packaging Institute, 1988). Few of these, however, include plastics among targeted recyclables, although the number of communities that collect plastics appears to be increasing steadily. For this reason, historical or current cost data on the inclusion of plastics in recycling programs are unavailable. Two recent analyses (IEc, 1988 and Center for Plastics Recycling Research, 1988) have estimated the incremental costs and revenues associated with adding plastics to curbside collection programs; results of these analyses are presented below (see Section 5.4.3.2).

A number of studies have provided templates to assist municipal officials in estimating program costs and benefits (e.g., Stevens, 1988, 1989a, 1989b; Glenn, 1988b; Glass Packaging Institute, 1988; Center for Plastics Recycling Research, 1988). Some of these studies have provided information on specific cost components (e.g., equipment costs), or have provided ranges of costs for major recycling program elements (e.g., labor, vehicle operation and maintenance), but none has provided comprehensive information on the costs and revenues of specific municipal programs. Nevertheless, a number of insights into the economics of curbside collection are emerging from the body of experience gathered by municipal and county recycling programs. Overall program costs and revenues are determined by the interaction of a large number of cost elements; some of these are influenced by the design of the recycling program, while others are more or less independent of the program setup. Table 5-17 reviews the effect of program design elements on the costs and revenues of curbside recycling. Because these program design elements often interact, and because changes in more than one element are often implemented simultaneously, it is difficult to isolate the impact of specific design elements on program costs and revenues. With this caveat, however, a number of general observations can be made:

Collection Strategy and Crew Size. Some studies have suggested that it is most cost-effective to collect MSW and recyclables simultaneously, using trailers on MSW collection vehicles (IEc, 1988). In practice, however, most communities have apparently chosen to operate independent recyclables collection crews. Few data are available to suggest which option, in practice, imposes the smaller cost for recyclables collection. If separate collection is implemented, both theoretical and practical evidence suggests that a one-man crew is most cost-effective.

Collection frequency -- Although increasing collection frequency (e.g., from bi-weekly to weekly collection) increases both capital and operating costs, it also tends to result in high participation rates and increased yields of recyclables. In a Plymouth, Minnesota, recycling program, tonnage collected rose from 40 tons/month to 240 tons/month when the town moved from monthly to weekly collection (Glenn, 1988a). In general, increasing collection frequency appears to generate a net economic benefit to the recycling program.

Providing containers for recyclables -- This option generates a capital cost. But the universal experience of program operators is that providing containers to residents is very important to generating high participation rates, and that incremental benefits far outweigh the costs of the containers.

Recyclables sorting -- Sorting of recyclables, if performed at all, may be carried out by residents or by the collection crew. If recyclables are not sorted, they will require additional processing before their sale. Requiring some sorting to be completed by residents reduces program costs (both for collection and for additional processing) and increases per-ton revenue from recyclable sales: on the other hand, this sorting option may increase collection costs (because more sophisticated collection vehicles are required), and may tend to reduce participation if sorting and storing a number of classes of recyclables becomes a burden on participants. In general, requiring residents to complete at least partial sorting (into two to (at most) four categories of recyclables) appears to be most cost-effective. In some cases, however, noneconomic goals may influence the selection of a sorting option; for example,

Table 5-17

SUMMARY OF PROGRAM DESIGN INFLUENCES
ON COSTS OF RECYCLING PROGRAMS

Program Parameter	Option	Impacts on Capital Costs	Impacts on Operating Costs	Impact on Quantity and Quality of Recyclables Collected	Impact on Revenues from Sale of Recyclables	Impact on Savings in Tipping Fees	Other Considerations
High collection frequency	--	Increase costs for collection equipment	Increase costs for labor, vehicle operation and maintenance, etc.	Increases quantity through increased participation rate	Increase in sales revenue	Increases savings	Increases participation
Provide containers for recyclables	--	Cost of containers	Small decrease in per-stop time for collection	Increases quantity through increased participation rate and increased collection per household	Increase in sales revenue	Increases savings	Increases participation
Recyclables sorting	No sorting	Least cost for collection equipment	Smallest per-stop time for collection crews; Requires fewest round trips for collection crews	Decreases quality (market value); Quantity collected may be greater than if sorting required of households	Decreased revenue unless extensive sorting done during processing	Increases savings if results in increase in quantity collected	May allow highest participation; Implies requirement for further processing

(cont.)

Table 5-17

SUMMARY OF PROGRAM DESIGN INFLUENCES
ON COSTS OF RECYCLING PROGRAMS

Program Parameter	Option	Impacts on Capital Costs	Impacts on Operating Costs	Impact on Quantity and Quality of Recyclables Collected	Impact on Revenues from Sale of Recyclables	Impact on Savings in Tipping Fees	Other Considerations
Recyclables sorting	By household	Collection vehicle must be compartmentalized (increases cost)	Increases per-stop time for collection crews Increases number of round trips for collection crews	Increases quality Tends to decrease quantity; decrease probably not significant if other steps taken to maintain participation	Appears generally to result in net increase in sales revenues	Reduces savings if quantity recycled is reduced; little or no impact if other steps taken to minimize reduction in quantity collected	May reduce participation Reduces requirement for processing
Recyclables sorting	By collection crew	Collection vehicle must be compartmentalized (increases cost)	Greatest per-stop time for collection crews Increases number of round trips for collection crews	Increases quality Quantity collected may be greater than if sorting required of households	Increase in sales revenue	Increases savings if results in increase in quantity collected	May increase participation Reduces requirement for processing

(cont.)

Table 5-17

SUMMARY OF PROGRAM DESIGN INFLUENCES
ON COSTS OF RECYCLING PROGRAMS

Program Parameter	Option	Impacts on Capital Costs	Impacts on Operating Costs	Impact on Quantity and Quality of Recyclables Collected	Impact on Revenues from Sale of Recyclables	Impact on Savings in Tipping Fees	Other Considerations
Promote program through mailings, articles, advertizing, personal contact	--	None	Cost of promotional programs	Increases both quantity and quality	Increase in sales revenue	Increases savings	Increases participation
Processing of recycled wastes	None -- recyclables sold as sorted by household or collection crew	No cost for processing facilities and equipment	No cost for processing labor or equipment maintenance	No impact on quantity; little impact on quality if sorting required of household or collection crews	Revenue less than if recyclables are processed	None	
	None -- unsorted recyclables sold to outside processor	No cost for processing facilities and equipment	No cost for processing labor or equipment maintenance	No impact on quantity; little impact on quality if sorting required of household or collection crews	Least revenue of all sorting/ processing combinations	None	

(cont.)

Table 5-17

SUMMARY OF PROGRAM DESIGN INFLUENCES
ON COSTS OF RECYCLING PROGRAMS

Program Parameter	Option	Impacts on Capital Costs	Impacts on Operating Costs	Impact on Quantity and Quality of Recyclables Collected	Impact on Revenues from Sale of Recyclables	Impact on Savings in Tipping Fees	Other Considerations
Processing of recycled wastes (Cont.)	Partial sorting and baling	Cost for processing facilities and equipment	Cost for labor, equipment maintenance, etc. Compared to no-process options, reduces cost for transport to markets	May increase quantity if reduced requirement for sorting by households increases participation	Revenue greater than for no-processing options, less than for complete processing option	Increases savings if results in increase in quantity collected	
	Complete	Highest cost for processing facilities and equipment	Highest cost for labor, equipment maintenance, etc.	May increase quantity if reduced requirement for sorting by households increases participation	Greatest revenue of all processing options	Increases savings if results in increase in quantity collected	

(cont.)

Table 5-17 (Cont.)

SUMMARY OF PROGRAM DESIGN INFLUENCES
ON COSTS OF RECYCLING PROGRAMS

Program Parameter	Option	Impacts on Capital Costs	Impacts on Operating Costs	Impact on Quantity and Quality of Recyclables Collected	Impact on Revenues from Sale of Recyclables	Impact on Savings in Tipping Fees	Other Considerations
Processing of recycled wastes (Cont.)	Regional or county processing center	Spreads cost over a number of communities	Spreads cost over a number of communities Increases labor and equipment costs because of need to transport to non-local center	May increase both If regional center can afford better processing	Increases revenues If regional center provides sophisticated processing Should increase revenues because allows coordinated marketing of large volumes of recyclables		Requires coordination and cooperation among communities
Size of collection crew	- -	None	One-man crew apparently most cost-effective	NA	NA	NA	

Source: Developed by Eastern Research Group.

Somerset County, New Jersey requires little sorting by participants, and employs handicapped citizens to collect and then sort mixed recyclables (Dittman, 1989).

Processing -- Processing includes a variety of activities, including final sorting (e.g., by color of glass), grinding or shredding, and baling of recyclables for sale. Processing imposes both capital and operating costs on a recycling program. Its primary economic benefit lies in the increased sales value of the recycled materials; another benefit may accrue if a minimal sorting requirement acts to increase participation rates and/or the volume recycled per participant.

Because of their high capital costs, processing facilities may be most economical if implemented as county or regional centers serving a number of municipalities; additional economies (expressed as increased sales revenues) may accrue because of the cooperative marketing and larger sales volumes allowed by a regional processing center. These economies will be reduced by increased operating expenses associated with the time and labor required to transport recyclables to a remote processing facility.

Promotion and Publicity for Recycling Programs -- Effective promotion can be critical to achieving and maintaining high participation rates in curbside recycling programs. Because many very effective promotional tactics can be implemented at low cost (e.g., bulk mailings, "doorknob" literature, articles in local papers), the net benefits of promotional campaigns appear almost universally to outweigh their costs.

Common to many of these program design parameters is their impact on participation rates in recycling programs. And participation rate appears to be the single variable most critical to determining the overall net economic cost or benefit of curbside collection. While the absolute value of operating costs (and potentially of capital costs as well) rises with increasing participation, the marginal capital and operating costs per ton collected fall. The marginal cost of processing also falls as participation (and tonnage collected) increase. On the revenues side, dollar-per-ton sales prices for recyclables are unaffected by increasing participation, and may actually increase if the additional tonnage allows a municipality to negotiate higher prices for its recycled materials.

The program parameters described above, and their associated cost and revenue impacts, are subject to control by recycling program operators. A number of additional cost and revenue elements are beyond such control, and may have a large impact on the economics of curbside collection. The most important of these are:

Tipping fees -- Avoided tipping fees represent a direct economic benefit of recycling. They vary from virtually nothing to as much as $200 per ton (Cook 1988).

Labor costs -- Labor costs are generally the largest single operating expense in curbside recycling programs, contributing as much as 85% to total annual program costs. They exhibit regional variation.

Prices obtained for recycled materials -- These prices are subject to wide variation, both over time and across geographic regions. Prices vary by resin type, resin mix, color, and degree of processing. For example, August 1988 prices for recycled polyethylenes were 15 to 29 cents per pound; a year before, cleaned and processed polyethylene resins sold for only 6 cents per pound (Brewer, 1988c). Across the country, there is significant variation in the value of recycled materials -- prices for the same grade of recycled HDPE or PET may vary by as much as a factor of two or more between regions (Recycling Times, 1989).

Given these many sources of variability in recycling program costs and revenues, it is not surprising that cost structures, per-ton costs and revenues, and net economic costs/benefits of curbside collection programs vary widely. Table 5-18 provides information on the costs and revenues generated by curbside recycling programs in eight municipalities across the country. This information has been gathered from a variety of sources, including published reports, internal reports generated by the municipalities, and contacts with local officials. Many of these reports appear to be incomplete (e.g., some lack any information on significant cost or revenue elements), and reporting format, accounting methods, and definitions of cost/revenue elements vary significantly. Much or most of the data are also self-reported, and as such have not been subjected to independent verification. For these reasons, the information presented in Table 5-18 cannot be used as a basis to make generalizations about the costs of curbside recycling. But the table does provide information on the range of costs and revenues associated with curbside programs, and the net economic impacts of these programs.

The reported revenue/cost ratio of these programs ranges from 0.34 (for a voluntary program in Austin, Texas) to 1.81 (for a mandatory, bi-weekly program in Montclair, New Jersey). Four of seven reporting programs calculate that the net revenues of curbside collection exceed program costs, while two other programs reported revenues nearly equal to program costs. The revenues reported from recyclable sales vary widely, from $12 per ton to a reported $47 per ton -- the highest per-ton revenue was generated by the one program that processes recyclables completely prior to their sale (Ann Arbor, Michigan).

Total annual costs per ton collected also varied widely, from approximately $40 per ton to nearly $170 per ton; the highest per-ton costs were again generated by the one program that processes recyclables. Operating costs contributed approximately 70% to 100% of the total costs associated with the recycling programs; the highest operating cost ($128 per ton) is reported by the Austin program, with a 25% participation rate. (Real costs for this program may actually be even higher, bacause the facilities and equipment were donated by the city at no explicit cost to the recycling program.) Avoided tipping fees are the primary contributor to program revenues in a number of these programs -- 65% of revenues in San Jose, California, 68% of revenues in Haddonfield, New Jersey, 72% of revenues in Ann Arbor, Michigan, 79% of revenues in Montclair, New Jersey, and 100% of revenues in East Lyme, Connecticut. (For programs that hire a contractor to implement recycling, it has been assumed that contract payments are approximately equal to the avoided cost of disposal of the recycled materials).

The Center for Plastics Recycling Research (1988) recently completed an extensive computer modeling study of the costs and benefits of curbside collection and multi-material recycling. Validated against the experience of five New Jersey recycling programs, this study confirms that curbside programs offer a net economic benefit under most plausible operating scenarios. The CPRR study also confirms that participation rate is the single most important variable affecting collection program economics, and demonstrates the importance of avoided tipping fees in determining the net economic impact of curbside recycling.

Table 5-18
COST INFORMATION AND PROGRAM CHARACTERISTICS FROM
EIGHT COMMUNITY RECYCLING PROGRAMS

Community	Ann Arbor, MI	Montclair, NJ	Austin, TX	San Jose, CA
Program Characteristics				
Type of program	voluntary	mandatory	voluntary	voluntary
Pick-up frequency	monthly	bi-weekly	weekly	weekly
Year program started	1978	1971	1982	1985
Materials recycled	a,b,c,d,e,g	a,b,c,d	a,b,c,d	a,b,c,d (a)
Required separation categories	4	2	3	3
Recycle and rubbish collect. crew	separate	separate	separate	separate
Recycle and rubbish collect. day	same	separate	same	same
Participation rate (%)	33	> 85	25	> 41
Number of households	20,000	14,500	90,000	20,000
Tons collected at curbside	2,500	4,980	7,200	6,500
Collector (public or private)	private	public	public	private
Processing method	complete	partial	none	none
Program costs				
Total capital expenditure	842,000	241,000	362,000	0
Processing facility	303,120	28,920	NA	NA
Processing equipment	143,140	19,280	NA	NA
Collection equipment	395,740	175,930	362,000	NA
Annualized capital costs				
(over 10 yr at 10%)	138,677	39,693	59,621	0
Total annual operating costs	146,323	442,500	924,000	222,124
Labor	--	261,000	615,000	--
Vehicle maintenance	--	14,500	234,000	--
Adminstration and overhead	--	167,000	75,000	27,418
Total annual costs	285,000	482,193	983,621	254,820
per household	14	33	11	13
per ton collected	114	97	137	39
Total revenue	417,500	691,960	363,400	278,524
Tipping fee savings	--	507,960	72,000	52,000
Recycled material sales	117,500	184,000	246,400	86,924
Collection contract fees	300,000	--	--	139,600
State grants/collection fees	--	--	45,000	--
Program cost summary				
Total revenue/total cost	1.46	1.44	0.37	1.09
Average sale price (per ton)	47	37	34	13
Total profits (costs)	132,500	209,767	(620,221)	23,704
Net profits (costs)/ton	53	42	(86)	4

(cont.)

Table 5-18 (cont.)
COST INFORMATION AND PROGRAM CHARACTERISTICS FROM
EIGHT COMMUNITY RECYCLING PROGRAMS

Community	East Lyme, CT	Haddonfield, NJ	Seattle, WA	Charlotte, NC
Program Characteristics				
Type of program	mandatory	mandatory	voluntary	voluntary
Pick-up frequency	weekly	weekly	weekly	weekly
Year program started	1974	1983	1988	1987
Materials recycled	a,b,c,d,e	a,b,c,d,g	a,b,c,d,e (a)	a,b,c,f,g
Required separation categories	4	3	1 or 3	1
Recycle and rubbish collect. crew	separate	separate	separate	separate
Recycle and rubbish collect. day	same	same	variable	same
Participation rate (%)	> 80	95	64	> 74
Number of households	5,000	3,000	94,000	9,100
Tons collected at curbside	2,100	1,703	23,985	1,329
Collector (public or private)	public	public	private	public
Processing method	none	none	partial	partial
Program costs				
Total capital expenditure	27,000	19,000	NA	591,108
Processing facility	NA	NA	NA	NA
Processing equipment	NA	NA	NA	NA
Collection equipment	27,000	19,000	NA	591,108
Annualized capital costs	4,447	3,129	NA	NA
(over 10 yr at 10%)				
Total annual operating costs	120,325	67,500	1,151,280	203,100
Labor	85,335	60,000	NA	147,100
Vehicle maintenance	6,150	7,000	NA	18,000
Administration and overhead	28,840	500	NA	38,000
Total annual costs	124,772	70,629	1,151,280	203,100
per household	25	24	12	22
per ton collected	59	41	48	153
Total revenue	168,000	100,900	1,319,175	113,449
Tipping fee savings	168,000	69,000	--	46,290
Recycled material sales	minimal	20,250	--	67,159
Collection contract fees	--	--	--	--
State grants/collection fees	--	11,650	--	--
Program cost summary				
Total revenue/total cost	1.35	1.43	1.15	0.56
Average sale price (per ton)	--	12	0	51
Total profits (costs)	43,228	30,271	167,895	(89,651)
Net profits (costs)/ton	21	18	7	(67)

Note: a: newspaper, b: aluminum, c: glass, d: metal, e: cardboard, f: plastic, g: misc.
(a): Figures do not reflect recently started plastics collection programs.
Sources: Seaman, 1989; Barger, 1989; San Jose, 1988; Battles, 1989; Clark, 1989; Watts, 1989; Schaub, 1989; IEc, 1988; EPA Journal, March, 1989.

One important gap in both actual and hypothetical cost estimates concerns recycling programs in urban areas containing a large proportion multi-family dwellings. Most of the curbside collection programs implemented in the United States to date have been in suburban or rural settings with very few multi-family units, and most estimates of curbside collection costs have focused on such settings. As pointed out in Section 5.4.2.1, there are a number of concerns specific to urban areas which may have a significant impact on the net cost of recyclables collection programs (e.g., lack of storage space in many apartments/condominiums, widespread use of dumpsters in urban settings, difficulty of access for collection vehicles). Given the large population residing in urban areas, and the critical shortage of MSW disposal capacity facing many of these areas, additional research into the costs of urban recycling programs is needed.

5.4.3.2 Costs of Adding Plastics to Curbside Collection Programs

Few curbside collection programs currently accept plastics for recycling. For example, of the eight programs described in Table 5-18, only three accept plastics. For this reason, few data have been collected on the costs of including plastics in curbside programs.

One study (IEc, 1988) has addressed a number of issues related to the addition of plastics to curbside programs. This study points out that the most significant cost impact of adding plastics to an established collection program is related to the fact that plastics have a very low density compared to other commonly collected materials -- the density of collected plastics is less than 30 pounds per cubic yard for uncrushed containers (40-50 pounds per cubic yard for hand-crushed PET containers), compared to 50-75 pounds per cubic yard for uncrushed aluminum cans (250 pounds per cubic yard for crushed aluminum cans), 145 pounds for mixed recycled metals, 500 pounds for newspaper, and 600-700 pounds for whole glass bottles (IEc, 1988; Center for Plastics Recycling Research, 1988). For a number of communities in Rhode Island, IEc has presented estimates of the increases in hauling time and cost associated with adding plastics to established collection programs (Table 5-19); the average increase among these communities was 67%.

Increased program costs are also associated with processing plastics and transporting them to a buyer. Baling plastics may require 10 to 12 times more baler strokes than baling a similar volume of newspaper. And when bales are transported, a 40 cubic yard trailer can hold only about $135 worth of PET plastics, compared to $240 worth of baled newspapers (IEc 1988, based on 1988 prices).

When these and other costs are totaled, IEc reports that the net cost of adding plastics to an established collection system is approximately 8 cents per pound recovered, or $160 per ton. Against these costs must be balanced the sales revenues generated by the recycled plastics, and the avoided cost of tipping fees. Table 5-20 presents a sensitivity analysis of the net cost or benefit of adding plastics to a curbside collection program as both per-ton sales revenue and tipping fee are allowed to vary. Under the assumptions governing the IEc analysis, adding plastics yields a net economic benefit if sales price is greater than approximately 8 cents per pound, or if tipping fees are greater than approximately $155 per ton. At lower sales prices or tipping fees, inclusion of plastics in collection programs may yield either a net cost or a net benefit; the realized net impact will depend on the combination of market prices and disposal

Table 5-19

COST IMPACTS OF ADDING PLASTIC TO
RHODE ISLAND CURBSIDE COLLECTION PROGRAMS

City/Town	Annual Round Trip Time Per Truck (Hrs.)				Annual Cost Per Truck ($)			
	No Plastic	With Plastic	Increase	Percent Increase	No Plastic	With Plastic	Increase	Percent Increase
Cranston	291	485	194	67%	8,046	13,410	5,364	67%
E. Greenwich	229	343	114	50%	6,919	10,378	3,459	50%
E. Providence	416	624	208	50%	11,132	16,698	5,566	50%
Johnston	153	267	114	75%	4,114	7,199	3,085	75%
Newport	607	970	363	60%	16,617	26,587	9,970	60%
N. Kingston	302	603	301	100%	8,508	17,016	8,508	100%
Warwick	286	515	229	80%	7,876	14,178	6,302	80%
W. Warwick	302	503	201	67%	7,305	12,175	4,870	67%
Woonsocket	425	667	242	57%	10,498	16,496	5,998	57%
MEAN	335	553	218	67%	9,002	14,904	5,902	67%

Source: IEc, 1988.

TABLE 5–20

ECONOMIC IMPACT OF ADDING PLASTICS TO A CURBSIDE COLLECTION PROGRAM
AT DIFFERENT TIPPING FEES AND PLASTICS PRICES

Sales Revenue from Recycled Plastics		Avoided Tipping Fee ($/Ton)							
$/Ton	$/Pound	$0	$25	$50	$75	$100	$125	$150	$175
$0	$0.00	(3,952)	(3,328)	(2,704)	(2,080)	(1,456)	(832)	(208)	416
$25	$0.01	(3,328)	(2,704)	(2,080)	(1,456)	(832)	(208)	416	1,040
$50	$0.03	(2,704)	(2,080)	(1,456)	(832)	(208)	416	1,040	1,664
$75	$0.04	(2,080)	(1,456)	(832)	(208)	416	1,040	1,664	2,288
$100	$0.05	(1,456)	(832)	(208)	416	1,040	1,664	2,288	2,912
$125	$0.06	(832)	(208)	416	1,040	1,664	2,288	2,912	3,536
$150	$0.08	(208)	416	1,040	1,664	2,288	2,912	3,536	4,160
$175	$0.09	416	1,040	1,664	2,288	2,912	3,536	4,160	4,784
$200	$0.10	1,040	1,664	2,288	2,912	3,536	4,160	4,784	5,408

Note: Each table entry represents the net annual (cost) or revenue associated with the addition of plastics to an established curbside recycling program at a given combination of sales price and tipping fee. For example, at a sales price of $100 per ton (5 cents per pound) of recycled plastics and a tipping fee of $75 per ton, the annual impact of adding plastics to a recycling program is estimated to be a net revenue gain of $416.

Source: IEc 1988

costs effective in a specific region. For example, if recycled plastics are sold for $125 per ton ($0.06 per pound), a community recycling program will realize a net revenue from plastics collection if tipping fees are greater than approximately $33 per ton.

The Center for Plastics Recycling Research (1988) has also calculated the cost of adding plastics to a curbside collection/multi-material recycling program. CPRR calculated that, under a plausible base case recycling scenario, the inclusion of plastics in a recycling program would increase the net economic benefit of the program by approximately 5%.

5.4.3.3   Costs of Rural Recycling Programs

"Rural recycling" here refers to recycling programs in communities that do not provide curbside MSW collection services. In such communities, MSW collection is typically carried out by one of two methods:

1. Residents may contract with a private hauler to collect and dispose of wastes.

2. Residents may bring their wastes to a central point (the community landfill or a transfer station), where it is accepted for disposal or for repacking and transport to a remote disposal site.

Until recently, rural localities have typically faced much lower MSW disposal costs than urban or suburban areas, and there has been little economic incentive to recycle wastes. Voluntary programs have been implemented in some areas, typically organized by environmentally conscious individuals or groups, but overall there has been very little recycling activity in rural settings. With the implementation of EPA's upcoming regulations for sanitary landfills under RCRA Subtitle D (expected in early 1990), and with increased concern nationwide regarding resource conservation and the environmental impacts of solid waste disposal, rural localities may experience increasing economic and citizen pressure to explore recycling alternatives.

With very few programs in place and little incentive for most rural communities to implement recycling efforts, very few data exist on the costs of recycling programs in rural areas. A recent study sponsored by the Ford Foundation (The Minnesota Project, 1987) has examined recycling programs in seven rural localities; the following discussion draws heavily upon this analysis.

In communities relying on private waste haulers, recycling might be implemented by a voluntary or mandatory requirement that residents separate recyclables from other wastes and that haulers collect the two classes independently. This option would require either that haulers make additional trips to each residential site, or that hauling vehicles include trailers for recyclables collection. The increased cost of recyclables collection would presumably be passed directly to residents in the form of higher waste collection contract costs. EPA knows of no communities that have attempted to implement such a recyclables collection program, nor of any studies that have attempted to determine the cost and/or feasibility of this recycling alternative. Very limited information suggests that a few communities have attempted to require private haulers to participate in such recycling schemes, but that resistance from haulers and residents has impeded their implementation (The Minnesota Project, 1987).

Where residents bring MSW to a landfill or transfer station, rural recycling programs may be implemented by requiring residents to separate recyclable articles from other wastes and to deposit them in segregated containers at the disposal/transfer site. Such programs could be made mandatory by instructing disposal/transfer site operators to refuse to accept wastes containing visible recyclable articles, although such an enforcement strategy might encourage illegal dumping of refused wastes. The Minnesota Project studied two municipalities that have implemented such programs (Table 5-21); one of these towns (Peterborough, NH) includes plastics in its recycling program. The mandatory program in Peterborough captured approximately 18% of total MSW tonnage at a collection center at the town landfill. Revenues included $12,000 ($22 per ton) from recyclable sales and avoided tipping fees of approximately $20,500, while expenses associated with the recycling program were approximately $29,500. The program had a net economic benefit of $3,500, or nearly 12% of program costs. The town reported only minimal problems associated with illegal "gate throwing" of waste by citizens who refused to separate recyclables. A voluntary recycling program in South Berwick, Maine, operating at the town transfer station, captures approximately 3% of the town's waste stream. Implemented as a centralized, unattended drop-off site, the program has virtually no expenses; therefore all of the $4,800 in revenues realized by the collection program ($1,500 in sales revenue plus $3,300 in avoided tipping fees) represents a net economic benefit to the town (The Minnesota Project, 1987).

Similar to the South Berwick program are a number of voluntary rural recycling programs using centralized or decentralized drop-off sites for recyclables. The Minnesota Project analyzed three such programs (Table 5-21), which diverted from 1.2% to 7.8% of MSW from disposal in affected localities. Two of these programs reported a ratio of revenues to costs of approximately 0.33; the third program reported a revenue/cost ratio of 1.1. As pointed out earlier, noneconomic considerations may influence program design and the outcome of any benefit-cost analysis of a recycling program. For instance, one of the three drop-off programs studied is operated by a county human services agency serving handicapped citizens, which considers net recycling program costs to be a reasonable expense as part of its commitment to its clients.

5.4.3.4 Costs of Container Deposit Legislation

As discussed in Section 5.4.2.1, container deposit legislation has been enacted in nine states and has successfully diverted millions of pounds of plastic soft drink bottles (and other bottles) from MSW disposal. Enactment and implementation of deposit legislation have frequently aroused controversy because of its purportedly significant economic impact on beverage distributors and retailers. In spite of often acrimonious economic debate, however, very little rigorous analysis of the economic impacts of deposit legislation has been completed -- most "bottle bill" analyses have borne unmistakable traces of their sponsors' political preferences.

As part of a review of proposed Federal container deposit legislation, EPA has initiated an analysis of the costs and benefits associated with such legislation (IEc, 1989). Preliminary results of this analysis are reported here; the most critical initial finding is that any costs

Table 5-21

SUMMARY OF COSTS AND OPERATING CHARACTERISTICS OF SEVEN RURAL RECYCLING PROGRAMS

Case Study Site/ Population	Target Materials	Type of Population Served	Collection Method	Approximate 1986 Tonnage	Recycling Program Expenses	Material Sales Revenues	Avoided Tip Fees/ Year	Avoided Tip Fees/ Year/Ton	Net Recycling Revenue (Cost)	Net Profit (Cost)/ Ton
Pierce Co, WI 32,126	a,b,c,g,n,o,t,Other	Residential	5 unattended drops Attended center	180	33,000	9,000	2,250	13	(21,750)	(121)
Morrison Co, MN 29,311	a,b,c,g,n	Residential Commercial	18 unattended drops Attended center Commercial pickup	722	150,000	36,500	11,900	16	(101,600)	(141)
Prairie du Sac, WI 2,145	a,b,c,g,n,o,p,t,other	Residential Commercial Industrial	Curbside	288	25,000	11,000	4,320	15	(9,680)	(34)
Ithaca, MI 2,950	a,b,c,g,n,p,s,t	Residential	Curbside Attended center	53 (a)	Not Avail.	Not Avail.	864 (a)	16	--	--
Arcata, CA 12,340	a,c,g,n,o,s,other	Residential Commercial Industrial	Unattended newspaper drops Attended center Commercial/industrial pickup	856	78,364	74,822	11,556	14	8,014	9
Peterborough, NH 4,893	a,c,g,n,o,p,s,t,other	Residential Minor Commercial	Town dump drop off	546	29,440	12,000	20,475	38	3,035	6
South Berwick, ME 5,600	a,g	Residential Minor Commercial	Transfer station drop off	83	minimal	1,500	3,320	40	4,820	--

Materials: a = aluminum, b = Bimetal cans, c = corrugated, g = glass, n = newspapers, o = grease and oil, p = plastic, s = scrap metal, t = tin cans and other.

Note: (a) Estimated from 4.38/month of May, 1987.

Source: Minnesota Project, 1987.

imposed on distributors and retailers are ultimately passed on to consumers (as increases in beverage prices), and that any such price increases have not had a significant impact on beverage markets or consumer purchasing patterns.

The costs of deposit legislation fall on three sectors: consumers, retailers, and distributors. Consumers bear a number of costs. Although deposits are redeemed when containers are returned to a collection center, consumers incur economic costs related to the time required to return containers and collect deposits. An economic cost may also be attributed to the time and inconvenience associated with container rinsing and storage prior to return. Consumers also ultimately reimburse retailers and distributors for the costs of their contribution to the collection program (see further discussion below).

Retailers also incur a number of costs, primarily in the labor required to provide deposit return services to consumers, the space required to store collected containers, and the administrative overhead associated with the collection/redemption program. Although retailers are typically compensated for their services by a per-container payment in excess of the consumer deposit, many retailers and their trade associations in "bottle bill" states claim that these payments do not cover their costs of participation in the deposit redemption program.

Beverage distributors are typically required, in effect, to run the container redemption system -- collecting containers from retailers, paying retailers a handling fee, and arranging to market (or dispose of) collected containers. If distributors cannot or choose not to sell collected containers to recycling processors (as they apparently sometimes have not, especially with plastic containers), they may also have to bear disposal costs. In some states unredeemed container deposits (which may amount to millions of dollars) are disbursed to distributors to compensate them for the costs of their contribution to collection/redemption programs. Even in these states, however, distributors frequently believe that they are not fully compensated for the costs of managing the deposit redemption system.

If retailers and/or distributors believe that they incur a net cost related to their participation in bottle deposit programs, they pass this cost back to consumers in the form of higher beverage prices. It has proven difficult to derive accurate estimates of the impact on consumer prices of container deposit legislation. A New York study calculated that consumer prices have increased an average of 2.4 cents per container for beer and approximately 1 cent per container for soft drinks as the result of deposit legislation. A similar study in Iowa suggested that retail prices of deposit beverages have increased approximately 2 to 3 cents per container (IEc, 1989).

5.4.4 Environmental, Human Health, Consumer, and Other Social Costs and Benefits Generated by Recycling Plastics

5.4.4.1 Environmental Issues

No known major environmental considerations impact the potential of plastics recycling as an alternative to reduce plastics disposal requirements. Collection and separation alternatives impose a variety of minor environmental costs, consisting primarily of energy use requirements related to recyclable collection, storage, and transportation (e.g., energy consumed by vehicles involved in a curbside recycling program).

Secondary processing alternatives employing homogeneous resin inputs generate environmental releases that are similar to those related to virgin plastics processing. Environmental impacts should be no greater than those associated with production of equal volumes of virgin plastics products and, because they employ existing resins as inputs, should be less than for virgin resin manufacturing.

Mixed resin secondary processing alternatives employ very mild conditions and produce minimal air and water pollution. Acid gas emissions are produced by some mixed resin processes, but these can be controlled with proven scrubbing technologies. One relevant long-term environmental consideration is that because they do not displace consumption of virgin resins and because they may not themselves be amenable to recycling, use of mixed resin secondary products may not eliminate the ultimate disposal requirement for their plastic constituents. Rather, use of mixed resin secondary processes delays that disposal requirement for the lifetime of the recycled product. For this reason, the environmental benefits of mixed resin processing should be measured in terms of deferring, rather than eliminating, plastics disposal and its associated environmental consequences (Curlee, 1986). Section 5.4.2 presented an analysis of these issues.

Mixed waste tertiary recycling processes produce a residual solid char (consisting primarily of carbon and ash) that must be disposed of; no toxicity testing has been performed on this substance. Tertiary processes employing homogeneous plastics with few additives produce little or no solid residue, however. Tertiary recycling products used as fuels produce emissions that should be compared to those of competing fossil fuels; no available evidence suggests that these emissions produce environmental impacts that are different from those associated with fossil fuel consumption.

5.4.4.2 Health and Consumer Issues

Increased recycling shows little potential of creating human health impacts. No serious concerns have been raised regarding potential health impacts of recycled plastics products. The act of recycling itself also has little potential for harming human health. Recycling does involve the storage of waste articles, some of which require washing to avoid odors or sanitation problems. Sanitation is therefore a potential concern both in households and institutions where initial recycling efforts are made, and in collection centers. This concern affects all MSW recycling, however, and is neither different nor more serious for plastics than for other recyclable MSW constituents. Data have not been developed on the significance of this concern.

5.4.4.3   Other Social Costs and Benefits

A number of policy considerations are related to plastics recycling alternatives. Many of these are implicit in the definition of these alternatives, and have been addressed in sections related to the four primary stages of plastics recycling.

A policy consideration related to mixed resin secondary recycling processes is that they may not eliminate the need to dispose of the recycled plastic, but defer that need for the lifetime of the recycled product (see Section 5.4.2). The benefits of mixed resin processing should not be understated -- because these technologies operate on mixed plastics wastes, they may promise the greatest diversion of plastics from MSW disposal. But balanced against these benefits are not only the longer-term requirement that mixed plastic recycled products be ultimately disposed of, but also the fact that markets for these products may remain problematical. This area of policy concern demands additional analysis as choices are defined between mixed plastic and other recycling options.

A possible policy conflict exists between recycling programs and the use of degradable plastics. Given the existing concerns about the purity of recycled resins, further contamination with degradable materials is problematic; identification and separation of these degradable plastics, however, may weaken the economic basis of recycling methods. Thus, policy makers may have to choose whether to emphasize recycling or use of degradable plastics, and they will also need to identify which strategies will be employed for which products. The chemistry of mixing degradable plastics with other plastics is discussed in Section 5.5.

Among the most important recycling alternatives, the major potential interaction appears to concern curbside collection and bottle deposit legislation, i.e., the potential of deposit legislation to remove the highest-value recyclables from the recycling stream and thus adversely affect the economics of curbside collection programs (see Section 5.4.2.1).

5.5   DEGRADABLE PLASTICS

Some of the environmental concerns regarding plastic wastes relate to the apparent indestructibility of these wastes when discarded to the ocean, or as litter. There is concern that plastics will accumulate in the environment indefinitely, leading to long-term environmental, aesthetic, and waste management problems. These environmental problems can potentially be ameliorated by the development and use of plastics that will degrade in the environment. This section outlines the types of degradable plastics that are being developed, their potential role in plastic product areas, and the present market status of degradable plastics.

Degradable products are not included in the integrated waste management system EPA prepared for its policy proposals in its "Agenda for Action," and thus do not have a defined role in current EPA policies. Further, degradable plastic products introduce a new range of environmental issues and their influence on current waste management concerns remains largely undefined. These uncertainties are described in the sections below.

5.5.1 Scope of the Analysis

This section summarizes available information about the current and potential development of degradable plastics and examines possible approaches to increasing the use of such materials. All types of degradable plastics intended for use in plastic product markets are considered here. Issues covered include types of degradation processes and the environmental implications of this waste management technique.

5.5.2 Types of Degradable Plastics and Degradation Processes

Six methods of enhancing or achieving degradation of plastic have been defined in the literature and are described below. The most important technologies, based on available data and apparent market potential, are photodegradation, biodegradation, and biodeterioration.

Photodegradation - Degradation caused through the action of sunlight on the polymer

Biodegradation - Degradation that occurs through the action of microorganisms such as bacteria, yeast, fungi, and algae

Biodeterioration - Degradation that occurs through the action of macroorganisms such as beetles, slugs, etc.

Autooxidation - Degradation caused by chemical reactions with oxygen

Hydrolysis - Degradation that occurs when water cleaves the backbone of a polymer, resulting in a decrease in molecular weight and a loss of physical properties

Solubilization - Dissolution of polymers that occurs when a water-soluble link is included in the polymer [Note: soluble polymers remain in polymeric form and do not actually "degrade." They are included here because they are sometimes mentioned in the literature on degradable plastics.]

Debate continues regarding the most appropriate definitions for these degradation processes as well as regarding the operational or performance standards for such processes. The absence of accepted definitions has been cited as a factor impeding the development of degradable plastics (U.S. GAO, 1988). The American Society for Testing and Materials (ASTM) has organized a committee to define terms for plastics degradation and to develop standards for testing and measuring "degradability."

The absence of accepted definitions for degradation also complicates the ensuing discussion. Plastics engineers have measured degradation according to changes in the tensile strength or the embrittlement of the material. Generally, degradation is considered to have occurred by the time the materials readily collapse or crumble (which is before they have completely disappeared). Field testing necessary to establish the final degradation products, however, has not been performed.

Photo- and biodegradation are discussed in detail below, but a general comment about the processes can be made here. First, the rate of degradation of plastic materials in the environment is a function of both the characteristics of the plastic product and the environmental conditions in which it is placed. The addition of characteristics that increase photodegradability, for example, is an effective waste management step only if the product is exposed to sunlight. Thus, degradable plastics must be matched with an eventual disposal practice (or with disposal problems that are to be mitigated) in order for intended effects to be produced.

In the subsections below, more information is provided about the mechanisms involved for the two primary degradation processes and the commercial activities that are being pursued. A summary of the degradation processes that have been introduced by manufacturers (although not necessarily commercially exploited) for plastic polymers is shown in Table 5-22.

5.5.2.1   Photodegradation

Photodegradation processes are based on the reactions of photosensitive substances that have absorbed energy from a specific spectrum of ultraviolet radiation, such as from sunlight. The reactions may cause a break in the linkages within the long polymer molecules. This shortening of the chains leads to a loss of certain physical properties.

Sunlight is the dominant source of the ultraviolet radiation that will produce photodegradation. Indoor lighting generally will not produce photodegradation both because window glass screens out most ultraviolet radiation from sunlight and because other indoor light sources do not produce much ultraviolet radiation. Because photodegradation is primarily an outdoor process, photodegradable plastic products used primarily indoors can therefore be given "controlled lifetimes." When the products are discarded outdoors -- as litter for example -- they will degrade more rapidly.

To enhance the photodegradation properties of a plastic, manufacturers have modified or developed new polymers that contain photosensitive substances in the polymer chain. Alternatively, they have used resin additives that are photosensitive and cause degradation of the plastic material. The principal technologies that have been developed for photodegradable plastics are described below.

MODIFICATION OF THE PLASTIC POLYMER -- Photodegradation may be accomplished by incorporating a photosensitive link in the polymer chain. The principal method used thus far has been the incorporation of carbon monoxide molecules, also referred to as carbonyl groups,

Table 5-22
DEGRADABLE PLASTICS TECHNOLOGIES

Degradation Mechanism	Developer	Product Sold	Current/ Potential Uses
Photodegradation	Ecoplastics, Willowdale, Ont.	Ketone carbonyl copolymers	Mulch film and trash bags
Photodegradation	Dow Chemical Midland, MI	Ethylene/carbon monoxide copolymer	6-pack yokes
Photodegradation	DuPont Co. Wilmington, DE	Ethylene/carbon monoxide copolymer	6-pack yokes
Photodegradation	Union Carbide Danbury, CT	Ethylene/carbon monoxide copolymer	6-pack yokes
Photodegradation	Ampacet Mt. Vernon, NY	Additive system	Trash bags
Photodegradation	Princeton Polymer Labs,Princeton, NJ	Additive system	Not available
Photodegradation	Ideamasters Miami, FL/Israel	Additive system	Mulch film
Biodegradation	ICI Americas Wilmington, DE	Aliphatic polyester copolymer	Bottles prod. planned
Biodegradation (a)	U.S. Dept. of Agric. Washington, DC	Starch additive	Blown film uses
Biodegradation/ Autooxidation	St. Lawrence Starch Mississauga, Ont.	Starch additive and metal compound	Trash bags and bottles
Biodegradation (a)	Epron Indus. Prod. United Kingdom	Starch additive	Not available
Solubility	Belland Switzerland	Soluble polymer	Not available

(a) For these products, only the additives undergo biodegradation;
   the polymer does not have exceptional degradation rates.
Source: Leaversuch, 1987 and Helmus, 1988.

into the polymers. If carbonyl groups absorb sufficient ultraviolet radiation, they undergo a reaction and break the linkage of the polymer chain. Copolymerization with carbon monoxide is the most common method of incorporating carbonyl groups into plastic.

The rate of photodegradation depends on the number of carbonyl groups added or incorporated, although most applications have used a 1% mixture. It has been postulated that if sufficient photodegradation occurs so as to substantially reduce the molecular weight of the plastic molecules, that biodegradation of the residual would be possible. Even if this postulate is true in some circumstances, photodegradation has not been accomplished to the degree necessary to allow subsequent biodegradation of lower-weight chemical molecules. For instance, polyethylene molecules may have molecular weights of 20,000 or higher. Photodegradation reduces this weight, but for biodegradation of the polymer to occur at significant rates the molecular weight must be reduced to approximately 500 (Potts, 1974 as referenced in Johnson, 1987). Such a reduction is not possible without more complete photodegradation than has been yet been achieved by polymer modification.

USE OF PLASTIC ADDITIVES -- Several types of additives have been commercially developed for enhancing photodegradability of plastics. One method uses a photosensitizing additive combined with a metallic compound to encourage degradation (Princeton Polymer Laboratory, as referenced in Johnson, 1987). Another method uses antioxidant additives (see Section 2 for a description of antioxidant additives). At low concentrations, antioxidant additives speed the rate of photodegradation.

5.5.2.2   Biodegradation

Manufacturers have developed potentially biodegradable products either by modifying the polymer or by incorporating selected additives. In the latter case, the plastic polymer left behind after degradation of the additive remains intact although it may no longer hold its original shape.

MODIFICATION OF THE PLASTIC POLYMER -- Most plastic resins, and all the commodity resins, are nonbiodegradable. More accurately, they are degradable at such a slow rate that they can be thought of as nonbiodegradable.

Some biodegradable resins exist, however, including selected polyesters and polyurethanes. These biodegradable resins were developed for low-volume specialty uses for which biodegradability is desirable, such as some agricultural applications (e.g., seedling pots for automatic reforestation machines). Some of these end products for biodegradable plastics are not materials that reach the MSW stream, so their uses have not represented decreases in the aggregate waste volumes.

As of a 1987 symposium on degradable plastics sponsored by SPI, biodegradable resins appropriate for use in packaging had not been developed (Johnson, 1987). One type of aliphatic polyester, polyester poly(3 hydroxybutyrate-3 hydroxyvalerate), or PHBV, has been developed by ICI Americas in England. It is biodegradable and reputed to have characteristics similar to polypropylene. It is not, however, currently price-competitive with nonbiodegradable plastics. Other techniques for enhancing biodegradation employ additives, as discussed below.

USE OF PLASTIC ADDITIVES -- Most development work on biodegradable additives has centered on the use of starch additives. Starch is highly biodegradable, and upon discard or burial it is consumed by microorganisms in the soil, if an active population of these organisms exists. The degradation of the plastic polymer that remains has not been enhanced by the addition of the starch. Starch may be employed in moderate amounts as a filler, i.e., at 5 to 10% relative to the resin. In some experimental work, it has been incorporated in amounts up to 60% of product volume. Autooxidants are also added to some products. One polymer manufacturer, St. Lawrence Starch, has claimed that on burial, the starch additive is consumed by microorganisms and the autooxidant reacts with metal salts in the soil to form peroxides. These help degrade the polymer itself until it is also biodegradable (Maddever and Chapman, 1987). The field research regarding this phenomenon, however, is extremely limited.

The U.S. Department of Agriculture has experimented with very high starch concentrations. In these volumes, the starch is gelatinized before incorporation into the polymer. Again, the starch in the product is biodegradable, and the remaining lattice of plastic polymer may be sufficiently porous (of low enough molecular weight) to be biodegraded as well (Budiansky, 1986).

5.5.2.3   Other Degradation Processes

Three other degradation mechanisms exist. As noted above, autooxidation operates by producing peroxide chemicals from plastic polymers that then degrade the polymers. Autooxidation additives are incorporated into the polymers and react with trace metals, such as those available in the soil after burial. Manufacturers of these systems assert that this process has been used, along with biodegradable processes, to provide a more complete degradation of polymers.

Hydrolysis occurs when water destroys links in the polymer chains, resulting in a decrease in the molecular weight of the polymers. Chemical groups that are susceptible to this type of attack must be present in the molecule for this to occur. Ester groups, which are present in a number of polymers, are an example of such a group.

Polymers have been developed that are water soluble under certain environmental conditions. Belland Co. has marketed a specialty resin that is soluble within specified pH ranges. This polymer actually washes away, but nevertheless the smaller pieces remain in polymeric form and are not chemically degraded. As a result, soluble polymers may not be considered biodegradable in the same sense as the other degradation mechanisms that are discussed. Outside the specified pH ranges, the material retains its physical properties.

5.5.3   Environmental, Health, and Consumer Issues and Other Costs and Benefits Generated by Use of Degradable Plastics

5.5.3.1   Environmental Issues

The production of new plastic materials with enhanced degradation characteristics raises several environmental questions about the disposal of the materials. In general, operating or field evidence about such issues is quite limited. The uncertainties about disposal of degradable plastics are noted in several sources (U.S. GAO, 1988; Leaversuch, 1987).

An important source of information about the behavior of degradable plastics in the environment could be the data submitted to the Food and Drug Administration (FDA) by manufacturers seeking approval for use of a polymer in food packaging or other food-contact uses. FDA requests data covering both consumer safety issues and environmental safety issues. In the latter category, FDA will require data concerning the following (U.S. GAO, 1988):

■   If the plastic polymer is itself degradable, under what conditions and over what timeframe

■   The potential for increased environmental introduction of degradation products and additives from a degrading polymer

■   The potential effects of small pieces of the degrading polymer on terrestrial and aquatic ecosystems

■   The effect of degradable polymers on recycling programs

To date, no food packaging manufacturer interested in utilizing degradable plastic technologies has submitted this information to FDA (Les Borodinsky, FDA, by telephone interview, March 31, 1989). Some companies, however, have initiated the FDA food additive petition process.

Photodegradable polymers -- Incorporating carbonyl groups into polymer chains does not appear to create a toxic compound in the polymer or a toxic degradation product. Among the tests performed to date are aquatic toxicity tests performed using the degradation products of the Ecoplastics polymer. All tests showed minimal toxicity (Dan, 1989). One manufacturer of a commercially available plastic secondary package, ITW HiCone, has submitted their six-pack rings and the product of its degradation to laboratory investigation. LD50 tests showed the ingestion of the carbonyl material to be nontoxic (Rosner-Hixson Laboratories, 1972) and an EP Toxicity Test showed an absence of hazardous materials in the degraded product (Allied Lab, 1988).

There are four general areas of concern regarding the potential environmental hazard of degradable plastics:

1)   Is the polymer itself more toxic due to its enhanced degradability?

2)   Are the byproducts of the degradation process toxic?

3)   Does the degradation process increase the leachability of additives from the polymers?

4)   Do the physical byproducts (i.e., the small pieces of undegraded plastic) pose a threat to wildlife?

The limited information for photodegradable and biodegradable polymers is described below. Manufacturers of both types of systems have agreed that their products do not create undesirable impacts. A summary of these assertions is presented in Table 5-23. Most of these descriptions were derived from the 1987 SPI Symposium on degradable plastics.

Photodegradable additives may cause some environmental concern. Autooxidizing metal salts are among the compounds being sold or developed for use. Essentially no field evidence about the use of these additives has been identified (E.A. Blair, Princeton Polymer Laboratories, by telephone interview, March 31, 1989). However, for the photodegradable plastic sold by Ideamasters, Gilead and Scott reported tests conducted at the University of Bologna, Italy, on the toxicity of decomposition products to plant life. These tests found no discernible uptake of metals by plant life (Gilead and Scott, 1987). Application of either the photodegradable resins or additives to a broad array of products that require pigments, plasticizers, or other additives, must be carefullly considered due to the increased potential for leaching of such additives as the material degrades. Leaching rate is related to the extent of surface exposure of the plastic. No investigations of this concern were identified. Further, EPA has found no investigations of whether the partially degraded materials present a greater concern for ingestion by wildlife than do normal plastics. A description of the severe injuries ingestion of plastic can cause is described in Chapter 3 (Section 3.4.1.2).

Biodegradable polymers -- Information on products of biodegradable plastics is similarly limited. As noted earlier, degradation is accomplished by modifying the polymer or by use of biodegradable additives. The biodegradable polymers (e.g., PHBV), by definition, can be entirely consumed by microorganisms and do not pose evident threats. The biodegradable additives are also environmentally benign; the plastic polymer they leave behind is not necessarily biodegradable itself, but it should not be inherently more toxic than normal polymers. As with the photodegradable resins, EPA has found no investigation of whether the physical byproducts of degradation (i.e., the small undegraded pieces of plastic material that remain) pose an ingestion threat to wildlife. In addition, no information on the leachability of additives (e.g., pigments) from these resins was identified.

Degradable plastics have been offered by some as a method for improving plastic waste management. However, current data do not indicate that any of the waste management options for plastics discussed in this report (i.e., source reduction, recycling, landfilling, and incineration) are benefitted by degradable plastics. Each waste management method is discussed below.

SOURCE REDUCTION. Currently available degradable plastics do not reduce the amount or the toxicity of the plastic waste that is generated. Thus, development of these materials is not considered a source reduction activity.

Table 5-23

SUMMARY OF REPORTED ENVIRONMENTAL
RESIDUALS FOR DEGRADABLE TECHNOLOGIES (a)

Developer	Product Sold	Apparent Environmental Impacts	Lit. Source
Ecoplastics, Willowdale, Ont.	Photodegradable ketone carbonyl copolymers	Accepted for food contact uses, Canada; no envir. impact	Guillet, 1987
Dow Chemical Midland, MI	Photodegradable ethylene/carbon monoxide copolymer	Not available	--
DuPont Co. Wilmington, DE	Photodegradable ethylene/carbon monoxide copolymer	Polymer approved for indirect food contact uses (adhesives only); no environmental impact	Statz and Dorris, 1987
Union Carbide Danbury, CT	Photodegradable ethylene/carbon monoxide copolymer	Degrad. products have much lower mol. wt.; no environ. impact	Harlan and Nicholas, 1987
Ampacet Mt. Vernon, NY	Photodegradable additive system	Not available	--
Princeton Polymer Labs,Princeton, NJ	Photodegradable additive system	Additives used are recognized as safe; no field test results	Blair, 1989
Ideamasters Miami, FL/Israel	Photodegradable additive system	Complete biodegradation to CO and water	Gilead and Ennis, 1987
ICI Americas Wilmington, DE	Biodegradable aliphatic poly-ester copolymer	Entirely biodegradable polymer; no negative environ. impact	Lloyd, 1987
U.S. Dept. of Agric. Washington, DC	Biodegradable starch additive	Not available	--
St. Lawrence Starch Mississauga, Ont.	Biodegrad./ auto-oxidant & starch additive	Entirely biodegradable materials; no negative environ. impact	Maddever and Chapman, 1987
Epron Indus. Prod. United Kingdom	Biodegradable starch additive	Not available	--
Belland Switzerland	Water-soluble polymer	Not available	--

(a) Results given are based on reports of authors, some of whom are employed by the manufacturers of the products. Additional test data were not identified. Not all of these technologies are currently available commercially.

Source: Compiled by Eastern Research Group from sources given.

RECYCLING. Recyclers have argued that use of degradable plastics will complicate recycling schemes by degrading the quality of recycled resins. They argue that a mix of degradable and nondegradable feedstock among recycled materials may invalidate some intended uses for reprocessed products.

Manufacturers of degradable plastics have argued that the addition of small amounts of degradable plastics will not have a significant effect on the quality of recycled products (Leaversuch, 1987). Further, manufacturers of photodegradable plastics argue that additives can be employed during the reprocessing stage so that the new products will not degrade (Dan, personal communication, February 22, 1989). Available data are not sufficient to indicate the resolution or possible magnitude of any problems of accommodating degradable plastics into recycling streams. There are no data indicating degradables could benefit plastic recycling systems.

The use of degradable plastics may benefit the recycling of *yard waste* (i.e., composting). Unlike regular plastic bags, degradable plastic bags that contain yard waste would not need to be removed before composting could begin. More information in the four areas described above is needed before this use should be promoted.

LANDFILLING. It has been claimed that degradable plastics will ease the capacity crisis facing some landfills in the United States. However, W.T. Rathje's work (see Section 4.2.1.3) indicates that degradation in a landfill occurs extremely slowly. In addition, more than half of the current MSW stream is composed of materials that are considered to be "degradable" (e.g., paper, yard wastes, food wastes), yet landfill capacity is still a concern. Therefore, development of degradable plastics is expected to have very little impact on current capacity concerns.

The increase in surface area produced by the loss of the starch additive or the breakdown of the plastic material by the photodegradation process also makes leaching of any additives more likely. Thus, some additives -- for example, colorants -- could be leached from waste in increased quantities after structural breakdown of the plastic.

INCINERATION. EPA is not aware of any information indicating that currently available degradable plastics will have any impact on incineration of MSW. Incineration will occur for the most part before any degradation can take place.

With regard to litter, the use of degradable plastics could encourage the careless discarding of wastes and aggravate the existing litter problem. No data were identified that could adequately address this question. It is noteworthy that public opinion polls have shown most people favoring efforts to substitute degradable products for nondegradable plastic products in order to reduce the durability of littered waste (Dan, 1989). The data may suggest that use of degradable plastics will not increase littering if the products are introduced simultaneously with programs that increase public concern and awareness of littering problems. Nevertheless, these data are not sufficient to forecast how littering rates may be affected by the more widespread use of degradable plastics.

The effectiveness of degradable plastics as a countermeasure to littering is also inherently uncertain. If degradable plastic is to reduce the unsightly nature of litter, it must degrade quite quickly after discard. Available information indicates that the most rapid photodegradation rates occur several weeks after disposal (see Table 5-24). Because much litter is discarded in urban areas where litter collection systems are in place, however, these wastes will probably not photodegrade quickly enough to disintegrate before they are collected by even a relatively infrequent cleanup cycle. Also, potentially biodegradable plastics, which require years to degrade, are not relevant to efforts to reduce litter.

Where no litter collection system is in place, photodegradable plastics may provide some benefit. Observations of littering tendencies, however, show that the presence of litter in an area tends to generate additional littering (Tobin, 1989). If this is the case, fresher discards will be repeatedly added to degrading plastics, and litter volume will never be observably reduced or eliminated (even if all litter were degradable).

5.5.3.2   Efficiency of Degradation Processes in the Marine Environment

The durability of plastic waste in the marine environment was identified in Section 3 as a particular environmental concern. Marine plastic wastes can be degraded by the same processes that affect wastes disposed of on land, but the rate of degradation usually differs between terrestrial and marine environments. The influences that change the relative rate of degradation in the marine environment are as follows (Andrady, 1988):

- Fouling of plastic reduces the rate of photodegradation. Materials exposed in the sea are initially covered, or fouled, by a biofilm and then by algal buildup and macrofoulants. These organisms reduce the solar ultraviolet radiation reaching the plastic.

- Seawater mitigates the heat buildup on the plastic, reducing the rate of degradation. Heat buildup from sunlight is transferred from the plastic to the surrounding environment more efficiently by water than by air. Thus, plastics floating in the sea are likely to show slower rates of oxidation and photodegradation. The significance of this differential, however, has not been well established.

- Coastal seawater is rich in microbial flora, increasing the rate of biodegradation. Plastics floating in coastal waters will be exposed to a greater variety of microbial actions.

- Moisture may increase the rate of degradation. High humidity is known to increase the rate of degradation of some types of plastics, possibly because small quantities of water increase the accessibility of the plastic molecule to atmospheric oxygen. Seawater may have the same effect on plastics, though any net change in degradation rate is probably small.

To test the relative rates of degradation on land and sea, Andrady exposed six types of plastic to terrestrial and marine environments (see Table 5-25). He defined degradation as a loss of tensile strength and extension for the materials. The degradation is not complete, i.e., a

Table 5-24

SUMMARY OF DEGRADATION RATES FOR AVAILABLE TECHNOLOGIES

Developer	Product Sold	Manuf.-Reported Time to Degradation (a)	Characteristics of Product Degraded
Ecoplastics, Willowdale, Ont.	Photodegradable ketone carbonyl copolymers	Not available	--
DuPont Co. Wilmington, DE	Photodegradable ethylene/carbon monoxide copolymer	4-5 days, Calif. in summer 60 days in Alaska in fall	LDPE polymer with 1% CO copolymer
Union Carbide Danbury, CT	Photodegradable ethylene/carbon monoxide copolymer	60 days in New Jersey in winter	LDPE polymer with 2.7% CO copolymer
Ampacet Mt. Vernon, NY	Photodegradable additive system	8 - 28 wk at varied U.S. locations and seasons	LDPE film with "Polygrade" masterbatch
Ideamasters Miami, FL/Israel	Photodegradable additive system	3 wks - Israel test; 48 wks - European test	--
ICI Americas Wilmington, DE	Biodegradable aliphatic poly-ester copolymer	Case 1 - In a matter of days in sewage treatment plant, Case 2 - 1 yr. (est.)	Case 1 - Thin film; Case 2 - Bottle
St. Lawrence Starch Mississauga, Ont.	Biodegradable starch additive	3-6 yr in sanitary landfill (est.)	Resin with 6% starch

Note: Not all of these technologies are currently available commercially.

(a) The meaning of "degradation" in the reports cited is varied.

Source: Compiled by Eastern Research Group from Society of the Plastics Industry, 1987.

Table 5–25

COMPARISONS OF DEGRADATION RATES
OF PLASTIC MATERIALS ON LAND AND
IN SEAWATER

		Percentage Decrease in the Mean Value of Tensile Property			
	Duration of Exposure (months)	Land		Sea Water	
Sample		Strength(a)	Extension	Strength(a)	Extension
Polyethylene film	6	6.6	95.1	no change	no change
Polypropylene tape	12	85	90.2	11	31.5
Latex balloons	6	98.6	93.6	83.5	38
Expanded (foam) polystyrene	10	32.9	18	82.3	65.2
Netting	12	no change	no change	no change	no change
Rapidly degradable polyethylene	1.2	46.2	98.6	27.1	88.9

(a) The strength measurements reported are based on the maximum load in the case of
netting and polypropylene tape materials.

Source: Andrady, 1988.

breakdown into elements does not occur. The loss of tensile strength may be adequate to prevent the degraded plastics from posing an entanglement threat to wildlife, although testing on this issue has not occurred. Andrady tested the terrestrial degradation rates by exposing samples on racks exposed to sunlight. He tested seawater conditions by tying samples to a pier and allowing the materials to float in the water.

Tensile strength is one indicator of the fragility of plastics. Andrady notes that measurements which replicate the stresses that plastic articles endure in the environment are difficult to generate and may not be accurately reflected by tensile strength measurements. As a result, more observation and experimentation with degradable plastics in the marine environment are needed.

Andrady's results indicate that for three out of five normal plastic samples, the materials degrade substantially more quickly on land than the equivalent sample in seawater. One of the samples did not degrade measurably in either environment. A final sample, expanded (foam) polystyrene plastic, degraded more quickly in sea water than on land.

Andrady also examined the performance of a degradable plastic, which was found to degrade more rapidly on land than in water. His data suggest, however, that the difference in rates is not as substantial as for the other plastics. Thus, Andrady's study indicates that degradable plastics may disintegrate sufficiently in the marine environment to achieve the desired aim of reducing the threat of entanglement.

The manufacturers of Ecolyte plastic have also tested their product's degradation rate under terrestrial and seawater conditions. They found that degradation in sea and fresh water is somewhat reduced relative to land, but is "still substantial" (Dan, 1989).

5.5.3.3   Human Health Issues

Degradable plastics raise a number of potential concerns for human health and the related issue of consumer product safety. Human health issues include 1) whether degradable plastics have a predictable lifespan, and 2) if not, whether they are toxic. Manufacturers of prototype degradable plastics have tried to achieve predictability for the shelf and useful life of their products. Premature degradation raises potential problems for human health -- e.g., the mixing of plastics materials with food -- and can result in a loss of consumer utility for the products.

According to the available literature, engineers have achieved substantial predictability in the durability of degradable polymers. By varying polymer mixes, particularly the amount of the degradable components in the product, engineers can predict the approximate rate of degradation given presumed conditions of environmental exposure. For example, a photodegradable material exposed to sunlight in a given region during a given season can be reasonably expected to retain its tensile strength for a specified length of time. Biodegradable plastics require an active microbial environment, such as might exist in soil, to achieve substantial degradation.

Degradation rates can only be engineered accurately, however (i.e., by the appropriate adjustments of polymer characteristics), if the relevant environmental exposure conditions for the products are known first. This prerequisite is particularly important for photodegradable plastics (Johnson, 1987; Harlan and Nicholas, 1987). Given the reduced ultraviolet light reaching products used and stored indoors, however, plastic engineers should be able to prevent most premature degradation. For biodegradable plastics, for which degradation rates are much slower, effective engineering of products should also be possible.

The data described above suggest that predictability of product lifespan is not a serious health concern with degradable plastics -- though thus far, consumer use of degradable plastics is so limited that only tentative conclusions can be developed. In the worst case, in which premature or unexpected product degradation occurs, the toxicity of the plastics could become an issue. Also, growth of surface microflora on biodegradable products in use (for example, growth on a biodegradable plastic razor) could raise health concerns.

The available data on prospects for use of degradable plastics in food-contact applications are limited. Currently, no direct food-contact use has been approved in the United States. One degradable plastic has been accepted for direct food-contact applications in Canada (Guillet, 1987).

5.5.3.4   Consumer Issues

Consumer utility is another factor that influences the feasibility of degradable plastics in wider commerce. Apart from considerations of environmental activism or concern, consumer willingness to purchase and use degradable plastics depends on their cost relative to conventional products, their convenience, and their quality for achieving the intended purpose. Present data indicate that the relative performance of degradable plastics is uncertain or unfavorable in each of these characteristics:

- Prices for degradable plastics are likely to be higher than for commodity resins because of the additional processing required, loss of important economies of scale in production (relative to those enjoyed by commodity resins), and the additional care needed during transport, delivery, and marketing to avoid premature exposure to degrading environmental elements.

- The convenience and quality of degradable products for consumers depends on manufacturers' ability to tailor product lifetimes to a length suitable for specific product uses as well as to ensure safe product storage in household use. At best, degradable plastics could equal the convenience and quality of nondegradable plastics. Some consumer markets may exist in which, as in medical applications, degradability is a distinct marketing advantage. The nature and size of such markets is probably modest. These factors suggest that market forces alone will not generate consumer support for degradable products, except for a few unusual market niches in which degradability itself is a valuable product attribute.

5.5.4 Cost of Degradable Plastics

The feasibility of increased use of degradable plastics is influenced by the cost of these polymers and the facility with which they can be processed. In general, degradable plastics will be sold at a premium to commodity resins, although the range of cost premiums cannot be exactly established with available information. At the SPI Symposium on degradable plastics, several presenters described their degradable resins as selling at only modest premiums to commodity resins. They also expressed confidence that premiums would decline as production levels increase. It must be presumed, however, that the addition of the degradability characteristic will require some additional processing and will thus generate some premium. A larger premium was described for sale of one resin, the biodegradable plastic polymer PHBV (Lloyd, 1987).

The ease of processing for degradable resins will also be a concern for product manufacturers, and could generate additional cost differentials that are not reflected in the market price of the resins. Resins are carefully engineered to optimize a variety of desirable characteristics, including ease of processing. The addition of the degradability characteristic to the resin is likely to be achieved only with some tradeoff of other resin features. Manufacturers are also concerned that waste or trim materials from processing cannot be reused with photodegradable resins, a limitation that increases raw material costs.

Further, storage and transportation of produced degradable products also require some additional controls. In general, manufacturers or shippers may need to institute controls on light exposure (for photodegradables) or moisture absorption and biological activity (for biodegradables).

5.5.5 Current Status of Efforts to Foster Manufacture and Use of Degradable Plastics

Future growth in the use of degradable plastics will depend on several factors, including market demand for and acceptance of degradable plastics and industry improvements in the technology for supplying degradable plastics. This section discusses the forces that have generated interest in as well as some existing uses for degradable plastics. These forces include various regulations and industrial research and development.

5.5.5.1 Regulations Requiring Use of Degradable Plastics

Government (at various levels) has passed legislation requiring the use of degradable plastics in selected applications. Table 5-26 shows a sample of the legislation that has been passed or proposed by states and localities. In general, the various bans have been fostered by concerns about the environmental impacts of nondegradable plastics. No investigations were noted of possible environmental concerns regarding degradable plastics.

Two communities, Berkeley, CA, and Suffolk County, NY, have passed resolutions restricting the use of nondegradable plastics in a number of applications. Suffolk County, for example, bans certain nonbiodegradable food packaging and nonbiodegradable plastic food utensils in take-out restaurants.

Table 5-26 SAMPLE OF RESTRICTIONS ON NONBIODEGRADABLE PLASTICS

State or Locality	Year	Description of Ratified Legislation
Alaska	1981	Bans nonbiodegradable six-pack carriers
Florida	1988	Bans nondegradable polystyrene foam and plastic-coated paper packaging for foods for human consumption. Requires all retail carry-out bags to be degradable
Maine	1988	Bans the use or sale of any polystyrene food or drink serving containers, whether or not manufactured with chlorofluorocarbons
Rhode Island	1988	Prohibits retailers from using plastic bags without offering consumers the choice of paper bags; exempts all biodegradable bags, boxes, and wrapping materials and all returnable containers from state sales taxes
Suffolk County, New York	1988	Bans the sale of certain nonbiodegradable food packaging, plastic grocery bags, certain PVC and PS packaging and utensils
West Virginia	1987	Taxes restaurants 5% of the wholesale value of nonbiodegradable and nonrecyclable plastics used

Sources: Wirka, 1988; EAF, 1988.

Wider mandates for use of degradable plastics have come from state and federal regulations. More than twenty states have passed laws mandating the use of degradable plastic holding devices (e.g., six-pack rings, contour-pak, film, etc.) for either beverages or beverages and other containers. Federal legislation in the form of the Degradable Plastic Ring Carriers law (Public Law 100-556) requires EPA to issue regulations by October 1990 specifying that regulated items are to be degradable (if feasible and as long as the degradable items do not pose a greater environmental threat than nondegradable items).

A product-oriented view of state and local regulations outlines the potential influence on plastic markets. Table 5-27 itemizes the major categories of plastic products that have come under regulation (either in the U.S. or internationally) and describes the national market size of each segment. The national sizes of the markets that have been regulated are also indicated. The largest market is for retail carryout bags, estimated at 760 million pounds in 1976 when these data were compiled. The aggregate size of the affected markets came to 1.8 billion pounds (using the 1986 data from the source material).

As also shown in the table, the total market for all product areas exceeded 50 billion pounds. Numerous plastic market areas are not currently being analyzed for use of biodegradables, including building and construction, furniture, transportation, and industrial uses of packaging.

5.5.5.2 Industry Initiatives on Degradable Plastics

Industry interest in degradable plastics has focused on only a few product types. The two primary markets for degradable plastics are regulation-induced markets of the type described above and special market niches in which degradable plastics outperform conventional products. Markets in the first category (see Section 5.5.5.1 above) are generated by public pressures and not from indigenous industry activities. These opportunities are likely to be the more important and more general area of interest for industry. For example, one executive of a large resin producer stated that the regulation-induced market for degradable plastics was the source of his company's interest in these plastics (Leaversuch, 1987). This section examines only the latter category of markets, the special market niches that industry will pursue without external encouragement.

These special market areas exist only among products in which these plastics outperform other materials and thus can capture a market share. Degradable agricultural mulch films are an example of a product that, when manufactured from degradable plastic, may outperform and thus be more valuable than nondegradable versions. Degradable films are spread on fields to provide mulch and then abandoned, while the nondegradable products must be eventually removed. By avoiding the removal step, farmers will accrue a cost savings that may exceed any cost differential between the conventional and degradable films. Similarly, degradable bags designed to hold materials (e.g., yard waste) destined for composting are also being produced. Unlike nondegradable bags, degradable bags can become part of the compost material. Several pilot programs using degradable bags are underway across the country. Industry has also pursued the development of biodegradables for medical applications, especially biodegradable or hydrolytically degradable surgical sutures. These products outperform conventional products by eliminating the need for the medical removal of the sutures.

Table 5-27

SIZE OF MARKETS
IN DEGRADABLE PRODUCT
AREAS

Market	Resin	Sample of Regulatory Coverage	U.S. Sales (million lb)
Beverage rings	Polyethylene	Numerous state laws prohibit nondegradable devices	125
Diaper backing	Polyethylene	Oregon bans nondegradable diapers	150
Retail carryout bags (a)	Polyethylene	Italy banned nondegradable bags in 1984	760
Disposable food service items(b)	Polystyrene Expandable poly- styrene	Suffolk, NY, banned nondegradable items	500
Egg cartons	Polystyrene	New Jersey proposed bans on these items	85
Industrial containers(c)	Polyethylene	Considered for ban in Oregon	200
Tampon applicators	Polyethylene	New Jersey proposed bans on these items	5
Total			1,825
Total U.S. resin sales - packaging (d)			13,200
Total U.S. resin sales - all market categories (d)			50,800

(a) Includes low- and high-density T-shirt, merchandise, trash, garment, and self-service bags.
(b) Includes thermoformed polystyrene and molded expanded polystyrene cups, plates and hinged containers and molded solid polystyrene cutlery, plates, cups, and bowls.
(c) Includes blow-molded high-density polyethylene drums, hand-held fuel tanks, and tight-head pails.
(d) Total resin sales for packaging are derived from Chem Systems and are based on 1985 data. Total sales data for resins are from Society of the Plastics Industry, 1988.

Source: Leaversuch, 1987 and additional materials as cited.

Other markets of this type may arise if a degradable product allows a task related to removal or disposal of a conventional product to be eliminated. Still, the overall significance of these markets is uncertain.

5.6 ADDITIONAL PROGRAMS TO MITIGATE THE EFFECTS OF PLASTIC WASTE

In addition to the waste management techniques discussed thus far, EPA and other Federal agencies can pursue several methods to mitigate the impacts of plastic wastes on the environment. These methods include 1) incremental controls on discharges of sewage into oceans and other navigable waters, 2) implementation of the MARPOL Annex V standards promulgated by the Coast Guard, and 3) programs to reduce litter. EPA can also undertake steps to mitigate problematical effects of plastic waste on the waste disposal methods currently used, namely incineration and landfilling.

5.6.1 Efforts to Control Discharges of Land-Generated Wastes from Sanitary Sewers, Stormwater Sewers, and Nonpoint Urban Runoff

Section 3 discusses the principal contributors of land-generated plastic wastes to the marine environment. These include:

- POTWs that cannot treat the capacity of normal "dry-weather flow" or POTWs that suffer downtime or breakdowns; at these facilities, untreated sewage may bypass the system and be released directly into receiving waters.

- Communities with combined sanitary and storm sewer overflows (CSOs); in these places, the volume of stormwater exceeds the capacity of the treatment plant during heavy rains, causing some of the effluent (consisting of both untreated sewage and stormwater with street litter) to be released directly to receiving waters.

- In communities with separate sewer and stormwater discharges, stormwater drains carry a variety of urban runoff including street litter.

Methods to correct these problems are discussed below.

EPA currently holds authority under the Clean Water Act (CWA) to regulate discharges from municipalities, including discharges from municipal waste water treatment facilities. EPA has made increasing use of its CWA authority by bringing legal action against cities that had failed to comply with regulatory requirements. Communities unable to treat all of the normal dry-weather flow due to treatment plant maintenance can construct backup holding tanks to retain excess flow for later treatment.

EPA is authorized to control CSOs under the Clean Water Act. EPA has developed a national control strategy to bring CSO discharges into compliance with the Act. The strategy presents three main objectives:

■ To ensure that all CSO discharges occur only as a result of wet weather

■ To bring all wet weather CSO discharge points into compliance with the technology-based requirements of the CWA and applicable state water quality standards

■ To minimize water quality, aquatic biota, and human health impacts from wet weather overflows that do occur

EPA will achieve these performance goals through a nationally consistent approach for developing and issuing NPDES permits for CSOs. The permits will require technology-based and water-quality based limitations for discharges; the control technology includes methods for controlling solid and floatable materials from CSOs, including plastic waste.

EPA is also studying the pollution contributions from stormwater discharges from communities with separate storm sewer systems. EPA is preparing a Report to Congress on this subject, and a portion of the report will assess the problems of floatable waste discharges. EPA has also proposed regulations controlling discharges associated with industrial activity from municipal separate storm sewer systems serving a population greater than 100,000 people.

5.6.2   Efforts to Implement the MARPOL Annex V Regulations

A substantial portion of plastic waste in the marine environment and on beaches is generated from vessels. Section 3 describes the quantities and types of materials discarded and their impacts on the marine environment.

The plastic waste generated from U.S.-flagged vessels, and from foreign-flagged vessels operating in U.S. waters should be substantially reduced as the result of the implementation of MARPOL (Marine Pollution) Annex V, an international treaty agreement for the protection of ocean resources. The U.S. legislation implementing this treaty is contained in the Marine Plastic Pollution Research and Control Act of 1987, which amends the Act to Prevent Pollution from Ships. The U.S. Coast Guard published interim final regulations on April 28, 1989. The regulations:

■ Prohibit the deliberate discard of plastic materials from vessels

■ Require ports to have "adequate reception facilities" to accept garbage that will be offloaded from ships

■ Restrict disposal of other garbage within various distances of shore

The only plastic wastes that are exempted from these regulations will be those materials that are lost in the course of normal commercial activities, such as nets or fishing line lost in fishing operations. Unfortunately, substantial amounts of netting can be lost during these operations; problems of "ghost fishing" by derelict nets, therefore, will not disappear even with perfect compliance with the regulations. Further, the reduction in waste disposal from vessels may also be less dramatic than desired because not all nations are signatories of the MARPOL treaty, and among those nations that are, compliance levels with their respective national regulations remain uncertain.

The authorizing legislation for MARPOL Annex V also requires EPA to examine the wider problems of plastic waste. Included among the EPA efforts is the preparation of this report on plastic waste problems. EPA is also directed to coordinate with the Department of Commerce and the Coast Guard to institute a program for encouraging the formation of citizens' groups to assist in the monitoring, reporting, cleanup, and prevention of ocean and shoreline pollution.

5.6.3 Efforts to Reduce Plastics Generated from Fishing Operations

Section 3 identified a number of problems caused by loss of fishing gear or other associated wastes to the marine environment. The National Oceanographic and Atmospheric Administration (NOAA) is currently evaluating methods to reduce the frequency of gear loss and the environmental impacts associated with such losses. Due to NOAA's ongoing effort, no attempt has been made to outline or analyze possible control methods for this report. The EPA will support NOAA in developing and implementing methods to control the loss and impacts of fishing nets, traps, and other gear.

5.6.4 Efforts to Control Discharges of Plastic Pellets

The only industrial waste stream of concern for this study is the plastic pellets that are found in the marine environment. These are frequently ingested by marine life and are also found in substantial quantities on beaches. The EPA initiatives and available options for control of this industrial waste stream are described here.

Section 3 describes the findings from the literature review and from original harbor sampling programs undertaken by the EPA Office of Water. Unfortunately, existing information is not adequate to characterize the point or nonpoint sources of plastic pellet wastes or to pinpoint the most effective control mechanism.

Any assessment of possible sources of plastic pellet waste requires a consideration of the flow of the pellets through the economy. Pellets are handled at several stages:

Plastic resin manufacturers - The manufacturers could lose some pellet materials during manufacturing, either to plant effluents or with plant solid waste, which then might be lost to the environment.

Plastic pellet transporters - Pellets are transported domestically primarily by rail or by truck in either large-quantity containers (e.g., tank cars) or small-quantity containers (e.g., fiber drums, paper bags). Some international shipments are transported by vessel. Transporters may lose some materials to the environment if their containers leak or are punctured. Cleaning of tank cars could also generate an effluent containing waste pellets.

Plastic product manufacturers - Plastic product manufacturers use pellets for a variety of molding and processing techniques. These firms have economic incentives to capture any waste pellets and reincorporate them into input streams. Nevertheless, some loss of pellets could occur because of spillage of pellet containers in the facilities or receipt of off-specification products. The lost pellets could be washed into sewer drains or discarded with facility solid waste.

Data on the relative contribution of these sources is almost entirely anecdotal and quite limited. According to some industry representatives, for example, resin manufacturers do not generate any significant pellet wastes. The volume of waste pellet material lost in product processing is unknown. Evidence is also not available concerning the loss of pellets in operations such as tank car cleaning.

The following are approaches for closing these possible points of release:

Reviewing terms of National Pollutant Discharge Elimination System (NPDES) permits - Effluents generated by either resin manufacturers or plastic processors could contain pellets. Such discharges may not be effectively controlled under the existing NPDES permits issued by EPA, partly because plastic pellets have not been recognized as an environmental problem until very recently. NPDES permits may also be held by plastic processing facilities that discharge to municipal sewer systems. In these cases also, EPA can review the permit terms to require more efficient control of pellet discharges. Note, however, that numerous plastic processing plants are quite small and may not have NPDES permits for either direct or indirect discharges.

Improving capture of plastic pellets in sewer or stormwater discharges - Issues associated with the capture of plastic materials from these discharge locations were described in Section 5.6.1 above. The same technologies can be used for pellets as for other plastic materials -- namely, skimming as well as screening of the wastewater effluents. Capturing plastic pellets, however, will be still more technically challenging because of their small size.

Improving the durability of pellet packaging - Pellets may be frequently released into the environment because of spills from damaged packaging. More durable packaging could reduce the rate of spillage or loss; currently, paper bags that can easily tear are used to ship large quantities of pellets.

Increasing educational initiatives - Efforts to educate the members of the plastics industry concerning the apparent damage caused by releases of plastic pellets could be broadly directed so as to help address all of the potential sources of this waste stream.

Pursuing further research on the sources of plastic pellet wastes - Sources of pellets are not well defined. Further field investigations would be helpful.

5.6.5 EPA Programs to Control Environmental Emissions from Incineration

EPA considers incineration as one of the waste management options in an integrated waste management system. Incineration and landfilling follow source reduction and recycling in the solid waste management hierarchy described in EPA's "Agenda for Action." EPA considers that properly operated and controlled incineration is a safe waste management option.

Section 4 noted a number of actual or potential emissions resulting from incineration of plastics found in municipal solid waste. EPA and the states regulate air pollution sources such as municipal solid waste incinerators under the Clean Air Act (CAA). The emission restrictions are directed at the total emissions from the combustors, not specifically emissions due to any component of the waste stream.

EPA regulates emissions directly and indirectly through several approaches under the CAA. Under the New Source Performance Standards (NSPS), EPA promulgated a limitation on the emission of particulate matter from municipal solid waste combustors. Additionally, EPA has promulgated general limitations on the pollutant levels under the National Ambient Air Quality Standards (NAAQS) Program. These are enforced through state level regulation and within the context of State Implementation Plans (SIPs). The latter describe the approach to be used by each state to achieve the national ambient limitations for each pollutant set by EPA. The pollutants covered by NAAQS, and thus by SIPS, include sulfur dioxide and particulates.

EPA is also planning to revise the NSPS regulations for new municipal solid waste incinerators and to provide guidance for controls for existing incinerators. The proposed regulations are expected to specify controls for acid gas emissions.

5.6.6 EPA Programs to Control Environmental Hazards Arising from the Landfilling of Plastic Wastes with Municipal Solid Waste

The final option for management of plastics in municipal solid waste is landfilling. Most plastic wastes in the MSW stream are landfilled. EPA programs for controlling the environmental effects of landfilled MSW, including any plastic waste, are summarized here.

Municipal solid waste landfills are regulated under Subtitle D of the Resource Conservation and Recovery Act (RCRA). This legislation establishes a framework for improvement of solid waste management systems, including:

- EPA's promulgation of general guidelines and minimum criteria for state solid waste management plans.

- EPA's promulgation of criteria for defining which facilities shall be considered "sanitary landfills" in the RCRA program. All other facilities are to be classified as open dumps. RCRA prohibits open dumping, and citizens or states can bring suit to enjoin such dumping.

These authorities allow EPA to establish the minimum criteria for operation of state solid waste management programs. If enforced, such criteria should provide for the protection of human health and the environment from the potential hazards of MSW disposal in landfills. Implementation and enforcement of an EPA-approved program, however, is the responsibility of individual state governments.

Further, EPA will be finalizing new MSW landfill criteria in the near future. The new rules. which were proposed in the Federal Register on August 30, 1988, will provide additional safeguards for protecting human health and the environment.

REFERENCES

Andrady, A.L. 1988. Experimental demonstration of controlled photodegradation of relevant plastic compositions under marine environmental conditions. Report prepared for the U.S. Department of Commerce, National Oceanic and Atmospheric Administration, Northwest and Alaskan Fisheries Center. Seattle, WA. 88-19. 68 p.

Barger, B. 1989. Telephone communication between Eastern Research Group and Brenda Barger, Resource Recovery Specialist, Mecklenburg County Engineering, Charlotte, NC. May 1.

Battles, P. 1989. Telephone communication between Eastern Research Group and Peter Battles, Director of Planning, East Lyme, CT. May 3.

Bennett, R.A. 1988. New Applications and Markets for Recycled Plastics. From the Recyclingplas III Conference: Plastics Recycling as a Future Business Opportunity (May 25-26). Plastics Institute of America, Inc. Technomic Publishing Company. Lancaster, PA.

Blair, E.A. 1989. Telephone communication between Eastern Research Group and Dr. E.A. Blair, Princeton Polymer Laboratory. March 17.

Borodinsky, L. 1989. Telephone communication between Eastern Research Group and Les Borodinsky, Division of Food Chemistry and Technology, Food and Drug Administration, Washington, DC. March 3.

Bree, W. 1989. Telephone communication between Eastern Research Group and William Bree, Recycling Department, Oregon Department of Environmental Quality. April 5.

Brewer, G.D. 1987. Mixed plastics recycling: Not a pipe dream. Waste Age. Nov 1987. pp. 153-160.

Brewer, G.D. 1988a. Recyclers: Cultivating New Growth for Packaging. Plastics Packaging. May/June.

Brewer, G.D. 1988b. Recycling Resources: A Plastics Industry Update. Plastic Packaging. Jan/Feb. pp. 40-46.

Brewer, G.D. 1988c. Pair Plan to Prevail Over Plastics. Waste Age. Aug 1988. p. 147.

Brewer, G.D. 1989. Comments submitted to the U.S. Environmental Protection Agency, Office of Solid Waste, on a draft Chapter 5 of OSW's Report to Congress, Methods to Manage and Control Plastic Wastes. Jun 13, 1989.

Budiansky, S. 1986. The world of crumbling plastics. U.S. News and World Report, Nov 24, 1986. p. 76.

California, State of. 1988. Annual Report of the Department of Conservation, Division of Recycling. Sacramento, CA.

Castro, L. 1988. Personal communication between Eastern Research Group and Lou Castro, Lykes Bros. Shipping Co. July 15.

Center for Plastics Recycling Research. 1988. Plastics Collection and Sorting: Including Plastics in a Multi-Material Recycling Program for Non-rural Single Family Homes. Rutgers University. Piscataway, NJ.

Clark, J. 1989. Telephone communication between Eastern Research Group and Jean Clark, Department of Public Works, Montclair, NJ. Apr 28.

Claus, P. 1987. Degradable Plastics in Europe. Proceedings of the Symposium on Degradable Plastics (Washington, June 1987). Society of the Plastics Industry. Washington, DC. p. 4.

Cook, J. 1988. Not in anybody's backyard. Forbes 142:172. Nov 28, 1988.

Council for Solid Waste Solutions. 1989. Fact sheets on plastics recycling initiatives: (1) Illinois and DuPont Create Recycling Partnership; (2) Recycling soda Bottles for Spic & Span Pine; (3) Nation's Largest Recycling Venture Launched. Washington, DC. May, 1989.

Curlee, T.R. 1986. The Economic Feasibility of Recycling: A Case Study of Plastic Wastes. Praeger. New York, NY.

Dan, E. 1989. Letter of Feb. 22, 1989 to Susan Mooney, U.S. Environmental Protection Agency, Municipal Solid Waste Program, from Erving Dan, Managing Director, Enviromer Enterprises, Leominster, MA.

Dipietro, R. 1989. Telephone communication between Eastern Research Group and Rich Dipietro, Manager of Packaging Management, Stanley Tools Corporation. New Britain, CT. May 12.

Dittman, F.W. 1989. Telephone communication between Eastern Research Group and Frank W. Dittman, Center for Plastics Recycling Research. May 1.

EAF. 1988. Environmental Action Foundation. Legislative Summary: Significant Packaging Initiatives Passed or Considered in 1988. Washington, DC. December.

EPA Journal. 1989. Five situation pieces. 15(2):35-40. March/April. U.S. Environmental Protection Agency. Washington, DC.

Franklin Associates. 1989. Comparative Energy and Environmental Impacts for Soft Drink Delivery Systems. Prepared for the National Association for Plastic Container Recovery (NAPCOR). Charlotte, NC. March.

Gehr, W. 1989. Telephone communication between Eastern Research Group and William Gehr, State of Vermont Department of Environmental Conservation. April 3.

Gilead, D. 1987. A New, Time-Controlled, Photodegradable Plastic. Proceedings of the Symposium on Degradable Plastics (Washington, June 1987). Society of the Plastics Industry. Washington, DC. p. 37.

Glass Packaging Institute. 1988. Comprehensive Curbside Recycling. Glass Packaging Institute. Washington, DC.

Glenn, J. 1988a. Junior, take out the recyclables. BioCycle. May/Jun:26.

Glenn, J. 1988b. Recycling Economics Benefit-Cost Analysis. BioCycle. Oct:44.

Guillet, J.E. 1987. Vinyl Ketone Photodegradable Plastics. Proceedings of Symposium on Degradable Plastics (Washington, June 1987). The Society of the Plastics Industry, Inc. Washington, DC. p. 33.

Harlan, G.M. and A. Nicholas. 1987. Degradable Ethylene Carbon Monoxide Copolymers. Proceeding of the Symposium on Degradable Plastics (Washington, June 1987). The Society of the Plastics Industry, Inc. Washington, DC. p. 14.

Helmus, M.N. 1988. The Outlook for Degradable Plastics. Spectrum. Arthur D. Little Decision Resources. Feb 1988.

IEc. 1988. Industrial Economics Inc. Plastics Recycling: Incentives, Barriers and Government Roles. Prepared for Water Economics Branch, Office of Policy Analysis, U.S. EPA. Industrial Economics Incorporated. Cambridge, MA. 152 pp.

IEc. 1989. Industrial Economics Inc. Potential Impacts of a National Bottle Bill on Plastics Recycling. Draft Report prepared for Water Economics Branch, Office of Policy Analysis, U.S. EPA. May 1989.

Johnson, R. 1987. An SPI Overview of Degradable Plastics. Proceedings of the Symposium on Degradable Plastics (Washington, June 1987). Society of the Plastics Industry. Washington, DC. p. 6.

Koser, W. 1989. Telephone communication between Eastern Research Group and Wayne Koser, Environmental Quality Specialist, Resource Recovery Section of the Waste Management Division of Michigan. Lansing, MI. April 3.

Leaversuch, R. 1987. Industry weighs need to make polymer degradable. Modern Plastics. Aug 1987. p. 52.

Lloyd, D.R. 1987. Poly(hydroxybutyrate-valerate) Biodegradable Plastic. Proceedings of the Symposium on Degradable Plastics (Washington, June 1987). Society of the Plastics Industry. Washington, DC. p. 19.

MacDonald, G. 1989. Telephone communication between Eastern Research Group and George MacDonald of the State of Maine Department of Economic and Community Development. Augusta, ME. April 7.

Maddever, W.J. and G.M. Chapman. 1987. Making Plastics Biodegradable Using Modified Starch Additions. Proceedings of the Symposium on Degradable Plastics (Washington, June 1987). The Society of the Plastics Industry, Inc.. Washington, DC. p. 41.

Maine DECD. 1988. Maine Department of Economic and Community Development. State of Maine Waste Reduction and Recycling Plan. Augusta, ME.

Massachusetts DEQE. 1988. Massachusetts Department of Environmental Quality Engineering. Plastics Recycling Action Plan for Massachusetts. Boston, MA.

Maczko, J. 1988. Personal communication between Dynamac Corporation and J. Maczko. Mid-Atlantic Plastic Systems. August.

Medeiros, S. 1989. Telephone communication between Eastern Research Group and Stephen Medeiros, Laser Fare LTD, Inc. Smithfield, RI. April 28.

Minnesota Project. 1987. Case Studies in Rural Solid Waste Recycling. Prepared for the Ford Foundation by the Minnesota Project.

MRI. 1974. Midwest Research Institute. Resource and Environmental Profile Analysis of Plastics and Non-plastic Containers. Prepared for the Society of the Plastics Industry.

NAPCOR. 1989. NAPCOR NEWSflash. National Association for Plastic Container Recovery. Mar 1989.

NOAA. 1988. National Oceanic and Atmospheric Administration. NWAFC Processed Report 88-16. Evaluation of Plastics Recycling Systems. Prepared by Cal Recovery Systems, Richmond, CA.

O'Sullivan, D. 1989. Telephone communication between Eastern Research Group and Denis O'Sullivan, Principal Packaging Engineer, Digital Equipment Corporation, Maynard, MA. May 12.

Phillips, J. 1989. Telephone communication between Eastern Research Group and Joseph Phillips, New York State Department of Environment and Solid Waste. April 2.

Plastics Recycling Foundation. 1988. Plastics Recycling: A Strategic Vision. Washington, DC.

Popkin, R. 1989. Source reduction: Its meaning and potential. EPA Journal. Mar/Apr.

Potts, J.E., R.A. Clendinning, W.B. Ackert and W.D. Niegisch. 1974. The Biodegradability of Synthetic Polymers. In: J. Guillet (ed). Polymers and Ecological Problems. Plenum Press. New York, NY. pp. 61-80.

Rattray, T. 1989. Telephone communication between Eastern Research Group and Tom Rattray, Associate Director of Corporate Packaging Development, Proctor and Gamble, May 12.

Recycling Times. 1989. March 28. p. 3.

San Jose. 1989. San Jose's Recycling Program Overview. San Jose, CA.

Schaub, R. 1989. Telephone communication between Eastern Research Group and Richard Schaub, Public Works Department, Haddonfield, NJ. April 28.

Schmitz, S. 1989. Telephone communication between Eastern Research Group and Stuart Schmitz of the Iowa Department of Natural Resources. Des Moines, IA. April 4.

Seaman, M. 1989. Telephone communication between Eastern Research Group and Martin Seaman.

Smith, N. 1989. Telephone communication between Eastern Research Group and Nora Smith, Senior Planner with the City of Seattle Solid Waste Utility. Seattle, WA. April 27.

Society for Research into the Packaging Market. 1987. Packaging Without Plastic - Ecological and Economic Consequences of a Packaging Market Free from Plastic. Research performed on behalf of The Association of the Plastics Producing Industry. Frankfurt am Main, W. Germany.

SPI. 1987. Society of the Plastics Industry. Plastic Bottle Recycling: Case Histories. The Society of the Plastics Industry, Inc. Washington, DC.

SPI. 1988a. Society of the Plastics Industry. Facts and Figures of the U.S. Plastics Industry. Washington, DC.

SPI. 1988b. Society of the Plastics Industry. Questions and Answers About Plastics Packaging and Degradability. Washington, DC.

SPI. 1989. Society of the Plastics Industry. States Requiring Plastic Container Coding. Council for Solid Waste Solutions. Washington, DC.

Statz, R.J. and M.C. Dorris. 1987. Photodegradable Polytheylene. Proceedings of the Symposium on Degradable Plastics (Washington, June 1987). The Society of the Plastics Industry, Inc. Washington, DC. p. 51.

Stevens, B.J. 1988. Cost analysis of curbside programs. BioCycle. May/June:37.

Stevens, B.J. 1989a. How to finance curbside recycling. BioCycle. Feb:31.

Stevens, B.J. 1989b. How to figure curbside's costs. Waste Age. Feb:52.

Stroika, M. 1988. Telephone communication between Eastern Research Group and Max Stroika, Manager of Purchasing, Freeman Chemical, Port Washington, WI. July 20.

Tobin, K. 1989. Telephone communication between Kit Tobin, Manager Network Services, Keep America Beautiful, Inc., Stamford, CT. and Eastern Research Group, Inc. April 5.

U.S. EPA. 1988. U.S. Environmental Protection Agency. The Solid Waste Dilemma: An Agenda for Action. Municipal Solid Waste Task Force. U.S. EPA/OSW. EPA-530-SW-88-052. Washington, DC.

U.S. GAO. 1988. U.S. General Accounting Office. Degradable Plastics - Standards, Research, and Development. Report to the Chairman, Committee on Governmental Affairs, U.S. Senate. GAO/RCED-88-208.

Washington State. 1988. Best Management Practices Analysis for Solid Waste: 1987 Recycling and Waste Stream Survey, Vol. 1. Prepared by Matrix Management Group for the Washington State Department of Ecology Office of Waste Reduction and Recycling. Olympia, WA.

Watts, A. 1989. Telephone communication between Eastern Research Group and Allan Watts, City of Austin, Department of Public Works. April 27.

West German Federal Office of the Environment, Berlin. 1988. Comparison of the Environmental Consequences of Polyethylene and Paper Carrier Bags. Translation by G. W. House. Environmental Plastics Group, Polysar International SA. Mar 1989.

Wienholt, L. 1989. Telephone communication between Eastern Research Group and Lissa Wienholt, Recycling Department of the Oregon Department of Environmental Quality. April 10.

Wirka, J. 1988. Wrapped in Plastics: The Environmental Case for Reducing Plastics Packaging. Environmental Action Foundation. Washington, DC.

Wirka, J. 1989. Design for a National Source Reduction Policy. Environmental Action Foundation. Washington, DC.

6. Objectives and Action Items

The previous five sections of this report described the production, use, and disposal of plastics (Section 2); concerns regarding plastic material and products in the marine environment (Section 3); impacts of plastics waste on the management of municipal solid waste (MSW) (Section 4); and available options for reducing the impacts of plastic waste (Section 5). This section presents the actions to be taken by EPA as well as recommended actions for industry and other groups to address the concerns identified in these earlier sections.

The objectives presented here are divided into two categories: those for improving the management of the MSW stream and those for addressing problems outside the MSW management system (e.g., improvements to the wastewater treatment and drainage systems). For each objective, action items are listed that represent what EPA believes are effective means of achieving that objective. In general, improvement in MSW management can play a substantial role in reducing the concerns presented by the plastic waste component of the MSW stream. Most of the objectives and action items given here, therefore, are aimed at promoting or improving management methods, such as source reduction, recycling, landfilling, and incineration. Improvements in landfilling and incineration will better the management of all MSW, not just plastics.

Section 3 identified several articles of concern in the marine environment based on their effects on marine life or public safety, or the aesthetic damage they cause. However, Section 3 highlighted the impacts that result from all types of marine debris, not just these articles of concern. Other debris, such as beverage bottles and food wrappers, is unsightly and offensive when found on beaches or in harbors. The objectives and actions provided here focus on the sources of all marine debris, not on the identified articles of concern.

Many studies and reports other than this document have assessed marine debris issues. These studies have been conducted by numerous organizations, including the National Oceanic and Atmospheric Administration (NOAA), the Center for Marine Conservation, and the Marine Debris Interagency Task Force on Persistent Marine Debris, which was created by President Reagan in 1987. The specific recommendations of these reports are not reproduced here, although EPA supports the general intent of these efforts.

313

6.1   OBJECTIVES FOR IMPROVING MUNICIPAL SOLID WASTE MANAGEMENT

6.1.1 Source Reduction

> **1.**   **ISSUE:** Material substitution efforts aimed at reducing plastic waste generation must not increase other environmental problems.
>
> **OBJECTIVE:** Develop a method for systematically analyzing source reduction efforts (for either volume or toxicity reduction) that involve substitution.

As stated in Section 5, decisions regarding material substitution must be carefully considered to avoid a mere shift of environmental impacts (e.g., using a substitute material to reduce the volume of MSW may increase the volume or toxicity of the industrial process waste produced). Before decisions are made regarding material substitution, therefore, the impact of proposed actions on the following factors should be assessed:

■   Energy use

■   Natural resource use

■   Production waste volume and toxicity

■   Product utility

■   Product safety

■   Management of the product once it becomes a waste

■   Costs of production and eventual costs to the consumer

■   International trade

Without this type of analysis, called a lifecycle analysis, determining the impact of source reduction efforts involving material substitution or other waste management options (e.g., recycling) is very difficult. In addition, a generic method may help determine the impact of new products or product changes on waste management (e.g., on incinerator operations, recycling) and the environment. Such an analysis can be used to support a voluntary or possibly a regulatory program for preventing the introduction of new products or packages that adversely affect MSW management or the environment.

ACTION ITEMS:

- EPA has issued a grant to the Conservation Foundation (CF) to evaluate strategies for MSW source reduction. CF has convened a steering committee of municipal solid waste source reduction experts representing a wide range of interests in government, industry, and public interest groups. The steering committee will examine policy and technical issues involved in conducting a lifecycle analysis. Determining when such an analysis will be needed will also be discussed. The steering committee will provide recommendations by the Fall of 1990.

- Building on the work conducted by CF (described above), EPA will develop a model for conducting a lifecycle analysis. The model will be evaluated by applying it to selected components of the waste stream. Work on model development will begin in late 1990. A preliminary model is expected by the end of 1991. Once testing of the model is complete, EPA will make it available to interested organizations.

---

2.   **ISSUE: Lead- and cadmium-based plastic additives contribute to the heavy metal content of incinerator ash.**

OBJECTIVE: Identify and evaluate substitutes for, and nonessential uses of, lead- and cadmium-based plastic additives.

Section 4 indicated that lead- and cadmium-based additives may contribute to ash toxicity. Because they are distributed in a combustible medium, these additives tend to contribute proportionately more to fly ash than to bottom ash.

ACTION ITEMS:

- EPA is continuing to evaluate the potential substitutes for lead- and cadmium-based plastic additives identified in Appendix C. Substitutes for lead and cadmium in other components of the waste stream are also being identified and evaluated. Findings of the study will be shared and discussed with manufacturers and users of identified products and additives, as well as with members of the public and Congress. The final report is expected by April 1990.

- EPA will evaluate options for regulating additive use in situations in which safe and effective substitutes are available or in products in which lead- and cadmium-based additives are not considered to be essential.

3.   ISSUE: Plastics represent a substantial -- and increasing --proportion of the volume of the MSW stream.

OBJECTIVE 1: Evaluate the potential for minimizing packaging as a means of source reduction.

OBJECTIVE 2: Encourage education and outreach programs on methods for and effectiveness of implementing source reduction at the consumer level.

OBJECTIVE 1: EVALUATE POTENTIAL FOR MINIMIZING PACKAGING

As shown in Section 2, plastic packaging waste represents a significant percentage of the total plastic waste stream. Therefore, source reduction considerations are appropriate in this area. Other components of the plastic waste stream may also be targeted for source reduction consideration in the future. Although some packaging appears to be excessive, it may serve such purposes as preventing pilferage, tampering, breakage, or food decay; or providing an area for labels. Thus, care must be taken in determining how to reduce packaging materials.

ACTION ITEMS:

- EPA issued a grant to the Coalition of Northeastern Governors (CONEG) in partial support of their Source Reduction Task Force. Under this grant, the CONEG Task Force worked with industry and the environmental community to develop specific regionally agreed-upon definitions for preferred packaging practices. A report was issued by the CONEG Task Force in September 1989 that provided preferred packaging guidelines and recommended that a Northeastern Source Reduction Council be formed. The Council will include representatives from the CONEG states (Connecticut, Maine, Massachusetts, New Hampshire, New Jersey, New York, Pennsylvania, and Rhode Island), industry, and the environmental community. The Council, which began meeting in October 1989, will develop long-range policy targeted at reducing packaging at the source and implementing the packaging guidelines. The Council will also develop educational materials for the general public. EPA is actively working with CONEG and the Council on these efforts.

- The steering committee convened by the Conservation Foundation to examine MSW source reduction issues (see p. 6-3) will develop recommendations for selection criteria and a framework for a corporate awards program. In such a program, corporations or other organizations would be recognized for their work in promoting and carrying out source reduction activities. Packaging may be one focus of this program.

- EPA is partially funding a research effort to analyze the production and disposal of six different packaging materials (glass, aluminum, steel, paper, plastic, and "composite" packaging) and the impact of public policies aimed at reducing or altering the mix of packaging materials. EPA is supporting part of this research effort on the economic policy issues raised in this packaging study. The economic basis for disposal fees, bans, or other policy options will be analyzed and the impacts of these measures will be evaluated. Results of this study are expected by early 1991.

- EPA has initiated a study of economic incentives and disincentives for source reduction efforts (not limited to plastic-related efforts) termed "Market Analysis of the Major Components of the Solid Waste Stream and Examples of Strategies for Promoting Source Reduction and Recycling." Incentives and disincentives that will be evaluated include:

    - volume-based waste charges
    - user charges
    - depletion allowances and freight rates

The findings of the study, which will be available in early 1990, will be reviewed for applicability to plastics packaging reduction. Recommendations for providing incentives or removing disincentives to plastics source reduction will be made at that time.

OBJECTIVE 2: EDUCATION AND OUTREACH ON SOURCE REDUCTION

Consumer preferences and habits are in part responsible for driving changes in packaging and material use. By encouraging consumers to change their habits, therefore, outreach programs could ultimately affect the types of plastic products on the market. These programs could show shoppers the importance of reusing household items as well as what items can be reused, and the advantages of buying bulk foods and durable items.

ACTION ITEMS:

- EPA has completed a study of interplay between consumers and industry in the purchase of products and packaging promoting source reduction and recyclability. This study, which was issued in September 1989, is entitled Promoting Source Reduction and Recyclability in the Marketplace. The study pulls together much of what has been written about the household consumer demand side of source reduction and recyclability. Areas covered include previous case studies, findings of recent surveys and opinion pools, research reports, current events, and consumer education materials. The study also addresses a relatively new area, but nonetheless of great importance to the future success of source reduction and recycling -- the interplay between consumers, manufacturers, and government in the marketplace.

- EPA will develop a series of brochures or fact sheets on source reduction. The first two will focus on how to achieve source reduction in the workplace and how consumers can buy source-reduced products. These will be available by mid-1990.

- Manufacturers and retailers should sponsor educational efforts. These efforts could include advertising packaging or products that incorporate source reduction efforts (e.g., were designed to produce less waste).

6.1.2   Recycling

ISSUE: Only 1% of post-consumer plastic waste is currently being recycled.

OBJECTIVE: Promote plastics recycling:

- Improve the recyclability of the waste stream
- Improve collection and separation of plastics
- Investigate processing technologies
- Enhance markets for recycled plastics
- Educate the public on plastics recycling

As explained in Section 2, approximately 10.3 million tons of plastics were discarded into the MSW stream in 1986, and that amount is expected to increase by more than 50% by the year 2000. Of the 10.3 million tons, only 1% is currently recycled. This percentage is growing because of the dedicated efforts of many organizations to implement plastic recycling programs. Many of the action items described below call for continuance of these efforts. In order to recycle a much greater percentage of plastic wastes, problems concerning collection/separation, processing, marketing, and public information must be resolved.

Efforts to promote the development of recyclable packaging have been proposed partly because of the trend in packaging to use multi-layer, composite plastics (e.g., squeezable bottles) and multi-material packaging (e.g., plastic cans with aluminum lids). These types of packaging pose recycling difficulties. Composite plastic packaging can only be recycled in a mixed plastic system (either by secondary or tertiary processing), and the collection infrastructure is currently limited (as are the markets for the products of mixed plastics recycling) and may continue to be limited. At this time, therefore, composite and multi-material packaging is usually disposed of, not recycled.

ACTION ITEMS:

Improving Recyclability of the Waste Stream

- EPA recommends that the steering committee convened for the Conservation Foundation's efforts (see p. 6-3) become a self-supporting and long-standing Waste Reduction Council. One role for this council could be to review packaging and products for their impacts on current waste management. Efforts to improve the recyclability of products and packages could be promoted.

Collection/Separation

- EPA is providing technical assistance to local communities and States for setting up recycling programs, including programs for plastics. This assistance takes various forms, including peer match programs and a clearinghouse for MSW management information.

- Industry should continue to support research on improving collection equipment and efficiency for plastics recycling.

- Industry should continue to provide assistance for community collection programs, including:
  - -financial assistance for the purchase of equipment for collecting all recyclables
  - -technical assistance for collection/separation methods
  - -technical assistance for creating an educational/outreach program for communities to increase volumes collected

- Industry, states, and local governments should work together to evaluate the efficacy of various labeling systems that:
  - -promote public awareness of recyclable plastics
  - -assist consumers with identification and separation of resins
  - -allow for mechanical separation of resins

More than one system may be required to achieve these goals.

Processing

- The Department of Energy, industry, and universities should support further analysis of the efficacy of tertiary recycling (converting plastic resins back into monomers).

- Industry should continue to sponsor research on secondary recycling technologies (i.e., for the conversion of post-consumer plastic items into new plastic products), particularly on improvements in product properties.

- Industry should continue to provide technical and financial assistance to communities for purchasing and operating appropriate processing equipment (e.g., balers, shredders).

Marketing

- The Department of Commerce should evaluate the current and potential future impacts of increased plastics recycling on virgin plastic production, markets, and on imports/exports of virgin and scrap plastic.

- EPA has initiated a study of incentives and disincentives to recycling: "Market Analysis of Major Components of the Solid Waste Stream and Examination of Strategies for Promoting Source Reduction and Recycling" (see p. 6-7). When the study is finalized, EPA will review the findings for their specific applicability to plastics. Recommendations for removing disincentives or instituting incentives will be made at that time.

- Given the uncertainties in the markets and supply levels for most recycled plastics, States and communities that have been involved with cooperative marketing strategies should share information regarding these strategies with other states and communities. This could be accomplished through EPA's MSW clearinghouse.

- Industry should continue its efforts to identify, establish, and expand markets for products of plastics recycling (e.g., plastic lumber).

Public Education

- EPA is making information on plastics recycling available to the public through the national information clearinghouse that provides information on all waste management options.

6.1.3   Landfilling and Incineration

ISSUE: Disposing of plastics through incineration or in landfills raises several environmental concerns.

OBJECTIVE 1: Further evaluate the toxicity and potential leachability of additives in plastic products.

OBJECTIVE 2: Carefully monitor the use of halogenated polymers (e.g., PVC) in consumer products.

OBJECTIVE 3: Improve the design and operation of both disposal options.

OBJECTIVE 1: FURTHER EVALUATE ADDITIVES

Current data (see Section 4) are extremely limited and do not allow a complete assessment of the contribution of plastic additives to landfill leachate or emissions from other waste management options (e.g., incinerators).

ACTION ITEM:

■ Industry should evaluate the toxicity of additives in plastic products when they are placed in a landfill environment or when incinerated.

OBJECTIVE 2: MONITOR PVC USE

Concern about the incineration and landfilling of plastics primarily involves the incineration of halogenated polymers and the use of certain additives. Some plastics, such as PVC, present both types of problems, i.e., incineration of PVC produces hydrogen chloride, an acid gas, and some PVC products require the use of additives that may leach from the product in a landfill.

Although the current level of PVC in the MSW stream is very small (approximately 0.6 - 0.11%), it may contribute a major portion of the chlorine present in MSW (see Section 4.3.2.4 for a discussion of PVC's contribution of chlorine to MSW). As the use of PVC increases, the cost of controlling acid gas emissions from combustion or leachate from landfills may increase. In addition, concern has been expressed about the role of incineration of halogenated polymers in the formation of specific toxic compounds like dioxins and furans (see Section 4).

FDA is considering several regulatory actions that, if approved, would increase the use of halogenated polymers in food packaging. For example, FDA has published a proposed regulation to provide for the safe use of certain vinyl chloride polymers in manufacturing bottles. FDA's proposed rule, if finalized, would allow for increased use of vinyl chloride polymers in consumer products and, therefore, increase the volume of these polymers in MSW.

Commenters on the FDA proposal questioned the impact of the vinyl chloride polymers on the disposal and management of MSW. In response, FDA has announced that it will prepare an environmental impact statement (EIS) on the effects of its proposed rule on vinyl chloride polymers and four food additive petitions involving chlorinated polymers. The EIS will indicate how carefully vinyl chloride polymer use should be monitored - and thus whether the FDA should approve additional uses of PVC and other halogenated polymers.

ACTION ITEM:

- EPA will work with FDA during the development of its EIS.

OBJECTIVE 3: IMPROVE DISPOSAL OPTIONS

As explained in "An Agenda for Action -- The Solid Waste Dilemma," EPA prefers source reduction and recycling as means of reducing the problems associated with disposal of MSW. Implementation of the source reduction and recycling action items discussed above may increase the viability of these management options, but there will always be plastics that cannot be reduced in usage or recycled. These plastics must be disposed of in landfills or incinerated. Information in Section 3 of this report indicates that stormwater discharges from landfills may provide a pathway to waterways for lightweight debris. Controls for landfills and incinerators must be adequate to protect human health and the environment.

ACTION ITEMS:

- EPA will finalize new MSW landfill criteria (proposed on August 30, 1988) by the Spring of 1990. These criteria will outline the controls necessary to protect human health and the environment. Control of stormwater discharges at these facilities was included in the proposed criteria.

- Under the authority of the Clean Air Act, EPA proposed regulations for new MSW incinerators and guidelines for existing incinerators in November 1989. The proposal identifies controls needed to reduce acid gas and dioxin emissions.

6.2   OBJECTIVES FOR HANDLING PROBLEMS OUTSIDE THE MSW
MANAGEMENT SYSTEM

6.2.1   Wastewater Treatment Systems/Combined Sewer Overflows/Stormwater Drainage
Systems

ISSUE: Wastewater treatment systems, combined sewer overflows, and stormwater
drainage systems contribute substantial volumes of plastics to the marine
environment.

OBJECTIVE: Improve regulation of these discharges and enforcement of
regulations, pursue research into control methods, and educate consumers about
proper disposal practices.

Plastic marine debris flows from many different sources (see Sections 3 and 5). Of these,
wastewater treatment systems, combined sewer overflows (CSOs), and stormwater discharges
have been identified as significant sources for many of the articles of concern identified in this
report as well as other marine pollution. Some articles originating from these sources may
include plastic pellets, tampon applicators, condoms, syringes, bags, and six-pack rings.

6.2.1.1 Wastewater Treatment Systems

When wastewater treatment systems experience failures and are completely or partially shut
down, untreated wastewater may be released to the environment. This wastewater may contain
floatable debris such as tampon applicators, condoms, syringes, and plastic pellets.
Inappropriate disposal of some of these items by consumers of plastic articles contributes to the
problem of plastic waste polluting the marine environment.

ACTION ITEM:

- Every community should assess its needs for improving wastewater treatment systems.
  Some options include use of back-up holding tanks during shut-down periods and
  better pre-treatment of wastewater by pellet manufacturers and plastic processors.

- Industry should implement labeling/educational programs that describe appropriate
  disposal methods for products and the impacts of inappropriate disposal. Currently, at
  least one manufacturer of plastic tampon applicators has initiated such a program, and
  has labeled its products with directions concerning appropriate disposal.

6.2.1.2 Combined Sewer Overflows

Most major municipal areas in the United States are served by a combination of sanitary sewers, separate storm sewers, and combined sanitary and storm sewers. CSOs are flows from combined sewers that occur when rain overfills the wastewater treatment system. EPA estimates that 15,000-20,000 CSO discharge points are capable of releasing floatable debris into the environment.

EPA developed "A National Control Strategy for CSOs" in January 1989. This document makes it clear that CSOs are point source discharges that require NPDES permits. Three objectives included in that document are:

- To ensure that all CSO discharges occur only as a result of wet weather

- To bring all wet weather CSO discharge points into compliance with the technology-based requirements of the Clean Water Act and applicable State water quality standards

- To minimize impacts from CSOs on water quality, aquatic biota, and human health

ACTION ITEMS:

- As stated in the National Strategy, all permits for CSO discharges must include technology-based limitations for the control of pollutants, including solid and floatable discharges. These permits will be issued by EPA or approved States. Enforcement actions will be taken for violations of these limitations.

- EPA is developing guidance for States and local communities on effective operation and control of a combined sewer system. Low-cost control mechanisms will be included.

- EPA will conduct a limited number of CSO sampling studies to pinpoint which articles are frequently released from CSO discharges. Results should be available by early 1990.

6.2.1.3 Stormwater Discharges

In many urban areas, runoff is discharged to the environment through separate storm sewers. Street litter can be transported to the marine environment through the storm water system.

ACTION ITEMS:

- EPA is developing a report to Congress on stormwater discharges, as required under Section 402(p)(5) of the Clean Water Act. The major objectives of the report are 1) to identify all stormwater discharges not covered by EPA's proposed regulations or by existing permits, and 2) to determine the nature and extent of pollutants in the discharges (floatables are only one type of discharged pollutant). This report will be completed by mid-1990.

- EPA will also prepare a second report to Congress by the end of 1991 on control mechanisms necessary to mitigate the water quality impacts of stormwater discharges that were identified in the first report.

- EPA will sample and study a limited number of stormwater discharges to pinpoint which articles are released from these sources. Samples will be taken by early 1990.

- As required by Section 402(p)(4) of the Clean Water Act, EPA has proposed regulations specifying permit application requirements for two categories of stormwater discharges: 1) stormwater discharges associated with industrial activity, and 2) discharges from municipal separate storm sewer systems serving a population greater than 100,000 people (see 53 FR 49416). Permit applications for these sources will be required one or two years after EPA completes its final regulations.

6.2.2   Other Sources of Marine Debris

6.2.2.1 Vessels

ISSUE: Plastics that are discarded or lost from vessels contribute to observed problems of entanglement and ingestion of plastics by marine wildlife.

OBJECTIVE 1: Implement Annex V of MARPOL, which prohibits the discharge of plastic waste at sea.

OBJECTIVE 2: Reduce the impact of fishing nets, traps, and lines in the marine environment.

OBJECTIVE 1: IMPLEMENT ANNEX V OF MARPOL

ACTION ITEMS:

- The Coast Guard has developed regulations that implement the requirements of Annex V of MARPOL.

- EPA supports NOAA's recent recommendation (Report on the Effects of Plastic Debris on the Marine Environment) that Federal and State agencies should enter agreements with the U.S. Coast Guard to enforce MARPOL Annex V.

- Port facilities, local communities, industry, and interested Federal Agencies (e.g., Navy, Coast Guard) should coordinate efforts to develop recycling programs for plastic waste that is brought to shore in compliance with MARPOL Annex V.

OBJECTIVE 2: REDUCE IMPACT OF FISHING GEAR

NOAA is currently conducting four feasibility studies on how to reduce the problems inherent in "ghost" fishing by lost nets or traps. These studies examine the following options:

- Use of degradable materials for nets or panels on traps

- Use of a bounty system to encourage retrieval of nets and traps

- Use of a marking system to assist in finding lost gear

- Negotiation with foreign-flagged vessels regarding proper disposal of nets and traps

ACTION ITEM:

- EPA will support NOAA in its efforts by sharing relevant information (e.g., research results on degradable plastics).

6.2.2.2 Plastic Manufacturers, Processors, and Transporters

ISSUE: Plastic pellets, which are identified as an article of concern in this report, may be released to the marine environment from plastic manufacturing plants, plastic processing facilities, or during transportation of plastic pellets.

OBJECTIVE: Determine specific sources of plastic pellets and evaluate control options.

ACTION ITEMS:

- EPA is assessing the sources of pellets through studies of CSO and stormwater discharges. Results are expected by early 1990.

- Industry should ensure that plastic pellets are transported in durable containers.

6.2.2.3 Garbage Barges

ISSUE: Garbage barges have been identified as sources of marine debris.

OBJECTIVE: Improve control of these sources.

ACTION ITEM:

- EPA will provide information to the Coast Guard on methods that owners or operators of garbage barges or other vessels that transport solid waste can use to reduce the loss of waste to the marine environment. The Coast Guard can incorporate these methods into the permits that these vessels must receive in order to operate.

6.2.2.4   Land- and Sea-Originated Litter

> ISSUE: Land- and sea-originated litter produce aesthetic, economic, and environmental impacts.
>
> OBJECTIVE 1:   Support current efforts to retrieve, monitor, and characterize beach marine litter.
>
> OBJECTIVE 2:   Support current litter prevention campaigns.

OBJECTIVE 1: SUPPORT LITTER RETRIEVAL AND CHARACTERIZATION

Litter prevention is the preferred option for reducing the aesthetic, economic, and environmental impacts of litter. Littering, however, is inevitable, and thus retrieval programs must be planned and implemented.

ACTION ITEMS:

- EPA will continue to sponsor a limited number of harbor surveys. These surveys help to remove unsightly floatable debris from the marine environment and provide data on the types of items in the marine environment.

- EPA will continue to work with NOAA in sponsoring beach clean-up activities. In addition to retrieving unsightly litter, these programs provide data regarding amounts, types of, and damage caused by plastic debris. These programs also educate participants on the marine debris issue.

OBJECTIVE 2: SUPPORT LITTER PREVENTION

Waste management methods such as source reduction and recycling will help reduce the volume of waste that is improperly disposed of as litter. The effectiveness of these methods for reducing litter on land or in the marine environment can be greatly enhanced by education and other litter prevention campaigns.

All levels of government, as well as environmental groups and industry, have supported and developed educational campaigns aimed at reducing littering behavior and encouraging clean-up campaigns. For example, the Department of the Interior has initiated an advertising campaign for litter reduction. EPA is working with NOAA, the U.S. Coast Guard, and other agencies to develop a public education program on marine debris. To date, this effort has included funding for the Second International Conference on Marine Debris, funding for the 1988 and 1989 beach cleanups conducted during COASTWEEKS '88 and '89 by the Center for Marine Conservation (CMC, formerly the Center for Environmental Education, or CEE), and sponsorship of an informal roundtable meeting to plan future national-level public education activities. Industry groups have also supported the efforts of CMC through magazine advertisements.

Some industries and commercial establishments that manufacture or use products which are often littered (e.g., fast food distributors) have printed reminders on the products to discourage littering. In addition, many State and local governments sponsor litter prevention programs and have instituted fines for littering. Keep America Beautiful (KAB) has been involved in the fight against litter since its inception in 1953.

ACTION ITEMS:

- EPA will continue to provide resources to distribute currently available educational materials on marine debris and its sources and effects.

- EPA is developing an educational program for consumers that describes the proper method for disposing of household medical wastes.

- Industry and public interest groups should continue their efforts to promote anti-litter behavior.

- Federal agencies should promote anti-litter behavior among their employees by displaying posters or other available materials throughout their buildings and grounds and by providing recycling opportunities.

6.2.3   Degradable Plastics

ISSUE: Degradable plastics have been proposed as a method to alleviate some litter or other environmental problems.

OBJECTIVE: Answer current questions regarding the performance and potential impacts of degradable plastics before promoting further applications.

Because of the many unanswered questions regarding the performance and potential impacts of degradable plastics in different environmental settings, EPA cannot at this time support or oppose the use of degradable plastics. EPA does not include degradables as part of its waste management strategy; however, there may be some useful applications of degradable plastics, such as for composting or agricultural mulch films. Additional data are needed in each of the following areas before an appropriate role for degradable plastics can be identified:

- *Degradable plastics in a landfill* -- Degradation in a landfill environment is primarily anaerobic and does not occur evenly throughout the landfill. While degradation does occur, it occurs very slowly. For example, even food wastes have been found in recognizable form after many years in a landfill. Use of degradable plastics in items disposed of in landfills, therefore, is very unlikely to reduce requirements for landfill capacity.

- *Degradable plastics as part of land-based litter and in the marine environment* -- Very little is known regarding how degradables perform in different environmental settings. The most critical unknowns are:

  - What is the rate of degradation in different environmental settings (e.g., on land in Alaska versus in water near Florida)?
  - What byproducts are formed and what are their environmental impacts?
  - Is leaching of additives greater in degradable plastics than in "nondegradable" plastics?
  - Will the degradation process have any adverse impact on aquatic life?
  - Will the use of degradables pose an ingestion threat to wildlife?
  - Will the use of degradables solve the problems of entanglement of wildlife?
  - Will the use of degradables increase littering?

  Until these questions are answered, EPA believes that (with the exception of plastic ring carriers that are discussed below) Federal, state, and local governments should refrain from promoting the use of degradable plastics.

- *Degradable plastics: recycling and source reduction* -- Many recyclers of plastics are concerned that degradables may seriously impair operations by disrupting the processing and/or adversely affecting the properties of the resulting products. The extent of this problem is difficult to determine at this time.

  Though use of degradables is sometimes labeled as a source reduction method, EPA does not consider degradables a source reduction technique because they are not expected to reduce the amount or toxicity of the waste generated. Degradable plastics must still be collected and managed as solid wastes. Degradation, if it occurs, would take place after the product is landfilled or incinerated; therefore the generation of waste is not affected.

ACTION ITEMS:

- EPA has initiated two major research efforts on degradable plastics. The first project is a multi-year effort that will provide information on degradation rates in different environmental settings, by-products formed during degradation, and ecological impacts of the degradation process. This project will be completed in late 1991. The second project will examine the impacts of degradable plastics on recycling. The effects of degradable plastics on a variety of recycling processes and products will be evaluated. Interim results should be available by mid-1990.

- Title I of the 1988 Plastic Pollution Control Act directs EPA to require that beverage container ring carrier devices be made of degradable material unless such production is not technically feasible or EPA determines that degradable rings are more harmful to marine life than non-degradable rings. The uncertainties regarding degradable plastics (discussed above) pose some difficulties for EPA's implementation of this Act; however, some specific information is known regarding ring carrier devices:

  - EPA has not identified any plastic recycling programs that currently accept or are considering accepting ring carriers. Therefore, degradable rings should not impair recycling efforts.

  - Ring carriers are usually not colored and therefore do not include metal-based pigments. Thus, concerns regarding leaching of pigments appear to be minimal for these devices.

  The research on degradable plastics (see above) now underway at EPA will help resolve remaining issues. EPA will initiate a rulemaking to implement the above legislation in 1990. A final rule is expected by late 1991.

- EPA will support and participate in ASTM's effort to develop standards for degradable plastics.

- Industry should demonstrate to EPA and the general public: 1) any benefits of degradable plastics, 2) cost-effectiveness of degradable plastics, and 3) whether degradable plastics pose less of an environmental risk than nondegradable plastics or disrupt other management practices (e.g., recycling). To do so, industry must generate and make available data such as the following:

  - Rates of degradation of plastic in different environments
  - Identification of chemical and physical degradation products and their impacts (including air emissions)
  - Possible impacts on recycling, assuming that use of this management technique expands
  - Leachability of additives from products during the degradation process.

Appendix A—Statutory and Regulatory Authorities Available to EPA and Other Federal Agencies

This appendix provides an overview of the legal authorities available to EPA and other Federal agencies for improving plastics waste management. Control of plastics and plastics waste could involve: 1) disposal of plastic wastes from vessels into the ocean; 2) disposal of plastic wastes from land sources to navigable waters (including the ocean); and 3) disposal of plastic waste from any source onto land.

This discussion covers laws which provide authority to EPA and other Federal agencies for regulatory action that could affect disposal of plastics. The laws covered are summarized in Table A-1.

A.1 SUMMARY OF FINDINGS

Following are the key findings of this appendix:

- Regarding the disposal of plastic wastes from vessels, the Act to Prevent Pollution from Ships, as amended, implements MARPOL Annex V international treaty prohibiting the disposal of plastic wastes from vessels in any navigable water. This law does not affect the accidental loss of nets or other gear during fishing operations.

- Regarding the disposal of plastic wastes into the ocean from land-based sources, the Ocean Dumping Act (formally called the Marine Protection, Research and Sanctuaries Act of 1972) prohibits the transporting of any material not associated with the normal operation of a vessel (such as plastic wastes from land) for the purpose of disposal in the ocean, except as permitted by EPA.

- Regarding the disposal of plastic wastes into navigable waters from land-based sources, the Clean Water Act could theoretically be used to restrict disposal of plastics in industrial or municipal effluents into navigable waters, but has not been so far: EPA has not considered plastic wastes an effluent of concern.

- Existing laws for the protection of fish and wildlife have limited effectiveness for controlling the disposal of plastic waste into navigable waters and thus for preventing marine entanglement .

- Regulation of non-hazardous solid wastes (such as plastics) on land is the responsibility of the states under Subtitle D of the Resource Conservation and Recovery Act (RCRA); the Federal government, however, has developed national performance standards for land disposal operations.

332

TABLE A-1

LAWS INFLUENCING METHODS OF PLASTICS DISPOSAL

Laws Affecting Disposal of Plastics by Vessels

    Marine Plastic Pollution Research and Control Act
    Refuse Act
    Greater Lakes Water Quality Agreement
    Outer Continental Shelf Act
    Federal Plant Pest Act
    Driftnet Impact Monitoring, Assessment, and Control Act

Laws Affecting Disposal of Plastics from Land to Sea

    Ocean Dumping Act (Marine Protection, Research and Sanctuaries Act of 1972, amended 1988)
    Clean Water Act
    Shore Protection Act
    Deepwater Port Act

Law Affecting Disposal of Plastics on Land

    Resource Conservation and Recovery Act
    Medical Waste Tracking Act (amends the Resource Conservation and Recovery Act)
    Clean Air Act

Laws Affecting Manufacture or Discard of Plastic Materials

    Toxic Substances Control Act
    Fishery Conservation and Management Act
    Endangered Species Act
    Marine Mammals Protection Act
    Migratory Bird Treaty Act
    Plastic Ring Legislation
    Food, Drug and Cosmetic Act
    National Environmental Policy Act

Source: Compiled by Eastern Research Group. Inc.

- Regarding the manufacture and disposal of plastics, EPA has the authority under Sections 5 and 6 of the Toxic Substances Control Act to regulate chemical substances that present or will present an "unreasonable risk of injury to health or the environment," but to date this authority has been applied only to substances of much greater toxicity than plastic products or plastic wastes.

- Recently passed laws (i.e., the Shore Protection Act; plastic ring legislation; and the Medical Waste Tracking Act, which amends RCRA) provide additional but specialized and limited authority to EPA for prevention of plastic waste management problems.

- A variety of other legislation provides some authority to EPA or other Federal agencies for control of actions that may generate or cause a release of plastic waste, but such legislation affects very few disposal activities.

A.2 LEGISLATION CONTROLLING THE DISPOSAL OF PLASTIC WASTES FROM VESSELS INTO NAVIGABLE WATERS

This section reviews the available legal authorities for controlling the disposal of plastic wastes from vessels. The most significant legislation for controlling plastic waste disposal from vessels -- the implementing legislation for MARPOL Annex V -- is discussed, as are a number of other laws that influence, to some degree, the disposal of wastes into the navigable waters. The discussion outlines the scope and coverage of these laws as well as, where appropriate, the actual extent of their influence over disposal practices.

A.2.1 The Marine Plastic Pollution Research and Control Act of 1987

The Marine Plastic Pollution Research and Control Act of 1987, which amends the Act to Prevent Pollution from Ships, implements the MARPOL (Marine Pollution) Annex V international treaty prohibiting the disposal of plastic wastes from vessels into any navigable water. The United States is one of the signatory nations to the treaty and is bound by the treaty to implement regulations that are consistent with MARPOL Annex V. The treaty and the U.S. legislation implementing the treaty domestically are expressly intended to eliminate vessels as major sources of plastic waste in the marine environment.

The regulations under MARPOL will be implemented and enforced by the Coast Guard, as have been previous Annexes covering oil and hazardous chemical wastes from vessels. On April 28, 1989, the Coast Guard issued interim final regulations. These regulations apply to all vessels of U.S. registration or nationality and any vessels (including foreign-flagged vessels) operating within the navigable waters of the United States, or within the exclusive economic zone of the United States (i.e., the 200-mile area affected by U.S. regulations on marine commercial activity). The maritime sectors that are regulated encompass merchant marine vessels, including passenger vessels, fishing boats, recreational vessels, offshore oil and gas platforms, and miscellaneous research, educational, and industrial vessels. Vessels operated by government agencies will be given five years to come into compliance with the MARPOL Annex V requirements.

MARPOL Annex V implementing regulations do not cover accidental loss of plastic materials (e.g., during commercial operations). Specifically, fishing nets lost accidentally at sea are not regulated under the MARPOL Annex V implementing regulations. Thus, some plastic materials will continue to be lost from vessels.

The MARPOL Annex V implementing regulations also require that ports have "adequate reception facilities" for receiving garbage that is brought ashore from incoming vessels. This term has not yet been defined by the Coast Guard. MARPOL Annex V provides authority for civil penalties for violation of disposal regulations.

The Marine Plastic Pollution Research and Control Act of 1987 also mandates several research studies of marine and other topics. The law directs EPA to perform this study of methods to reduce plastic waste disposal issues. Similarly, the Commerce Department is directed to submit a Report to Congress on the effect of plastic materials on the marine environment. The EPA in conjunction with other Federal, state and interstate agencies, is charged with preparing a restoration plan for the New York Bight. This plan shall identify and address pollutant inputs and their impact on marine resources of the Bight, and also analyze and recommend appropriate mitigation technologies.

A.2.2 Additional Legislation

Refuse Act: MARPOL Annex V will make the Refuse Act -- the only remaining vestige of the 1899 Rivers and Harbors Act and the only legislation that had previously affected the disposal of wastes from vessels -- largely irrelevant for controlling the disposal of plastic waste. The Refuse Act regulates the disposal of wastes into the ocean from vessels operating within the three-mile limit from land. The Refuse Act prohibits all garbage disposal, including disposal of plastics, in U.S. coastal waters or inland waterways. The law does allow for specific ocean discharges if permitted by the U.S. Army Corps of Engineers.

The Refuse Act was intended to prevent problems with navigation and fouling of rivers by debris. It could be interpreted broadly to prohibit virtually all waste disposal from vessels. Nevertheless, the law has been described by the Coast Guard as inadequate for the purposes of controlling waste disposal to navigable waters (Kime, 1987) because it imposes no civil and only modest criminal penalties for violations. As a result, the Coast Guard cannot impose administrative penalties; instead, it can only bring labor-intensive, time-consuming judicial actions against violators.

Great Lakes Water Quality Agreement (GLWQA): A regional restriction on disposal of vessel wastes is in effect for the Great Lakes. The authority for this prohibition comes from Annex V to the 1978 Great Lakes Water Quality Agreement (GLWQA). This agreement is a product of the International Joint Commission, a body concerned with cross-boundary issues between Canada and the United States. Under Annex V to the agreement, the discharge of garbage, including "all kinds of victual, domestic, and operational wastes" is prohibited and subject to appropriate penalties. According to Great Lakes-based Coast Guard officials, most vessels comply with this treaty. Most Great Lakes vessels have been equipped with compactors to handle collected wastes (Hall, 1988).

Outer Continental Shelf Act: This act, which restricts disposal from offshore activities, is designed to support the exploration, development, and production of minerals from the Outer Continental Shelf while protecting the character of waters in this area. Under this authorizing legislation, the Minerals Management Service (MMS) has developed orders prohibiting the disposal of all solid wastes, including plastics, from operating offshore structures and from associated support vessels. As a result of the MMS regulations, the new regulations being implemented under MARPOL Annex V will generate few incremental requirements for offshore oil and gas operations; nearly all waste disposal from these structures is already restricted.

Federal Plant Pest Act (1957): This law influences ocean dumping because it increases the obstacles to disposal of shipboard wastes in port. Under the authority of this Act, the Department of Agriculture established the Animal and Plant Health Inspection Service (APHIS). This service requires that all garbage off-loaded in a U.S. port from vessels coming from foreign countries be treated to prevent infestations. Off-loading is generally allowed only after garbage has been steam-sterilized or incinerated. A network of APHIS inspectors is on call nationwide and routinely boards ships that have entered U.S. ports. Recent data indicate that close to 50,000 ship inspections are made annually (Caffey, 1987).

Owners of ships coming from foreign ports are charged for use of APHIS-approved facilities, which provide for incineration of wastes. To avoid this expense, wastes are sometimes dumped overboard before the vessels reach port; the regulations have thus become an incentive for illegal ocean disposal. Further, as of January 1989, a number of ports, including some major ports, did not have local APHIS facilities for receiving shipboard wastes.

Driftnet Impact Monitoring, Assessment, and Control Act of 1987: This law also provides some protection for ocean resources. This legislation requires the Department of Commerce to collect information and analyze the impacts of driftnet fishing by foreign vessels operating beyond the exclusive economic zone of any nation. It also authorizes international agreements with other nations that have fishermen operating in the North Pacific and affecting U.S. marine resources. The legislation also requires the study and creation of a driftnet marking, registry, and identification system to determine the source of abandoned fishnets and fragments.

A.3   LEGISLATION CONTROLLING THE DISPOSAL OF PLASTIC WASTES FROM LAND SOURCES TO NAVIGABLE WATERS

Plastic debris can also enter navigable waters from land-based sources such as industry, sanitary and stormwater sewer systems, and municipal solid waste handling facilities. This section describes the legislation for controlling the disposal of plastic wastes from land sources to navigable waters.

A.3.1 The Ocean Dumping Act

National concern for the environmental threat from ocean dumping led to passage of the Marine Protection, Research and Sanctuaries Act in 1972, now termed the "Ocean Dumping Act." Under this Act, EPA and the U.S. Army Corps of Engineers are responsible for regulating the transportation and dumping of wastes in the ocean. This legislation prohibits the transportation of waste from land-based sources (such as plastic wastes from the land) for the purpose of ocean disposal, except as permitted by EPA.

A.3.2 The Clean Water Act

The stated objective of the Clean Water Act of 1972 is to "restore and maintain the chemical, physical, and biological integrity of the nation's waters." To achieve that objective, the act establishes two national goals: 1) to reach a level of water quality that provides for the protection and propagation of fish, shellfish, and wildlife and for recreation in and on the water; and 2) to eliminate the discharge of pollutants into U.S. waters. The principal means to achieve these goals is a system that imposes effluent limitations on, or otherwise prevents, discharges of pollutants into any U.S. waters from any point source.

Under this law, EPA has prepared effluent limitations guidelines and standards for numerous categories of industrial facilities that discharge pollutants into the nations waters either directly or through publicly owned treatment works (POTWs). Regulations include those for process water discharges from manufacturers and fabricators of plastics. Existing regulations for this industrial category, however, control total suspended solids, biochemical oxygen demand, and toxic discharges but do not explicitly regulate outflows of raw plastic materials (although permit writers could do so). Such wastes may be washing into navigable waters. Similarly, for other industries, the Clean Water Act regulations restrict end-of-pipe discharges of toxic chemicals to the environment. The Clean Water Act could theoretically be used to restrict disposal of plastics in industrial effluents into navigable waters, but to date this has not occurred in Federal regulations; EPA has not considered plastic wastes an effluent of concern.

Untreated sewage from POTWs can also be discharged into surface waters when the volumes of incoming waste are larger than the treatment capacity of the facility or when a facility is malfunctioning or undergoing maintenance. These untreated wastes may contain various amounts of plastic debris that normally would be removed by screens and skimmers during treatment. EPA has used its authority under the Clean Water Act to bring legal action against cities that have failed to meet Federal standards for treatment of wastewater.

In some communities where storm sewers are combined with municipal sewage systems, intense storms can cause sewer overflows and result in both untreated sewage and stormwater being discharged directly into receiving waters. These combined sewer overflows (CSOs) may contain various kinds of sewage-associated plastic debris (e.g., disposable diapers, tampon applicators, condoms, and other sanitary items) as well as street litter collected by stormwater runoff.

EPA has identified its authorities under the Clean Water Act as a means of supplementing its regulations in this area. The Agency is in the process of revising regulations under the National Permit Discharge Elimination System (NPDES) to add new permit application requirements for stormwater discharges from cities with large (over 250,000) and medium-sized (between 100,000 and 250,000) populations.

The Clean Water Act can also be used to regulate discharges of plastic wastes from stormwater discharges where separate sanitary and stormwater sewers exist. EPA is currently studying its options for regulating this source of waste and is preparing a report to Congress on this topic.

A.3.3 Shore Protection Act of 1988

Under the Shore Protection Act, waste handlers must minimize the release of municipal or commercial waste into coastal waters during the loading or unloading of wastes from vessels or during the transport of wastes by vessel. In addition, the owner or operator of any waste source or receiving facility must provide adequate control measures to clean up any municipal or commercial waste that is deposited into coastal waters.

A.3.4 Deepwater Port Act

This legislation includes provisions to protect marine and coastal environments from any adverse effects due to the development of deepwater ports. The law authorizes regulations that could prevent marine pollution by requiring clean up of pollutants generated and by defining proper land disposal methods for any synthetic materials related to the construction of deepwater ports. To date only one facility, the Louisiana Offshore Oil Port, is licensed under this legislation (Serig, 1989).

A.4 DISPOSAL OF PLASTIC WASTE FROM ANY SOURCE ONTO LAND

Plastic debris is deposited on land mainly in the course of normal land disposal of municipal solid waste. The most important existing legislation in this area, the Resource Conservation and Recovery Act (RCRA), regulates the operation of municipal solid waste landfills. Additionally, the discussion below describes the role of the Clean Air Act, which regulates solid waste combustion.

A.4.1 Resource Conservation and Recovery Act (RCRA)

Regulation of non-hazardous solid waste (as defined by regulation) is the responsibility of states pursuant to Subtitle D of RCRA. EPA has developed national performance standards for the land disposal of non-hazardous solid wastes. The Federal regulations do not consider components of the solid waste separately and thus do not single out plastic wastes for special consideration. States are responsible for developing, implementing, and enforcing their own regulations, which must be at least as protective as the federal standards. The Federal

Appendix A: Statutory and Regulatory Authorities 339

government can influence plastics disposal by providing information on how to recycle these materials and by helping states to implement waste reduction efforts.

Subtitle C of RCRA regulates the disposal of hazardous wastes. Under RCRA, a waste is considered hazardous if it is a solid waste that

...because of its quantity, concentration, or physical, chemical, or infectious characteristics...may pose a substantial present or potential hazard to human health or the environment when improperly treated, stored, transported, or disposed of.

EPA has prepared a listing of the materials that are considered to be hazardous based on evidence of health or environmental dangers they pose. For materials that are not listed, EPA has defined test procedures to determine whether they are hazardous. Most plastics do not have the characteristics defined as hazardous (ignitability, corrosivity, reactivity, or toxicity) and are not listed as hazardous and, thus, are not regulated as hazardous wastes.

A related question concerning the applicability of RCRA to the plastics disposal problem concerns the treatment of the ash and residues that result from the incineration of plastics and other materials in MSW. As noted in Chapter 4 of this report plastics may contribute to the heavy metal content of ash. Legislation regarding the proper handling of incinerator ash is currently pending.

Under the recently passed Subtitle J of RCRA (the Medical Waste Tracking Act of 1988), EPA has set up a demonstration program in several states for tracking medical wastes from their generation to disposal. Based on the results of the program, EPA will evaluate the need for regulations to ensure that medical wastes are handled and disposed of properly. Medical wastes may contain plastic tubing and syringes, as well as certain other plastic medical debris from hospitals, doctors' offices, clinics, and laboratories.

A.4.2 Clean Air Act (CAA)

Under the authority of Section 111 of the Clean Air Act (CAA), EPA is currently developing regulations and guidelines governing air emissions from new and existing municipal waste combustors. These requirements could affect the emissions of various gases into the atmosphere which come from incineration of plastics. In addition, under CAA authority, EPA could control whether certain plastics are combusted or whether they must be separated by the source before incineration. Such source separation requirements could affect the amount of plastic recycled or the method of disposal. EPA is also considering regulation of air emissions from municipal solid waste landfills.

A.5   OTHER LEGISLATION THAT INFLUENCES THE MANUFACTURE OR DISCARD OF PLASTIC MATERIALS

A number of additional laws govern activities that are relevant to plastic materials and problems generated by their disposal. These general-coverage laws could be employed to regulate the manufacture, use, and disposal of waste materials.

A.5.1   Toxic Substances Control Act (TSCA)

The Toxic Substances Control Act (TSCA) of 1976 provides EPA with authority to require testing of new and existing chemical substances entering the environment and to regulate them where necessary. Under Sections 5 and 6 of TSCA, EPA is given broad authority to take whatever regulatory measures are deemed necessary to restrict chemicals suspected of posing harm to human health or the environment.

One of the most severe impacts of plastics disposal in water bodies is the injury to and death of fish, marine mammals, and birds that become entangled in plastic or mistake it for food. To date, however, EPA has applied its authority under Section 5 of TSCA for new chemicals and Section 6 of TSCA for existing chemicals only to substances of much greater toxicity (such as polychlorinated biphenyls, chlorofluorocarbons, and asbestos) than plastic products or plastic wastes. The law has never been employed for general solid waste problems such as management of plastic wastes. Review of a chemical under these authorities tends to focus on the toxicity of the chemical itself rather than the products it will be used in (e.g., plastics) or potential disposal problems associated with those products.

A.5.2   Food, Drug and Cosmetic Act

The Food, Drug and Cosmetic Act (FDCA) authorizes regulations to ensure the safety of food and medical products. This regulatory authority extends over the products that come in contact with food or drugs or that are used for medical purposes (e.g., blood bags, artificial joints, or valves for placement in the body), including plastic materials. By regulating the chemical and physical nature of plastic materials for certain uses, FDCA influences the nature of a portion of the plastic materials that are disposed. This influence is significant because substantial amounts of plastic materials are used in packaging, containers, health supplies, and other products that fall under the jurisdiction of the U.S. Food and Drug Administration (FDA) (although no quantitative estimates are available.)

FDA's regulations are promulgated to ensure the safe use of food-contact materials and medical products under the stated conditions of use. For example, as part of the determination of safety, FDA requires that the sponsor of food additive petitions provide data on the potential migration of components of the food-packaging material to food. These data are typically obtained from extraction experiments using food-simulating liquids. Migration data are considered in conjunction with toxicological data to determine whether the proposed use of the packaging materials is safe.

FDA is required by the National Environmental Policy Act (see Section A.5.5) to consider the environmental impact of its actions, including actions involving plastics. For example, FDA recently announced that it will prepare an environmental impact statement (EIS) on the effects of its proposed action on vinyl chloride polymers. The EIS was prompted due to concerns over the impact of vinyl chloride polymers on MSW management.

FDA has had a substantial influence on the types of plastic products manufactured, particularly with regard to additives used in these products. As a result of FDA's safety reviews, the presence of additives in the waste stream and the aggregate toxicity of the plastic materials discarded have been reduced. In general, the scope and reach of FDA's influence on the plastic products it has regulated or will regulate is greater than restrictions on plastic products as envisioned under TSCA or other EPA status.

However, FDA does not regulate the plastics used in building and construction materials except for special cases such as pipes used in food processing plants, nor does it regulate plastics used in automobiles. Some of these products, particularly polyvinyl chloride, may employ toxic additives that are not approved for use in contact with food and drugs.

A.5.3 Fish and Wildlife Conservation Laws

Several fish and wildlife conservation laws (the Fishery Conservation and Management Act, the Endangered Species Act, the Marine Mammals Protection Act, and the Migratory Bird Treaty Act) have some potential influence over plastics entering the environment. To date, however, they have not been used for this purpose.

The Fishery Conservation and Management Act, also called the Magnuson Act, prohibits the disposal of fishing gear (such as plastic fishing nets) overboard. This rule is enforced against foreign ships that operate within the 200-mile Exclusive Economic Zone through the foreign vessel observer program. There are, however, no counterpart regulations for domestic ships. This statute does not prohibit accidental loss of plastic fishing gear at sea.

The Endangered Species Act and Marine Mammals Protection Act prohibit the "taking" of animals of protected species by any means. Entanglement of mammals or birds is prohibited under a strict interpretation of these rules, but in most cases the entanglement of an individual fish or mammal cannot be linked to a specific act of disposal. Thus, no attempt has been made to enforce these regulations against maritime industries.

The United States has entered into four separate treaties (with Canada, Mexico, Japan, and the U.S.S.R.) to protect migratory bird species. The Migratory Bird Treaty Act (MBTA) provides the domestic framework for satisfying the international obligations under this treaty. This legislation prohibits the unpermitted capture or killing of migratory birds.

The MBTA has been successfully enforced against killers of migratory birds even when there was no intent by the individuals involved. Such success implies that the MBTA could be enforced against fishermen who find entangled migratory birds in their nets (Gosliner, 1985). Nevertheless, this law has not been widely used to penalize fishermen who capture or kill migratory birds incidental to their fishing operations.

A.5.4 Degradable Plastic Ring Carrier Law

A law recently passed by Congress, termed An Act to Study, Control, and Reduce the Pollution of Aquatic Environments from Plastic Materials and For Other Purposes (PL 100-556), directs EPA to develop regulations that require that any plastic ring carriers used for packaging, transporting, or carrying multipackage cans or bottles be made of degradable materials. Regulations are not required if EPA determines that the risks posed by degradable ring carriers are greater that those posed by nondegradable carriers. To reduce litter and to protect fish and wildlife, many states have already enacted laws requiring that plastic ring carriers be made from degradable material.

A.5.5 National Environmental Policy Act

The National Environmental Policy Act (NEPA) directs Federal agencies to consider environmental factors in planning their projects and activities. NEPA directs all agencies to prepare an Environmental Impact Statement (EIS) for any major Federal action that will significantly affect the quality of the environment. An EIS must identify and discuss the environmental effects of the proposed action and identify, analyze, and compare options. Further, EPA must review any EIS prepared by other agencies.

The NEPA process ensures that relevant environmental issues are considered in an agency's decision making process. While this process, including the EPA review, does not provide any specific regulatory authority it does permit agencies to base their decisions on environmental considerations, when balanced with other factors. In the course of this review, plastic waste management issues could be addressed.

REFERENCES

Caffey, R.B. 1987. Hearings before the Subcommittee on Coast Guard and Navigation and the Subcommittee on Fisheries and Wildlife Conservation and the Environment of the Committee on Merchant Marine and Fisheries, House of Representatives. Statement of Dr. Ronald B. Caffey, Assistant to the Deputy Administrator, Plant Protection and Quarantine Program, Department of Agriculture. July 23, 1987.

Gosliner, M. 1985. Legal authorities pertinent to entanglement by marine debris. In Proceedings of the Workshop on the Fate and Impact of Marine Debris. November 27-29, 1984. Honolulu, Hawaii.

Hall, G. 1988. Telephone communication between Jeff Cantin of ERG and Gordon Hall of the Lake Carriers Association. March 2, 1988.

Kime, J.W. 1987. Hearings before the Subcommittee on Coast Guard and Navigation of the Committee on Merchant Marine and Fisheries, House of Representatives, on HR 940. Testimony of Rear Admiral J. William Kime, Chief, Office of Marine Safety, Security and Environmental Protection, U.S. Coast Guard.

Serig, H. 1989. Telephone communication between Jeff Cantin of ERG and Howard Serig, Office of Secretary, Department of Transportation, May 3, 1989.

Appendix B—State and Local Recycling Efforts

Most recycling programs operate today at the local level as part of either city or county programs. As landfill space has dwindled and dumping costs have risen, many states have moved to encourage or mandate municipal solid waste recycling through legislation. State recycling initiatives include laws that require communities, counties, or regions to develop recycling programs, bottle bill container laws, and measures that encourage, but do not mandate recycling. The discussion below and Table B-1 summarize a number of state recycling programs.

Currently eleven states have some type of manatory municipal solid waste recycling legislation. Although these laws vary widely in scope and content, defining state waste reduction or recycling goals is the initial step in developing a state waste management strategy. Most of these states have set statewide waste reduction goals which usually require or encourage some type of community curbside or drop-off recycling. Some states (Connecticut, Pennsylvania, Rhode Island, New Jersey, New York, Maryland, and Florida) require communities, counties, or regions to develop curbside collection programs. Many of these laws contain some unique initiatives. Connecticut law prohibits the landfilling or incineration of twelve specified materials after 1991. In Oregon and Wisconsin curbside recycling is not specified, but residents must be provided with some opportunity to recyle. Illinois has set a goal of twenty-five percent waste reduction (as have a number of other states) and requires communities to meet this figure locally. Other states, like Washington and Minnesota, where many successful local recycling programs exist, have passed more general legislation which set source reduction and recycling as the state waste management priorities.

Container deposit laws are another method for states to encourage recycling. Bottle bills have existed in nine states since 1987, and have been considered by many others. Oregon, a state with high return figures, has required a deposit on some beverage containers since 1971. Eight of these programs set a five-cent deposit on certain beverage containers; one (Michigan) has set a ten-cent deposit. The specific containers included vary from program to program. All include carbonated beverage containers. Most recently, Florida passed an alternative bottle bill requiring that an advanced disposal fee be added to the price of all containers by 1992. This law is unique in that it not only applies to beverage containers, but also to other containerized products.

In addition to the legislation mentioned above, industry or non-profit recycling collection networks exist in most states. These operations vary in sponsor, size, and materials collected. In many states, programs include established drop off centers and promotion campaigns which could act as foundations for statewide collection programs.

It is estimated that more than 600 curbside collection programs are currently operating in communities across the country. Pilot programs to discover the most effective means of collecting the most recyclable material are now common. The participation rate and the amount of waste set out by each household are indicators frequently cited when evaluating the success of curbside collection programs. There are many program variables that influence these evalation criteria. Some potentially significant factors include: collection frequency; whether collection is on the day of MSW pick-up or on a separate day; whether home storage containers are provided and, if so,

Table B-1

HIGHLIGHTS OF STATE RECYCLING LEGISLATION AND PROGRAMS

State	Year	Program Description
Alaska	1983	Alaskans for Litter Prevention and Recycling, statewide litter prevention program and recycling; 10 recycling centers and 2 mobile units.
California	1972	California Waste Management Board, solid waste management agency, and other public and private recycling organizations; 3,672 multi-material recycling centers; special events: "Recycle Week".
	1986	California Beverage Container Recycling and Litter Reduction Act (Recycling Act) targeting 65% redemption of all container types in 1989; requires "convenience zone" redemption centers.
	1986	The AB 2020 legislation requires the Department of Conservation to determine (by region) which materials can be recovered economically. Manufacturers must pay a "processing fee" to ensure a reasonable return to recycles.
Colorado	1983	Recycle Now!, multi-material recycling program; over 400 recycling centers; collected tonnage to date: beverage containers 100,000, newsprint 242,000; special events: "Recycle Month" and "Clean-Up Week".
Connecticut	1980	Bottle bill legislation.
	1987	Legislation set a state-wide goal of 25 percent reduction of solid waste by 1991 (yard waste is included in the goal). The effort is coordinated through a number of established regions. Municipalities are required to recycle twelve materials, including PET and HDPE plastic containers. None of the materials are to be knowingly accepted at any landfill or waste-to-energy facility.

(Cont.)

Table B-1 (cont.)

HIGHLIGHTS OF STATE RECYCLING LEGISLATION AND PROGRAMS

State	Year	Program Description
Florida	1984	Florida Business and Industry Recycling Program, multi-material recycling; 190 recycling centers; collected tonnage 1984 to 1986: 73,900 tons of aluminum, 71,250 tons of glass.
	1988	Senate Bill 1192 provides waste minimization incentives, measures to reduce non-biodegradable material production and increase recycling, an alternative bottle-bill program.
	1988	State Law established the goal to reduce solid waste by 30 percent by 1994. All counties and cities with populations greater than 50,000 must develop recycling programs by July 1, 1989 and to separate a majority of specified materials, including plastic bottles from the waste stream.
	1988	Requires the Department of Environmental Regulation to include any conditions in solid waste facility permits that are necessary to reach the state's goal of 30 percent recycling.
Illinois	1981	Illinois Association of Recycling Centers, multi-material recycling, works with the Department of Energy and Natural resources as well as the Illinois Environmental Protection Agency; 200 recycling centers (including mobile units).
	1986	Office of Illinois Solid Waste and Renewal Resources, technical and financial assistance provide on recycling efforts; 138 recycling centers.
	1988	Requires communities of over 100,000 and the City of Chicago to develop waste management plans that emphasize recycling and alternatives to landfills. Also set a twenty-five percent statewide recycling goal.
Iowa	1979	Bottle bill legislation passed, retailer redemption centers.

(Cont.)

Table B-1 (cont.)

HIGHLIGHTS OF STATE RECYCLING LEGISLATION AND PROGRAMS

State	Year	Program Description
Kansas	1983	Kansas Beverage Industry Recycling Program, multi-material recycling; 16 recycling centers; collected tonnage in 1986: 1770; special events: "Recycle Month".
Kentucky	1980	Kentucky Beverage Industry Recycling Program, multi-material recycling; 35 recycling centers; collected tonnage in 1986: 31,846.
Louisiana	1982	Keep Louisiana Beautiful, Litter control/ recycling/beautification; 155 recycling centers.
Maine	1975	Bottle bill legislation passed, redemption centers, diverts roughly 5.5% of the total waste stream.
	1987	An amendment to the Solid Waste Law an established waste recovery system before issuing permits for incineration or landfill facilities.
	1988	The State of Maine Waste Reduction and Recycling Plan sets a municipal recycling goal of 25% recycling by January 1, 1994. There are thirty existing public recycling programs in the state of Maine. Most recycling programs collect separated materials at a drop-off center. The City of Brunswick offers residents curbside collection and services centers.
Maryland	1984	Maryland Beverage Industry Recycling Program, multi-material recycling and litter control; 130 recycling centers; special events: "Recycle Week".
	1988	Legislation established a statewide mandatory recycling program aiming to recycle 15-20 percent of the county solid waste stream, depending upon the population of the county, in five years.

(Cont.)

Table B-1 (cont.)

HIGHLIGHTS OF STATE RECYCLING LEGISLATION AND PROGRAMS

State	Year	Program Description
Massachusetts	1983	Bottle bill legislation passed, retailer redemption centers.
	1987	The Massachusetts Solid Waste Act established five regional recycling programs to coordinate construction of facilities, material collection and sales, and the distribution of financial incentives. Municipalities must agree to pass mandatory recycling ordinances to receive assistance for recycling costs (public education, technical or equipment costs).
Michigan	1978	Bottle bill legislation passed, retailer redemption centers.
	1986	The Clean Michigan Fund established to lessen the state's dependence on landfills by supporting resource recovery programs and organizations through direct assistance (in the form of grants).
Minnesota	1980	Recycle Minnesota Resources, beverage container and multi-material recycling; 125 recycling centers; collected tonnage in 1986: 6,500.
	1980	The Minnesota Waste Management Act. A 1984 amendment forbids any waste disposal facility supported, directly or indirectly, by public funds to accept "recyclable material" except for transfer to recycler.
Montana	1971	Associated Recycles of Montana, household and industrial multi-material recycling; 50 recycling centers; special events: "Recycle Month".

(Cont.)

Table B-1 (cont.)

HIGHLIGHTS OF STATE RECYCLING LEGISLATION AND PROGRAMS

State	Year	Program Description
Nebraska	1979	Nebraska Litter Reduction and Recycling Programs, grants and technical assistance; 260 recycling centers; collected tonnage in 1986: 665,866.
	1980	Nebraska State Recycling Association, Statewide recycling coalition for promotion and assistance; more than 100 recycling centers including community drop-off centers.
New Hampshire	1983	New Hampshire Beautiful, litter control/litter pickup/public education/recycling grants to municipalities; 11 private and 77 municipal recycling centers; collected tonnage in 1986: 890.
New Jersey	1987	New Jersey Mandatory Source Separation and Recycling Act (statewide voluntary began in June, 1982) requires counties to recycle 15 percent of the previous year's total municipal solid waste in the first full year and 25 percent by the end of two years. Counties have six months to determine three recyclable materials, besides leaves, which are economically recoverable. The Act calls for the establishment of collection (curbside and collection center) and marketing systems and separation ordinances for residents, businesses, and industry; 500 recycling centers; collected tonnage in 1985: 890,000.
	1987	The Recycling Act also requires that the establishment of county waste management planning goals be part of the waste-to-energy facility permit process. No "designated recyclables" are permitted on the tipping floor of such a facility.
North Carolina	1983	Keep North Carolina Clean and Beautiful, Department of Transportation Branch, focuses on litter prevention/reduction, recycling and beautification; 24 local programs throughout the state; special programs: "Clean-Up Week".

(Cont.)

Table B-1 (cont.)

HIGHLIGHTS OF STATE RECYCLING LEGISLATION AND PROGRAMS

State	Year	Program Description
New York	1983	Bottle bill legislation passed, retail redemption centers.
	1988	State Solid Waste Management Act, requires each municipality to implement a source separation plan by Sept 1, 1992, "where economically feasible." The goal is to reduce/reuse/recycle fifty percent by weight of the State's solid waste. The law establishes a number of measures and standards affecting waste producers and processors.
	1988	Applicants for landfill permits must submit analyses of recycling potential and a plan for implementing a recycling program.
Ohio	1980	Ohio Litter Prevention and Recycling, litter prevention/recycling of household items; 28 recycling centers; collected tonnage in 1986: 13,546, special events include "Ohio Recycling Month".
Oklahoma	1982	Oklahoma Beverage Industry Recycling Program, multi-material recycling; 46 recycling centers; collected tonnage in 1986: 29,276, special events: "Recycling Month".
Oregon	1971	Bottle bill legislation passed, retail redemption centers.
	1983	The State Recycling Opportunity Act requires local governments to provide citizens with the opportunity to recycle through curbside collection or drop-off centers. Cities of more than 4,000 must provide at least monthly collection. 106 towns and cities practice curbside separation and five of these collect plastics. Roughly ten depot collection programs are in place in major cities. The Act defines recyclable materials as "any material or group of materials which can be collected and sold for recycling at a net cost equal to or less than the cost of collection and disposal of the same materials."

(Cont.)

Table B-1 (cont.)

HIGHLIGHTS OF STATE RECYCLING LEGISLATION AND PROGRAMS

State	Year	Program Description
Pennsylvania	1974	Recycling and Energy Recovery Section, Bureau of Solid Waste Management, multi-material recycling/energy recovery; 500 non-profit community collection centers, 125-150 scrap dealers; 120 curbside recycling programs; special events: "Recycle Month".
	1983	Pennsylvania Recycling Network, multi-material recycling; 440 recycling centers; 105 curbside collection programs (both municipal and private); collected tonnage in 1986: 110,000.
	1988	Passed legislation requiring communities larger than 10,000 to start recycling programs by September, 1990. Smaller communities have until 1991. The responsibility for solid waste management is shifted from municipalities to counties. A statewide landfill surcharge is used to finance the local recycling collection programs.
Rhode Island	1984	Ocean State Cleanup and Recycling Program, litter control/multi-material recycling; 35 recycling centers.
	1986	A comprehensive recycling law requires each city or town to separate solid waste into recyclable and non-recyclable material prior to disposal in a state-owned facility. Municipalities must divert fifteen percent of their waste stream in within three years. Much of this effort will be focussed on curbside separation programs for which residents will be supplied with a plastic recycling bin. The law also requires all commercial generators and managers of multi-unit housing to submit a plan for recycling and waste reduction.
South Carolina	1987	South Carolina Governor's Task Force on Litter, litter reduction and public recycling awareness, funded by the private sector.

(Cont.)

Table B-1 (cont.)

HIGHLIGHTS OF STATE RECYCLING LEGISLATION AND PROGRAMS

State	Year	Program Description
Texas	1967	Keep Texas Beautiful, Inc., nonprofit educational/coordination organization serving a growing base of community-litter prevention programs; 350 recycling centers; collected tonnage in 1986: 13,827; special events: "Recycle Week".
	1983	Texas Recycles Association, multi-material recycling, education, community relations; at least 300 recycling centers and 4 theme parks.
Vermont	1973	Bottle bill legislation passed, redemption centers, has diverted approximately 6 percent of the state's solid waste stream.
	1987	State Solid Waste Act encourages local recycling collection programs and regional waste management plans by providing technical and financial assistance. The plan stresses reduction, reuse, and recycling. Currently, there are more than 55 collection programs or collection drives operating in the state.
Washington	1962	Committee for Litter Control and Recycling, industry coalition; more than 1,000 recycling centers.
	1970	Washington State Recycling Association, multi-material recycling; approximately 100 recycling members.
	1971	Anti-litter Law established funding, through a special tax, for public education, waste recepticals, and litter policing. A 66 percent litter reduction has resulted. Currently recycles 1,177,400 tons of material equalling 22.4 of the states total waste stream. This is accomplished through curbside and drop off center methods. The State Department of Ecology lists only HDPE bottles as the only plastic material being recycled in significant quantities. It is estimated that 1,700 tons (12.7 percent of HDPE bottles) were recycled in 1987.

(Cont.)

Table B-1 (cont.)

HIGHLIGHTS OF STATE RECYCLING LEGISLATION AND PROGRAMS

State	Year	Program Description
Washington (Cont.)	1971	Solid Waste Management Act established the following solid waste priorities: 1) waste reduction, 2) recycling, 3) energy recovery and incineration, and 4) landfilling. Analysis of the state's waste disposal practices and to decide how to achieve this agenda have been renewed annually.
	1989	Renewed Solid Waste Management Act establishes a statewide goal of 50 percent recylcing of municipal waste by the year 1995. The legislation does not require the use of specific recycling methods in obtaining this goal.
West Virginia	1982	West Virginia Beverage Industry Recycling Program, multi-material recycling/liter control/education; 23 recycling centers; collected tonnage in 1986: 7,335.
Wisconsin	1984	Wisconsin Recycles, recycling awareness program; collected tonnage in 1986: 25,000.
	1984	Bureau of Solid Waste Management, waste reduction and recovery; approximately 600 community recycling programs and 650 companies; special events: "Recycle Week".
	1984	Recycling Act requires municipalities to provide citizens with recycling drop-off centers. Owners and operators of solid waste disposal sites and transfer stations must provide recycling collection centers if none exists. A number of specified materials must be accepted at these centers.

Sources: IEc, 1988; EDF, 1987; EAF, 1989; EPA Journal March/April, 1989; National Softdrink Association, 1988; Vermont, 1989; California, 1988; Maine, 1988; Minnesota, 1988; Washington State, 1988.

the type of containers; the number of curbside separation categories; and mandatory vs. voluntary recycling. Table B-2 provides information on the participation rates, collection quantities, and the program specifics listed above from twenty-two municipal curbside collection programs.

It is difficult to measure the success of a collection program. The accuracy of participation figures is often uncertain, as not all cities have defined tracking systems. Participation rates vary from week to week. Different seasons, for example, and occasional holidays create fluctuations in waste quantities and set-out rates. Most households do not recycle every week. In the communities represented in Table B-2, weekly participation is roughly half the rate of overall participation. This indicates that residents set out recyclable on every other collection period. This holds true regardless of the collection frequency. The ratio of overall to daily participation rates of programs with bi-weekly collection, such as Ann Arbor, Minneapolis, and Montclair, is comparable to that of similar programs with weekly collection schedules.

In an effort to increase participation rates, most programs are now coordinating rubbish collection and recycling pick-up on the same day. The aim is to encourage residents to include recyclable in their existing waste set-out routine. There are, however, successful programs that collect recyclables and rubbish on different days. The six programs in Table B-2 that have different waste and recyclable pick-up days do not reflect lower success rates. Montclair, NJ, one such program, estimates participation at eighty-five percent and collection amounts at 686 pounds per household per year. Same day waste and recycling collection may be more efficient for collection workers and residents, but it is not vital to obtaining high collection rates.

Making recycling convenient to participants appears to maximize resident participation. Supplying households with storage containers for recyclables is one common means of making home storage and sorting as effortless as possible. Distributing recognizable containers that appear at the curbside on collection day also helps promote recycling programs. There are many different container types currently available. While there are some advantages to supplying containers, cities that do not provide containers appear to have participation rates only slightly lower than those that do.

Before they can be sold as raw materials, recyclable must be separated from the waste stream by either residents, collection crews, or processing facilities. It is most efficient to separate waste prior to set-out, rather than commingling and then separating. It has been argued that requiring this extra effort of residents may decrease participation rates. The examples in Table B-2 support this contention, showing higher participation rates for programs accepting commingled waste. Although it requires more extensive processing methods, many communities are now accepting commingled waste or waste separated into very few categories.

As mentioned in the previous section, a few states have developed waste management legislation that mandates curbside recycling programs. While these laws help establish programs, it is unclear to what degree they raise household participation rates above those of voluntary programs. Quantities of recycled materials also are not noticeably higher for mandatory programs than for voluntary programs. Participation rates are on average 10-20 percent higher for mandatory programs, but a number of very successful voluntary programs do exist (e.g., Montclair, New Jersey). It should be considered that programs in communities or states mandating curbside recycling require properly organized, promoted, and implemented programs. These factors may be responsible for elevated success rates, and not simply the fact that the program is mandatory.

Table B-2

A SUMMARY OF CURBSIDE RECYCLING PROGRAM CHARACTERISTICS AND PARTICIPATION RATES

Community	Population	Households Served	Tons/Year	Pounds/House/Year	Partici-pation(b)	Participation Overall %	Participation Collection Day %	Collection Frequency	Materials Collected(a)	Same Day As Trash	Provide Home Storage Containers	Household Set-Out Requirement
Austin, TX	450,000	90,000	7,200	160.00	V	20-25	10-12	Weekly	N,G,T,A	Mostly	20,000	Separate
Cheltenham, PA	35,500	9,500	N.Avail.	N.Avail.	V	40	30-35	Weekly	N,G,A	Yes	Yes	Separate
Davis, CA	47,000	11,000	3,200	581.82	V	60	50	Weekly	N,G,C,A	Yes	No	Separate
East Lyme, CT	N.Avail.	5,000	2,100	840.00	M	80	N.Avail.	Weekly	N,A,G,T	Yes	N.Avail.	Separate
Evesham Twp., NJ	36,000	8,500	2,995	704.71	M	85-90	50	Weekly	G,T,A,MP	No	Yes	Commingled
Groton, CT	10,000	1,900	626	658.95	M	75-85	50	Weekly	N,C,G,T,A	No	No	Commingled
Haddonfield, NJ	12,500	4,400	1,703	774.09	M	95	66	Weekly	N,G,T,A	Yes	Yes	Commingled
Hamburg, NY	10,500	3,350	N.Avail.	N.Avail.	M	98	N.Avail.	Weekly	N,C,G,T,MO	Yes	No	Commingled
Marin Co., CA	N.Avail.	44,000	12,500	568.18	V	60	35-40	Weekly	N,G,T,A,P	Yes	Yes	Separate
Mississauga, ONT.	400,000	90,000	14,000	311.11	V	80	40	Weekly	N,G,T,A	Yes	Yes	Commingled
Mecklenburg Co., NC	460,000	9,100	2,336	513.41	V	71	37	Weekly	N,G,A,P	Yes	Yes	Commingled
Niagara Falls, ONT.	70,000	19,500	2,307	236.62	V	75-80	45	Weekly	N,G,T,A,P	Yes	Yes	Commingled
Plymouth, MN	43,000	12,500	2,800	448.00	V	N.Avail.	53-56	Weekly	N,C,G,T,A	No	Yes	Separate
San Jose, CA	720,000	180,000	6,500	72.22	V	58	25	Weekly	N,G,T,A,P	Yes	Yes	Separate

(cont.)

Table B-2 (cont.)

A SUMMARY OF CURBSIDE RECYCLING PROGRAM CHARACTERISTICS AND PARTICIPATION RATES

Community	Population	Households Served	Tons/Year	Tons/House/Year	Partici-pation(b)	Participation Overall %	Participation Collection Day %	Collection Frequency	Materials Collected(a)	Same Day As Trash	Provide Home Storage Containers	Household Set-Out Requirement
Seattle, WA	500,000	94,000	23,985	510.32	V	48	29	Weekly	G,T,A,MP	Yes	No	Commingled
Springfield Twp., PA	22,000	6,800	1,972	580.00	V	70	60	Weekly	N,G,A	Yes	Yes	Commingled
Sunnyvale, CA	116,000	28,000	4,078	291.29	V	50-60	21	Weekly	N,G,T,A,P,MO	Yes	Yes	Separate
Upper Moreland Twp., PA	28,000	6,200	N.Avail.	N.Avail.	M	62	50	Weekly	N,G	Yes	Yes	Separate
Ann Arbor, MI	108,000	20,000	1,700	170.00	V	50	25	2/month	N,C,G,T,A,MO	Yes	No	Separate
Minneapolis, MN	360,000	120,000	7,600	126.67	V	25-30	15	2/month	G,T,A,MP	No	No	Separate
Montclair, NJ	38,500	14,500	4,980	686.90	M	85	50	2/month	N,G,A	No	No	Commingled
St. Cloud, MN	44,000	9,770	403	82.50	M	N.Avail.	30	Monthly	N,G,A	No	No	Separate

(a) Materials: A = aluminum, G = glass, P = plastic, N = newspaper, C = corrugated cardboard, MO = motor oil, MP = mixed paper

(b) V = voluntary, M = mandatory

Sources: Biocycle, May/June, 1989.

REFERENCES

California, State of. 1988. Annual Report of the Department of Conservation, Division of Recycling. Sacramento, CA.

EAF. 1989. Environmental Action Foundation. Mandatory Statewide Recycling Laws. Feb 1989. Washington, DC.

EDF. 1988. Environmental Defense Fund. Coming Full Circle: Successful Recycling Today. New York, NY.

EPA Journal. 1989. Five situation pieces. 15(2):35-40. March/April. U.S. Environmental Protection Agency. Washington, DC.

IEc. 1988. Industrial Economics Inc. Plastics Recycling: Incentives, Barriers and Government Roles. Prepared for Water Economics Branch, Office of Policy Analysis, U.S. EPA. Industrial Economics Incorporated. Cambridge, MA. 152 pp.

Maine DECD. 1988. Maine Department of Economic and Community Development. State of Maine Waste Reduction and Recycling Plan. Augusta, ME.

Minnesota. 1988. Research Provided to the Members of the Governor's Select Committee on Recycling and the Environment. State of Minnesota. St. Paul, MN. Oct 1988.

National Softdrink Association. 1988. Promoting Recycling to the Public.

Vermont. 1989. Solid Waste Management Plan. Vermont Agency of Natural Resources. Waterbury, VT. Feb 1.

Washington State. 1988. Best Management Practices Analysis for Solid Waste: 1987 Recycling and Waste Stream Survey, Vol. 1. Prepared by Matrix Management Group for the Washington State Department of Ecology Office of Waste Reduction and Recycling. Olympia, WA.

Appendix C—Substitutes for Lead and Cadmium Additives for Plastics

C.1. INTRODUCTION

There is a wide range of cadmium- and lead-based products that are used in a variety of plastic applications. The selection of a particular lead or cadmium pigment and/or heat stabilizer is dependent on processing requirements, resin characteristics, and end-product uses. Identification of substitutes is, therefore, a complicated and application-specific task. There are, however, some general classes of substitutes available for each cadmium and lead-based additive type. This does not indicate, however, that there is a substitute for every application. The following sections describe the performance, cost, feasibility, and other considerations that dictate how substitutes can replace traditional lead and cadmium products. The toxicity of potential substitutes is not considered in this report. Evaluation of toxicity is important because it varies by substitute. To fully characterize substitutes, however, the toxicity of individual substitutes would have to be considered.

C.2. SUBSTITUTE COLORANTS AND THEIR PROPERTIES

Although there is some agreement on the performance characteristics (i.e., chemical compatibility, light and heat fastness, hue, and intensity) which are of concern to most consumers (plastics manufacturers), the selection of a substitute will depend in large part upon the individual consumer's ranking of the importance of these attributes. For example, a manufacturer of beach balls may worry mostly about hue and intensity and not be concerned about heat fastness. On the other hand, a manufacturer of high-performance

358

automotive polymers may find a less vibrant color acceptable if the substitute is heat-resistant and performs suitably in the other aspects.

Particular colorant-polymer combinations may also be ruled out for a variety of other reasons. The first problem that might be encountered is a chemical incompatibility between the colorant and the polymer solvent, resin, or manufacturing by-products. For example, some pigments may not be used in PVC because of their sensitivity to acid (Kirk Othmer 1983). In addition to a chemical resistance problem, the colorant may not be able to survive harsh processing conditions for certain polymer resins. For example, although the end use of a plastic polymer may not require high temperatures, extensive heating during processing is often required to melt and mold a plastic. In addition to these obstacles, a colorant-polymer combination may be ruled out because of the end use of the plastic product. A combination that is sufficiently lightfast and suitable for indoor use may be ruled out for exterior applications. Table C-1 provides the performance properties both for lead- and cadmium-based pigments and for their potential substitutes. The specific processing conditions and color requirements for polymer applications determine the substitutes that may be appropriate for replacing each cadmium- or lead-containing pigment.

For purposes of this analysis, substitutes for lead and cadmium pigments are chosen based only on the characteristic of having a similar hue as reported in the Plastic Additives Handbook (1987). In specific applications, many factors influence pigment choice, including lightfastness and heat stability, but these considerations have not been evaluated for this analysis. Although not all substitutes work in all applications, some suitable substitute usually can be found for individual applications. One exception to this general availability of substitutes is in the area of pigments for high

Table C-1. Comparison of Performance Properties: Lead and Cadmium Pigments and Their Potential Substitutes

	Hues	Chemical Formula	Heat Stability[a] (°C)	Lightfastness[b]	Chemical Resistance	Opacity/ Hiding Power	Remarks on Performance (Compatible Polymer/ Resin)
Lead/Cadmium Pigment							
Lead:		Pb					
Chromate	Orange	PbCrO$_4$	230-250	6-8	Sensitive to acids/bases	Opaque	PVC, LDPE +; HDPE, PS
Sulfate	Yellow-Orange	PbSO$_4$	230-250	6-8	Sensitive to acids/bases		PVC, LDPE +; HDPE, PS
Molybdate	Reddish Orange	PbMoO$_4$	220-250	6-8	Sensitive to acids/bases		PVC, LDPE +; HDPE, PS
Chromate + Iron Blue	Greenish Yellow to Medium Shades of Olive Green	---	-	-	-		
Cadmium:		Cd				Opaque	
Sulfide	Orange shade of Yellow	CdS	300	8	Sensitive to acids		General suitability
Sulfide + Zinc	Greenish Yellow	Cd + Zn	-	-	-		
Sulfide + Selenium	Red and Maroon	CdS + Se	300	8	Sensitive to acids		General suitability
Sulfide + Mercury	Reddish Orange to Bluish Red	CdS + Hg	-	-	-		
Substitute Colorant							
Inorganic							
Nickel titanium	Yellow	NiTiO$_2$	300	8	--	Opaque	General suitability, but greatly reduced tinting strength
Iron Oxide	Red	Fe$_2$O$_3$	300	8	--	Opaque	General suitability H-PVC =
Organic							
Monoazo	Yellow	--	260	7-8	--	Transparent	LDPE, PS +; PVC -
Monoazo naphthol	Red	--	280	5-7	--	Transparent	PVC, PS, LDPE +; HDPE (++)
Quinacridone	Red	--	240-280	7-8	--	Transparent	PS, PVC, LDPE +; HDPE (++)
Perylene	Red	--	220-300	7-8	--	Transparent	PS, PVC, LDPE +; HDPE (++)

Table C-1 (Continued)

Hues	Chemical formula	Heat Stability[a] (°C)	Lightfastness[b]	Chemical Resistance	Opacity/ Hiding Power	Remarks on Performance (Compatible Polymer/ Resin)
Dyes						
Pyrazolone derivative Yellow	--	300[d]	7-8[d]	Poor	--	PMMA, N-PVC, PS +
Azo Dye Red	--	260[d]	2-5[d]	Poor	--	PMMA, N-PVC, PS +

[a] For heat stability, the temperature is stated at which no coloristic changes occurred during normal dwell times (approximately 5 minutes) in processing machines.

[b] Determination of lightfastness is carried out in accordance with DIN 53 389. 1 is the lowest value; 4 has 8 times the fastness level of 1; the highest value is 8, testing not normally being carried out beyond this level. See Plastic Additives Handbook (1987) for more information.

[c] Polymer Codes: PS = polystyrene; LDPE = low density polyethylene; PMMA = polymethyl methacrylate; HDPE = high density polyethylene; PVC = polyvinyl chloride; N-PVC = Unplasticized PVC.

Polymer Performance: + denotes suitable/recommended; ■ denotes limited suitability/recommended; - denotes not suitable/recommended; ( ) denotes a qualification of the statement; * denotes caution is needed in the case of HDPE articles sensitive to distortion.

[d] for dyes, lightfastness and heat stability depends to an especially high degree on the plastic to be colored.

Source: Plastic Additives Handbook 1987, Brannon 1988.

temperature resins. In these areas, the vibrant colors provided by cadmium pigments are difficult to replace because even though substitutes may match in hue, they cannot withstand the high temperature of the processing and use environments in which they would be required.

Because application-specific considerations are often critical for establishing exact substitution patterns, lead- and cadmium-based pigments are grouped together (i.e., substitutes are considered for the color yellow; not chrome yellow (lead) and cadmium yellow). A substitute for each major hue (red and yellow) was chosen from the chemical families that are comprised of inorganic and organic pigments and dyes. In addition, two organic substitutes, quinacridone and perylene pigments, were chosen based on information from Mobay (1989) that reported they could be used with high performance polymer systems (e.g., nylon, polyesters). Table C-2 identifies the substitutes that can replace specific lead and cadmium pigments based on color considerations.

C.2.1. Costs of Lead- and Cadmium-Based Pigments and Their Substitutes

For most of the lead-based and cadmium-based colorants, many acceptable potential substitutes are available, although as mentioned before, the costs of the substitutes may be significantly higher. The price of cadmium has increased as other markets, such as that for nickel-cadmium batteries, increase the demand for cadmium. On the other hand, lead pigments have remained inexpensive. Table C-3 shows recent prices for the most common lead- and cadmium-based pigments, as well as prices for some pigments (mostly organic) that could be used as substitutes in various applications.

According to one industry contact, as a rule of thumb, the performance of organic pigments increases as the price does (Hoechst-Celanese 1989). Certain pigments may not perform as well as less expensive organics in a few measures

Table C-2. Substitute Products That Replace Lead- and
Cadmium-Based Products by Color

Substitute Colorant	Hue	Possible Substitute For[a]
Inorganic		
Nickel titanium	Yellow	Lead Chromate; Cadmium Sulfide Cadmium sulfide + zinc
Iron Oxide	Red	Lead Molybdate; Cadmium/sulfide selenide
Organic		
Monoazo	Yellow	Lead Chromate; Cadmium Sulfide Cadmium/sulfide selenide
Monoazo naphthol	Red	Lead Molybdate; Cadmium/sulfide selenide
Quinacridone	Red	Lead Molybdate; Cadmium/sulfide selenide
Perylene	Red	Lead Molybdate; Cadmium/sulfide selenide
Dyes		
Pyrazolone derivative	Yellow	Lead Chromate; Cadmium Sulfide Cadmium sulfide + zinc
Azo Dye	Red	Lead Molybdate; Cadmium/sulfide selenide

[a] Possible substitutes are based primarily on colorants reported in the
Plastic Additives Handbook (1987).

Sources: Plastic Additives Handbook 1987, Brannon 1988.

Table C-3.   Costs: Lead and Cadmium-Based Pigments
and Their Substitutes

Hue		Cost ($/lb.)[b]
Yellow	**Lead/Cadmium Pigments**	
	Lead Chromate	$ 1.55
	Cadmium Sulfide	$14.60
	Cadmium Sulfide + Zinc	$14.60
	Substitute Colorants[a]	
	Nickel Titanium (Inorganic)	$ 3.50
	Monoazo (Organic)	$17.95
	Pyrazolone Derivative (Dye)	$20.32
Red	**Lead/Cadmium Pigments**	
	Lead Molybdate	$ 2.25
	Cadmium/Sulfide Selenide	$18.15
	Substitute Colorants[a]	
	Iron Oxide (Inorganic)	$ 0.79
	Monoazo naphthol (Organic)	$24.25
	Quinacridone (Organic)	$32.00
	Perylene (Organic)	$41.20
	Azo Dye (Dye)	$10.30

[a] The substitute colorants listed are based on
having a similar hue (i.e., red or yellow) as
reported in the Plastic Additives Handbook (1987).
In addition to cost, specific selection of a substitute is dependent on a diverse set of performance properties. For a comparison of these
properties, see Table C-1.

[b] Costs were determined by contacting chemical
companies and requesting prices on given pigments
(from Plastic Additives Handbook). The costs are
based on the most commonly used (standard)
packaging sizes (40-60 lb. containers) reported by
the chemical companies.

Sources: Plastic Additives Handbook 1987,
Bayer-Mobay 1989, BASF 1989.

of performance, but the quality must be higher in at least one aspect of performance or there would be no demand for the product. For example, prices generally increase as the acceptable processing temperatures rise. This consideration, combined with the smaller production scales of specialty and high-performance plastics help maintain the price differential.

In addition, the amount of product used for an application is dependent mainly on the shade and brilliance required. For example, if a faint yellow is required, less lead chromate is used than would be required for a darker yellow. In general, for similar hues, 25 percent less organic pigment is required compared to an inorganic pigment (Hoechst-Celanese 1989). Estimates are not available as to the amount of dye required for coloring.

C.2.2. Other Factors Affecting Selection of Substitutes and Substitute Costs

If lead and cadmium pigments were not available, chemical companies may be more willing to invest in the research and development of substitutes, because there would be less low-cost competition for any new substitutes developed. From Table C-1, it is evident that there are many potential substitutes available for lead and cadmium pigments; however, it may be difficult to find adequate replacements for some pigments (such as some of the very high-performance cadmiums used in nylons) at any cost. In those cases, it may be necessary for the plastic manufacturer to sacrifice one aspect of performance, such as hue or brilliance, in exchange for another, such as heat or lightfastness.

Several of the companies contacted (Harshaw 1989, Heubach 1989) are still committed to the manufacturing and/or distribution of lead-based pigments, but the majority of the companies contacted have ceased to supply them. The same cannot be said for some of the cadmium-containing colorants; their performance characteristics are more difficult to produce using either organic or other inorganic pigments. Although there are organic pigments that match cadmium pigments in brilliance, hue, and lightfastness, they generally are not adequate in high-temperature situations. On the other hand, although many of the inorganic compounds are quite heat resistant, they are tinctorially weak or colorless (white).

The toxicity of heavy metals is well known, and therefore many manufacturers have shied away from using compounds containing metals such as lead and cadmium. For example, General Electric stopped using lead pigments 10-12 years ago and ceased production of cadmium pigments at the beginning of 1989 (General Electric 1989a, General Electric 1989b).

Research by many other manufacturers continues in order to find substitutes that are compatible with high-performance engineering resins such as polycarbonate, nylon, and other polyesters (Modern Plastics 1987). Combinations of substances are sometimes used to complement each other and improve the overall qualities of products. For example, in an application requiring some hiding power and resistance to heat, a mix of a high performance organic colorant and an inorganic compound may work. The inorganic pigment lends its hiding power while the organic pigment provides intense color. In some instances the inorganic compound may also provide some color, thereby reducing the need for large quantities of organic colorant, which would be required if the inorganic pigment were colorless (white) like titanium dioxide.

Although dyes can often be used in place of pigments, they currently do not hold a large share of the market for colorants. Because dyes must be soluble in the resins they are used in, each variety of dye may be compatible with only a few types of resins. In addition, the lead- and cadmium-based pigments are generally applied in a solid form, either as color concentrates or dry powders. Although dyes are also found in solid form, they are usually used as liquids which would require some changes in the plant equipment and result in additional costs to the plastic manufacturer.

C.3. SUBSTITUTE STABILIZERS AND THEIR PROPERTIES

Substitute products that can replace lead- and cadmium-containing heat stabilizer products have been developed and continue to be investigated for several reasons. The toxicity of lead and cadmium compounds, availability of an increasing number of technically superior alternate products, the lower costs of substitutes, and increasing costs associated with using lead and especially cadmium products have all been influencing factors. This section addresses substitute products, compares costs to existing lead and cadmium products when possible, and discusses important factors that affect substitution.

C.3.1. Substitutes for Lead-Containing Heat Stabilizers

The majority of lead-based heat stabilizers have been replaced in applications where substitution is possible. For example, organotin stabilizers (e.g., alkyltin mercaptides, alkyltin carboxylates, and estertin mercaptides) barium/zinc, and metal-free stabilizers can be used in rigid applications, pipes and fittings, pigmented profiles,[1] foamed profiles, and records that previously used lead-based products (Argus 1989c). Table C-4 identifies PVC articles and stabilizer systems that can replace lead-based products in rigid PVC applications (Plastics Engineering Handbook 1976).[2]

---

[1] Profiles are typified by such products as channels, gaskets, decorative trim, siding panels, window frames, and other rigid structures used in indoor and outdoor construction and building.

[2] It is important to note that the toxicity of these substitutes has not yet been evaluated.

Table C-4. Potential Substitutes[a] for Lead-Based Heat Stabilizers
in Rigid PVC Products

PVC Item	Lead Stabilizers	Methyl, Butyl and/or Octyltin Mercaptides	Butyltin Esters	Metal-Free Stabilizers
Pipes and Fittings	X	X		
Pigmented Sheets and Profiles	X	X	X	
Foamed Profiles	X	X		
Phonograph Records	X			X

[a] Ba/Cd stabilizers are not considered to be substitutes for lead stabilizers due to toxicity considerations and the requirements of this analysis (i.e.. lead and cadmium products are both under investigation and therefore, are not considered to be substitutes for one another), although they may be technically and economically feasible in some applications.

[b] Phonograph records may be stabilized with certain metal-free stabilizers (see discussion on page C-15).

Source: Plastics Additives Handbook 1987.

Lead stabilizers also have been used in flexible PVC applications. These applications can be split into those for which cost-effective and reliable substitutes have been developed (shoes, sandals, and soles), and those for which substitution has lagged (electrical insulation and jacketing). Electrical insulation applications use the majority of lead-based stabilizers because of the critical non-conducting nature of lead products (Vinyl Institute 1989b).

The use of electrical cable insulation and jacketing can be divided among three major use areas:

- power wiring,
- telephone cable, and
- cords and connectors for appliances and other consumer items.

The critical properties of weathering, humidity resistance, and thickness of the jacket in the power wiring and telephone cable applications have made substitution difficult given that lead imparts these properties. Lead PVC heat stabilizers for these applications provide outstanding use characteristics and there are currently no products available on the market that can replace these lead-based products (BF Goodrich 1989).

The power and telephone cable uses account for about 50 percent of lead stabilizer usage for jacketing and insulation, while the cord/connector applications account for the remaining 50 percent. It is believed that these lower performance cord/connector applications can be replaced with alternate stabilizers (Argus 1989c, BF Goodrich 1989). These reformulated products were, however, too experimental or could not be identified at this time. Table C-5 identifies the substitutes for lead-based stabilizers used in flexible PVC applications.

Table C-5. Potential Substitutes[a] for Lead-Based Heat Stabilizers
in Flexible PVC Products

PVC Item	Lead Stabilizers	Ba/Zn Stabilizers	Butyltin Esters	Teflon[b]
Cable Insulation and Jacketing	X			X
Shoes, Sandals, Soles	X	X	X	

[a] Ba/Cd stabilizers are not considered to be substitutes for lead stabilizers due to toxicity considerations and the requirements of this analysis (i.e., lead and cadmium products are both under investigation and therefore, are not considered to be substitutes for one another), although they may be technically and economically feasible in some applications.

[b] Teflon is a technically feasible substitute for PVC coatings, but is not a one-for-one substitute in that heat stabilizers are not replaced, but rather reformulation is required (i.e., teflon replaces PVC). The use of teflon has not been examined by the industry as a viable replacement for cable insulation and is expected to be on the order of five to ten times more expensive (Bedford Chemical 1989). Teflon also may not possess sufficient flexibility for many applications (Vinyl Institute 1989b).

Sources: Plastic Additives Handbook 1987, Bedford Chemical 1989, Argus 1989c.

C.3.2. Substitutes for Cadmium-Containing Heat Stabilizers

Barium/Cadmium heat stabilizers have a wide range of applicability in rigid and flexible PVC applications. The products and substitutes are identified in Tables C-6 and C-7. There is some overlap with lead-based stabilizers, but these are not considered to be substitutes for cadmium-containing products and vice-versa. The processing techniques (e.g., calendaring, extrusion, injection molding, blow molding, pressing, coating), the processing conditions (temperature, mixing, alkalinity) and a host of other reasons including the presence of other additives, the end-use of the product (indoor/outdoor), the costs of substitutes, and the toxicity of substitutes influence which products are ultimately considered to be substitutes. In general, there are potential substitutes for cadmium-containing stabilizers, including barium/zinc, calcium/zinc, and tin-based stabilizers.

C.3.3. Costs of Lead- and Cadmium-Based Heat Stabilizers and Their Substitutes

Lead and cadmium heat stabilizers have seen widespread use because of their relatively low cost compared to newer substitute products. As concern has mounted regarding the toxicity of lead and cadmium products and substitute products have been perfected, costs have declined (Argus 1989b). Table C-3 presents the relative costs of lead, cadmium, and substitute products.

It must be noted that the actual substitution pattern for lead and cadmium stabilizers is very complicated. The wide range of applications, processing considerations, and other factors that affect substitute development and entry into the market make it difficult to distinguish exact substitution patterns. The costs presented in Table C-8 should be considered, therefore, rough approximations.

Table C-6.   Potential Substitutes[a] for Cadmium-Based Heat Stabilizers
in Rigid PVC Products

PVC Item	Ba/Cd Powders	Butyltin Mercaptides	Butyltin Esters	Barium/ Zinc Solids
Films for Non-Food Applications	X	X		X
Pigmented Profiles				
-- Indoor	X	X		X
-- Outdoor	X		X	X
Foamed Profiles	X	X		X

[a] Lead stabilizers are not considered to be substitutes for Ba/Cd stabilizers
due to toxicity considerations and the requirements of this analysis (i.e.,
lead and cadmium products are both under investigation and therefore, are not
considered to be substitutes for one another), although they may be
economically and technically feasible in some applications.

Source: Plastic Additives Handbook 1987.

Table C-7. Potential Substitutes[a] for Cadmium-Based Heat Stabilizers
in Flexible PVC Products

PVC Item	Ba/Cd Powders	Ba/Cd[b] Liquids	Butyltin Mercaptides or Esters	Ba/Zn Stabilizers	Ca/Zn Stabilizers
Films for Non-Food Applications	X	X	X	X	X
Profiles and Flexible Tubes for Non-Food Applications	X	X		X	X
Shoes, Sandals and Soles	X	X		X	
Artificial Leather Coatings		X		X	X
Dippings		X		X	X

Note: Zn - zinc; Ca - Calcium

[a] Lead stabilizers are not considered to be substitutes for Ba/Cd stabilizers
due to toxicity considerations and the requirements of this analysis (i.e.,
lead and cadmium products are both under investigation and therefore, are not
considered to be substitutes for one another), although they may be
economically and technically feasible in some applications.

[b] The relative amount of cadmium in liquid stabilizers can be reduced by the
addition of zinc fatty acid salts that replace the corresponding cadmium salts
(Modern Plastics 1987).

Source: Plastic Additives Handbook 1987, Vinyl Institute 1989b.

Table C-8.   Costs of Lead, Cadmium, and Potential Substitute Heat Stabilizers

Stabilizer Class	Approximate Cost Range ($/lb.)	Amount Used Per Hundred Parts of Resin
Lead		
Lead Compounds	0.50-1.00	1.0-3.0
Organo-tin Stabilizers	1.00-3.00	1.5-2.5
Barium/Zinc Stabilizers	2.00-4.00	1.5-3.0
Teflon	*	Not Applicable
Cadmium		
Barium/Cadmium Liquids	0.95-1.75	1.0-4.0
Barium/Cadmium Solids	1.85-2.75	1.5-3.0
Barium/Cadmium/Zinc Products	0.95-1.70	1.0-3.0
Zinc/Calcium Products	1.00-3.00	1.0-3.0
Liquid Organo-tin Stabilizers	3.00-4.50	2.0-4.0
Solid Organo-tin Stabilizers	8.00-10.00	1.5-3.0
Barium/Zinc Liquids	1.25-2.50	1.0-4.0
Barium/Zinc Solids	2.00-4.00	1.5-3.0

* Teflon is a different class of substitute in that it would replace the end-product, PVC coatings, used for wire and cable insulation. It is not currently considered a stable substitute for economic reasons. The cost of PVC coatings is roughly $0.50 to $1.00/lb. and for teflon >$5/lb.

Sources: Argus 1989a, 1989b, 1989c; Bedford Chemical 1989.

C.3.4. Other Factors Affecting Selection of Substitutes and Substitute Costs

There are a number of considerations that must be included in the selection of stabilizer substitutes and estimation of costs for comparison to costs for lead and cadmium stabilizers. It has not been possible to characterize each of these considerations, but they are provided for completeness:

- Substitute stabilizer packages that can replace lead or cadmium products may be required in quantities greater or lesser than the products they replace. They may be cheaper or more expensive, at the concentration level required, or they may be viable only for some applications.

- The addition of co-stabilizing products may reduce substitute costs, improve performance to a level above that of the lead or cadmium product, or increase product service life.

- It may be possible to combine stabilizers so that a synergistic effect is achieved, thereby improving performance and/or reducing costs.

- New substitutes are constantly being developed and made available. Some of these, based on antimony and metal-free stabilizer systems (e.g., diphenylthioureas, and ß-aminocrotonates) have not been widely accepted, but may influence the stabilizer market over the next few years.

REFERENCES

Argus. M. Croce. 1989a (May 10). New York, NY. Transcribed telephone conversation with Mark Wagner, ICF Incorporated, Fairfax, VA.

Argus. D. Stimpfl. 1989b (May 11). New York, NY. Transcribed telephone conversation with Mark Wagner, ICF Incorporated, Fairfax, VA.

Argus. D. Brilliant. 1989 (May 17). New York, NY. Transcribed telephone conversation with Mark Wagner, ICF Incorporated, Fairfax, VA.

BASF-Basic Organics Group. 1989 (May 16). Transcribed telephone conversation with Tanya Yudleman, ICF Incorporated, Fairfax, VA.

Bayer-Mobay Corporation. 1989 (May 16). Transcribed telephone conversation with Tanya Yudleman, ICF Incorporated, Fairfax, VA.

Bedford Chemical. D. Gauw. 1989 (May 4). Bedford Chemical, Division of Ferro Corporation. Bedford, OH. Transcribed telephone conversation with Mark Wagner, ICF Incorporated, Fairfax, VA.

Brannon, SM. 1988 (February 1-5). 43rd Annual Conference, Composites Institute, the Society of the Plastics Industry. Colorants for Composites -- A Review.

General Electric Color Lab. D. Bryant. 1989a (May 4). Technician, General Electric Color Lab, Mt. Vernon, IN. Transcribed telephone conversation with Peter Weisberg, ICF Incorporated, Fairfax, VA.

General Electric Color Lab. D. Bryant. 1989b (December 19). Technician, General Electric Color Lab, Mt. Vernon, IN. Transcribed telephone conversation with Thomas Hok, ICF Incorporated, Fairfax, VA.

B.F. Goodrich. G. Lefebvre. 1989 (May 17). Cleveland, OH. Transcribed telephone conversation with Mark Wagner, ICF Incorporated, Fairfax, VA.

Harshaw Colors. M. DiLorenzo. 1989 (May 12). Division of Eagelhard Corp. Transcribed telephone conversation with Don Yee, ICF Incorporated, Fairfax, VA.

Heubach, Inc. 1989 (May 12). Transcribed telephone conversation with Don Yee, ICF Incorporated, Fairfax, VA.

Hoechst-Celanese. D. wave. 1989 (May 9). Transcribed telephone conversation with Don Yee, ICF Incorporated, Fairfax, VA.

Kirk-Othmer. 1983. Encyclopedia of Chemical Technology. John Wiley and Sons Publishing Co., Inc. Vol. 18, pp. 184-206; Vol. 14, pp. 168-183; Vol. 6.

Mobay. J. Graff. 1989 (May 3). Transcribed telephone conversation with Don Yee, ICF Incorporated, Fairfax, VA.

Modern Plastics. 1987 (September). Colorants -- pp. 68-71; Heat Stabilizers -- pp. 70-71. McGraw-Hill Publishing Co.

Plastics Engineering Handbook. 1976. 4th Edition. Van Nostrand Reinhold Publishing Company.

Plastic Additives Handbook. 1987. Hanser Publishers. Munich, Germany.

Vinyl Institute. R. Gottesman. 1989a (May 15). Little Falls, NJ. Transcribed telephone conversation with Mark Wagner, ICF Incorporated, Fairfax, VA.

Vinyl Institute. R. Gottesman. 1989b (July 11). Little Falls, NJ. Comments received on the Draft Plastics Section.

Part II

Recycling in the Industrial Sector

The information in Part II is from *Plastics Recycling in the Industrial Sector: An Assessment of the Opportunities and Constraints,* prepared by T. Randall Curlee and Sujit Das of Oak Ridge National Laboratory for the U.S. Department of Energy, November 1989.

Executive Summary

This report addresses numerous issues related to plastics recycling in the industrial sector: manufacturing and post-consumer plastic waste projections, the estimated energy content of plastic wastes, the costs of available recycling processes, institutional changes that promote additional recycling, legislative and regulatory trends, the potential quantities of plastics that could be diverted from the municipal waste stream and recycled in the industrial sector, and the perspectives of current firms in the plastics recycling business.

Post-consumer wastes are projected to increase from 35.8 billion pounds in 1990 to 48.7 billion pounds in 2000. Plastic packaging is expected to account for about 46% of all post-consumer plastics during the coming decade and remain the by-far largest single contributor to the waste stream. The product category with the largest projected percentage increase is the building and construction sector, which is projected to increase from 2.0% of the total in 1990 to 8.0% in 2000. Manufacturing nuisance plastics are projected to account for about 4% of total plastic wastes.

The total energy inputs required to manufacture the resins that are projected to appear in the solid waste stream in 1990 are estimated to exceed $1,378 \times 10^{12}$ BTUs. If all manufacturing nuisance plastics and post-consumer plastics are incinerated, the product heat of combustion from those plastics is projected to be about 644×10^{12} BTUs in 1990. The estimate for total energy inputs is equivalent to about 1.8% of what total U.S. energy consumption was in 1985. The estimate for product heat of combustion is equivalent to about 0.8% of the 1985 estimate for total energy consumption.

Recent estimates of the costs of recycling plastic wastes suggest that data are not sufficiently detailed or validated to draw definitive conclusions. Given the caveats, two recently introduced secondary processes appear to be economically viable if the appropriate materials are available. Although the costs of secondary technologies appear favorable, no good information is available to suggest the potential size of the market for the lumber-like materials being produced. Neither is there information about how the prices of lumber substitutes may decrease as the production of those products increases.

Significant changes have occurred recently in terms of the institutional structures and regulations that impact on plastics recycling. In the majority of cases, these changes promote additional recycling. For example, mandatory moves to curbside collection of plastics and subsidies for the development and use of new recycling technologies have made plastics recycling more economically attractive. In some cases, however, the institutional and regulatory trends have

380

questionable implications for plastics recycling and are reflective of the continuing uncertainty about plastics in general.

This report presents several scenarios in which specific technical, economic, institutional, and regulatory conditions are assumed. Given the most probable scenario, the quantity of plastics available for recycling will grow tremendously. If curbside collection of segregated household plastics becomes the norm and if economically viable technologies are available to recycle commingled plastics containing 50% contaminants, most manufacturing nuisance plastics will become recyclable. In addition, about 50% of the post-consumer plastic waste stream -- i.e., about 24 million pounds in 2000 -- will be a candidate for recycling.

This assessment concludes that there are significant opportunities for the industrial sector to recycle plastic wastes during the coming decade. Most technical, economic, institutional, and regulatory trends point toward more possibilities and greater incentives for additional recycling activities. There are, however, potential barriers. Technologies that can accommodate dirty commingled plastics require further development, especially tertiary technologies that could produce high-valued pre-polymers. Separation technologies also require further development if some of the larger waste streams are to be recycled in a secondary or tertiary sense. Market constraints may limit significantly the potential for secondary products, especially those products that compete with wood or concrete. Finally, regulatory programs, which will impact on the collection of plastics and the economic viability of recycling operations, are currently being developed at the federal, state, and local levels. Unfortunately, the specifics of these regulations are highly uncertain at this time.

The lack of information about potential secondary markets is one of the most severe handicaps faced by government and private-sector decision makers. If, on the one hand, lumber-like materials have a large potential market, regulators may select to divert plastics away from the municipal waste stream in a form that will be appropriate to these technologies. In that case, alternative technologies, such as tertiary processes may be de-emphasized. On the other hand, if the potential market for these lumber-like materials is small, the development of alternative recycling technologies becomes more important.

1. Introduction

The industrial sector has historically recycled in excess of 75% of its waste plastics in either a primary or secondary sense -- i.e., clean scrap has been used in place of virgin resin or the scrap has been used to manufacture products with less demanding physical and chemical properties. More recently, a limited quantity of industrial waste has been recycled in a tertiary sense. Tertiary recycling refers to processes such as pyrolysis and hydrolysis that convert the waste polymers to useful chemicals or fuels. A fourth alternative, quaternary recycling, refers to incineration with heat recovery and has not been used extensively with respect to the recycling of separated plastic wastes.

The portion of manufacturing waste that has not historically been suited for recycling has been disposed of by landfill or incineration. These plastic wastes, which have commonly been referred to as manufacturing nuisance wastes, may be heavily contaminated with other materials or may consist of hard-to-recycle thermosets. Thermosets differ from the more popular thermoplastics in that their interlinking bonds prevent melting and reforming into new products. Although the size and composition of manufacturing nuisance plastics have been estimated, those numbers are now somewhat dated. Further, the technical, economic, institutional, and regulatory environment within which the industrial recycling of these wastes may occur is changing rapidly.

While the additional recycling of industrial waste may offer additional economic and environmental payoffs, an even greater potential exists with the recycling of post-consumer wastes. Recent trends toward more source separation of plastics, improved technologies to separate plastics from other materials in the municipal waste stream, and government mandated recycling programs all suggest that the quantity of plastics available for recycling as a segregated waste will grow. In addition, the total use of plastics continues to grow at a rapid rate. The identification of the quantities, qualities, and sources of post-consumer wastes, and in particular the divertable portion of the post-consumer stream -- which in this report will be referred to as divertable plastic waste (DPW) -- is important for planning in the industrial sector. Particular R&D areas may be shown to be most effective in promoting additional recycling of this portion of the waste stream.

The following section of this report gives updated estimates of the quantities and qualities of manufacturing nuisance plastics. Also presented is information on the current uses of plastics and how the post-consumer plastic waste stream is expected to change between now and the year 2000. Section 3 gives estimates of the energy contents of manufacturing and post-consumer plastic wastes during the coming decade. Section 4 presents recent information on the costs of recycling as compared to the costs of disposal. Recent institutional changes that have helped to promote plastics recycling in the industrial sector are summarized in Section 5. Legislative and regulatory trends that directly or indirectly promote additional recycling are discussed in Section 6. Several scenarios depicting probable future technical, economic, institutional, and regulatory conditions are discussed in Section 7. Those scenarios are then overlaid on the information presented in Section 2 on post-consumer waste to suggest the quantities, qualities, and sources of DPW in future years. Section 8 focuses on the current plastics recycling industry and discusses the problems and opportunities for additional recycling as currently perceived by the industry. Conclusions are summarized in the final section.

2. Plastic Waste Projections

Curlee (1986) presents projections of the future uses of plastics in various product categories, manufacturing nuisance plastics, and post-consumer wastes for the years 1984, 1990, and 1995. This section updates those projections and extends the projection to the year 2000. In addition, some additional disaggregation of the projections is given for two important product sectors -- transportation and packaging. For the most part, the methodology used in this section is the same as detailed in Curlee (1986, Chapter 4 and Appendix B). The interested reader is referred to that document for details of the methodology. A general overview of the methodology and specific changes made in this update are documented herein. The objective of this section is to present information about the sources, quantities, and likely qualities of future plastic waste streams. This information will be helpful in the later assessment of the applicability of different recycling technologies to different plastic waste streams.

2.1. PROJECTED U.S. RESIN PRODUCTION AND USE

2.1.1. Methodology

Projections of U.S. resin production and use of specific resins in major product categories were made using time series analysis. Historical production of all resins and the use of resins in nine major product categories were regressed against time and a constant. The relationships were then used to project the future production and use of plastics in the United States.

2.1.2. Results

Figure 2.1 illustrates the historical growth of total resin production (including polyurethane) in the U.S. for the years 1974 through 1987 and gives projections of future production through the

384

year 2000.[1] While there have been significant downturns in production during this historical period, the general trend is up sharply. For example, total resin production increased from about 37 billion pounds in 1980 to more than 55 billion pounds in 1987. According to the projections, total resin production will exceed 62 billion pounds in 1995 and approach 72 billion pounds by the year 2000.

Figures 2.2a and 2.2b illustrate the historical use of plastics in several major product areas during the years 1974 to 1987 and give projections for those product areas through the year 2000. [Note that because the Society of the Plastics Industry's <u>Facts and Figures of the U.S. Plastics Industry</u> (from which the historical data were obtained) excludes the use of polyurethanes in their data series, Figures 2.2a and 2.2b do not include polyurethanes. Polyurethanes are considered separately in a later figure.] The packaging and the building and construction sectors are by far the largest of the sectors considered. Resin usage in the packaging sector has increased from 10.0 billion pounds in 1980 to 15.2 billion pounds in 1987. By 1995 resin usage in that sector is projected to increase to 19.5 billion pounds and by 2000 to 22.6 billion pounds.[2] Resin usage in the building and construction sector increased from 6.4 billion pounds in 1980 to 11.3 billion pounds in 1987. The use of plastics in building and construction applications is projected to increase to 15.2 billion pounds in 1995 and to 17.8 billion pounds in 2000. The use of plastics in the other major product areas has been relatively flat in recent years.

---

[1]See Appendix A for the numbers corresponding to the graphical presentations given in this section. Also see Appendix A for explanation of the data sources and results of the regression runs on which the projections are based.

[2]Note that the projections given in this section assume that no significant structural changes will occur, such as changes in technology or changes in regulations that might impact on the use of plastics in specific product categories.

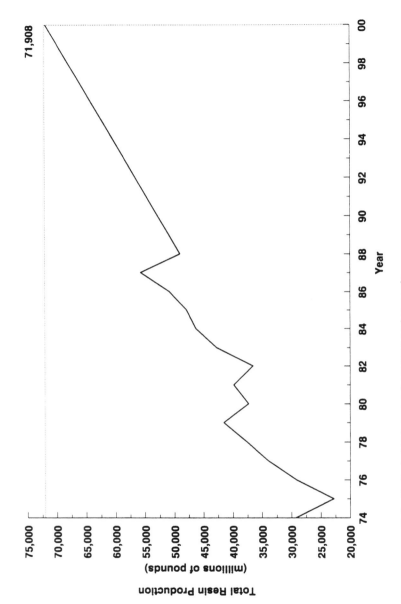

FIGURE 2.1. TOTAL U.S. RESIN PRODUCTION

Note: Projected Values are from 1988 and onwards

FIGURE 2.2a. RESIN USAGE BY PRODUCT TYPE

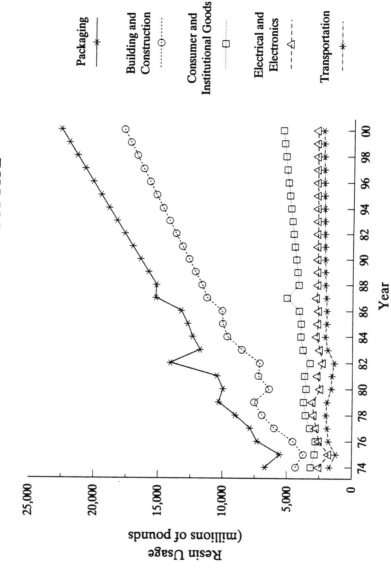

Packaging

Building and Construction

Consumer and Institutional Goods

Electrical and Electronics

Transportation

Resin Usage (millions of pounds)

Year

Note: Projected Values are from 1988 and onwards.
Excludes the use of polyurethane

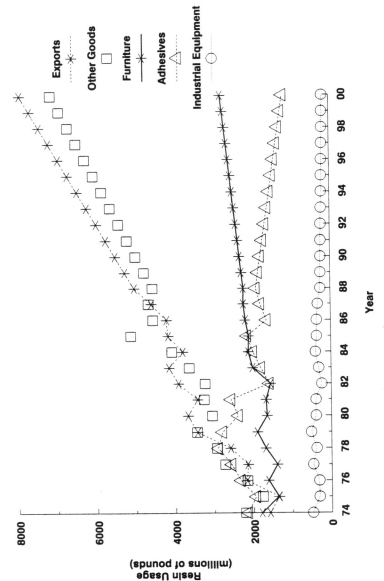

FIGURE 2.2b. RESIN USAGE BY PRODUCT TYPE

Exports

Other Goods

Furniture

Adhesives

Industrial Equipment

Resin Usage
(millions of pounds)

Year

Note: Projected Values are from 1988 and onwards
Excludes the Use of Polyurethane

Figure 2.3 and Table 2.1 further disaggregate the use of plastics in the large and controversial packaging sector. Data on the historical usage of plastics by packaging type is taken from <u>Modern Plastics</u>, and the projections are made using the same methodology used to make the above mentioned projections of total usage of plastics in the polyurethane sector. The projections of total usage of plastics in the polyurethane sector are lower than in Figure 2.2a. In the latter case a different source, SPI's <u>Facts and Figures of the Plastics Industry</u>, was used which reports higher historical usage in the packaging sector.

Containers and lids account for the largest percentage of plastics used in the sector and have increased from 4.2 billion pounds in 1981 to 6.4 billion pounds in 1987. About 9.3 billion and 11.2 billion pounds are projected to be used for containers and lids in 1995 and 2000, respectively. Films account for the second largest percentage, increasing from 3.3 billion pounds in 1981 to 4.4 billion pounds in 1987. The film segment of the packaging sector is projected to increase to 5.9 billion pounds in 1995 and to 6.9 billion pounds in 2000. The packaging sector as a whole is expected to consume more than 21 billion pounds of plastics in 2000.

Figure 2.4 summarizes the actual and projected usages of polyurethanes in different sectors over the same time frame. Total production of polyurethanes is only about 4% of total resin production. However, polyurethanes pose special problems for recycling and therefore are important to consider as a separate resin stream. The two largest uses of polyurethanes are in the furniture and fixtures sector and the building and construction sector. The furniture and fixtures sector accounted for 0.8 billion pounds in 1985 and is projected to account for 1.3 billion pounds by 2000. The building and construction sector consumed 0.4 billion pounds in 1985 and is

FIGURE 2.3. RESIN USAGE IN PACKAGING APPLICATIONS

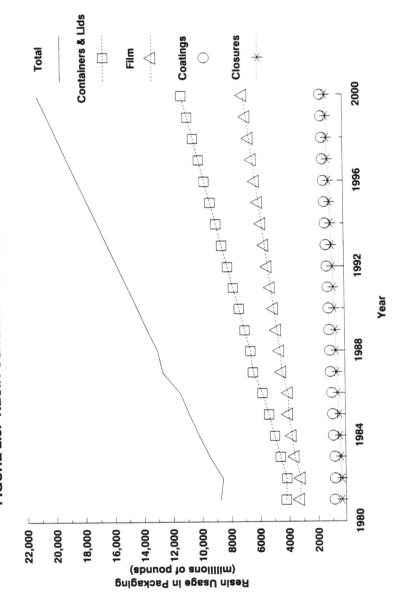

Note: Projected Values are from 1988 and onwards

TABLE 21. HISTORICAL AND PROJECTED USE OF PLASTICS IN SELECTED PRODUCT CATEGORIES IN PACKAGING

YEAR	CLOSURES	COATINGS	CONTAINERS & LIDS	FILM	TOTAL
1981	373	889	4214	3257	8733
1982	364	847	4159	3189	8559
1983	417	860	4580	3585	9442
1984	472	955	4965	3764	10156
1985	503	988	5355	4000	10846
1986	612	1035	5801	4004	11452
1987	676	1114	6433	4426	12649
1988*	701	1130	6595	4548	12974
1989	754	1172	6977	4746	13650
1990	808	1214	7360	4945	14326
1991	861	1256	7743	5143	15003
1992	914	1298	8126	5341	15679
1993	967	1340	8508	5540	16355
1994	1021	1382	8891	5738	17032
1995	1074	1424	9274	5936	17708
1996	1127	1466	9656	6134	18384
1997	1180	1509	10039	6333	19061
1998	1234	1551	10422	6531	19737
1999	1287	1593	10805	6729	20414
2000	1340	1635	11187	6928	21090

Note: *Starting year for projections

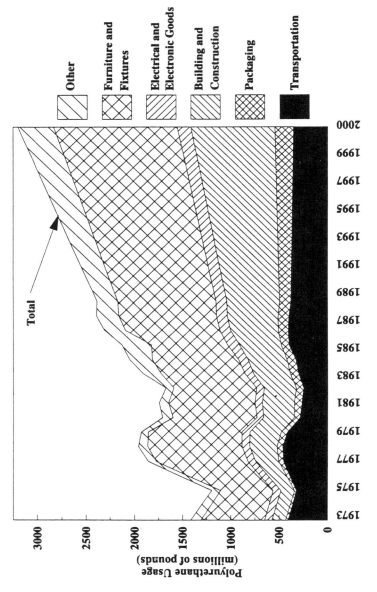

FIGURE 2.4. POLYURETHANE USAGE IN SELECTED MARKETS

projected to consume 0.9 billion pounds by 2000. Most of the projected increase in total polyurethane usage is in the building and construction sector.

2.2 PROJECTIONS OF MANUFACTURING NUISANCE PLASTICS

Projections of manufacturing nuisance plastics are given most recently in Curlee (1986). Those projections are, however, based on assumptions about the percentage of throughput that becomes a nuisance plastic from Leidner (1981). Leidner's work is, in turn, based on research published by Milgrom in 1971. The underlying assumptions on which the projections are based are therefore somewhat dated. Although manufacturing nuisance plastics have been estimated to compose only 8% to 9% of the total plastic waste potentially available for recycling, manufacturing waste is important because the segregated collection of manufacturing waste is easier to accomplish than is the case with post-consumer wastes.

This subsection addresses the question of what percentage of throughput becomes a nuisance plastic. That information is then used in combination with the projections of total resin production discussed in the previous subsection to formulate updated projections of manufacturing nuisance plastics. The update of the relevant percentages followed discussions with individuals familiar with the technological changes that have occurred with respect to plastics manufacturing equipment. Although this was not a formal survey of such individuals, the information gathered can be assumed to reflect a relevant range of estimates.

2.2.1. Methodology

A three step process was used. First, estimates of the percentage of total plastic throughput that has historically become a nuisance plastic at various stages of production -- i.e., resin producer; fabricator; converter; and packager, assembler, and distributor -- were updated. Second, the percentages derived from the first step were applied to the projections of total future U.S. resin production discussed in the previous subsection. This step gives projections of the total quantities of nuisance plastic to be produced by the different manufacturing processes or stages of production. Third, the estimates and projections derived from the second step were disaggregated by resin type according to the percentages of each resin produced in the U.S. in 1987 (as given in <u>Facts and Figures of the U.S. Plastics Industry</u>, 1988 edition).

2.2.2 Results

Figure 2.5 and Table 2.2 give estimates of the percentages of throughput that become nuisance plastics at different stages of production. The values in the first column of Table 2.2 indicate the percentages of resin content at different stages of production that would be affected. Note that five different sets of numbers are given, reflecting different opinions about technological changes that have occurred since Milgrom's 1971 publication. The second column in Table 2.2 reflects the numbers given in Leidner (1981) and subsequently in Curlee (1986). The third column reflects the conclusions of a follow-on assessment by Milgrom published in 1979. The fourth column assumes that Leidner's 1981 estimates are reduced by 50% and reflect the opinion of Albert Spaak, Technical Director of the Plastics Institute of America [Spaak (1988b)]. The fourth column assumes a 75% reduction in Leidner's 1981 numbers, but does not directly reflect the opinion of any particular expert. The final column reflects the opinion of Pearson (1988). Figure 2.5 presents this information in graphical form.

Table 2.3 gives estimates and projections of manufacturing nuisance plastics and disaggregates those estimates by stages of production. The numbers include the production of polyurethane. Figure 2.6 presents information about total nuisance plastics in graphical form. Note that while all the experts agree that technology has improved such that nuisance plastics are less today than in 1971, there is significant disagreement about the extent to which those wastes have been reduced.

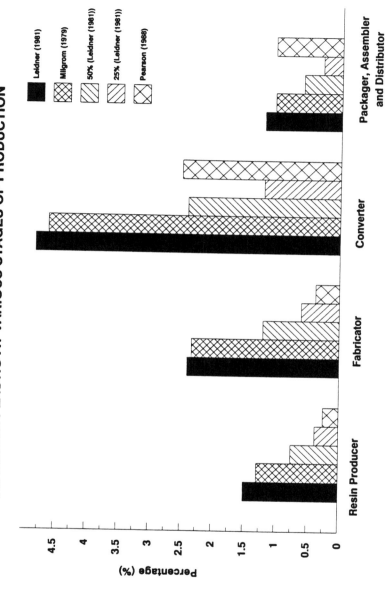

FIGURE 2.5. PERCENTAGE OF THROUGHPUT THAT BECOMES A
NUISANCE PLASTIC AT VARIOUS STAGES OF PRODUCTION

TABLE 22   PERCENTAGE OF THROUGHPUT THAT BECOMES A NUISANCE PLASTIC AT VARIOUS STAGES OF PRODUCTION

	Percentage of Commodity Resin Affected	Leidner (1981)	Milgrom (1979)	Percentage of Throughput that Becomes a Nuisance Plastic		Pearson (1988)
				50% Leidner(1981)	25% Leidner(1981)	
Resin Producer	100.0	1.5	1.29	0.75	0.375	0.250
Fabricator	84.2	2.4	2.33	1.20	0.600	0.375
Converter	40.1	4.8	4.60	2.40	1.200	2.500
Packager, Assembler and Distributor	68.0	1.2	1.04	0.60	0.300	1.040

TABLE 23. PROJECTIONS OF MANUFACTURING NUISANCE PLASTICS DISAGGREGATED BY STAGE OF PRODUCTION (IN MILLIONS OF POUNDS)

1990 Estimates

	Leidner (1981)	Milgrom (1979)	50% Leidner (1981)	25% Leidner (1981)	Pearson (1988)
Resin Producer	793	682	396	198	132
Fabricator	1068	1037	534	267	167
Converter	1017	975	509	254	530
Packager, Assembler and Distributor	431	374	216	108	374
Total	3309	3067	1655	827	1203

1995 Estimates

	Leidner (1981)	Milgrom (1979)	50% Leidner (1981)	25% Leidner (1981)	Pearson (1988)
Resin Producer	936	805	468	234	156
Fabricator	1261	1224	630	315	197
Converter	1201	1151	600	300	625
Packager, Assembler and Distributor	509	441	255	127	441
Total	3906	3620	1953	976	1419

2000 Estimates

	Leidner (1981)	Milgrom (1979)	50% Leidner (1981)	25% Leidner (1981)	Pearson (1988)
Resin Producer	1079	928	539	270	180
Fabricator	1453	1411	727	363	227
Converter	1384	1326	692	346	721
Packager, Assembler and Distributor	587	509	293	147	509
Total	4503	4173	2251	1126	1636

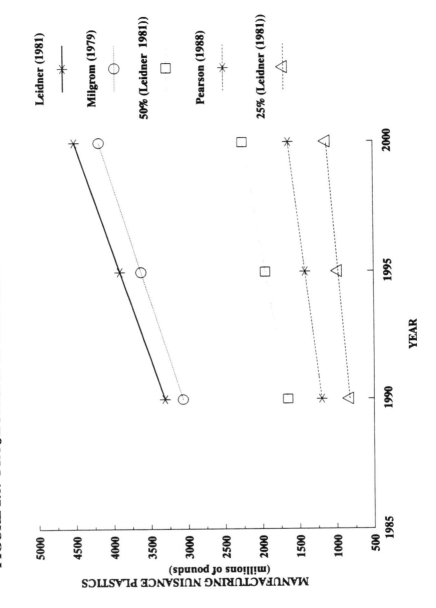

FIGURE 2.6. PROJECTIONS OF MANUFACTURING NUISANCE PLASTICS

Table 2.4 gives projections of manufacturing nuisance plastics by resin type for the years 1990, 1995, and 2000, given the assumptions about waste generation rates made in Leidner's assessment. Table 2.5 provides similar information, given the "50% Leidner" case. The total quantities of nuisance plastics have been disaggregated according to the percentage composition of total 1987 plastic resins produced in the United States.

Table 2.6 gives information from another source on estimated manufacturing nuisance plastics. The estimates from a recent presentation by Albert Spaak (1988a) are disaggregated by a combination of resin types and products and are the result of a mini-survey conducted by the Plastics Institute of America (PIA). It is interesting to note that the findings of the survey place estimates and projections of nuisance plastics significantly higher than any of the cases considered above. The 1990 projected quantity of nuisance plastics at 5.89 billion pounds exceeds the 3.3 billion pounds projected for our highest case -- the Leidner (1981) assumptions.

2.3. PROJECTIONS OF POST-CONSUMER PLASTIC WASTES

Given that plastic wastes are difficult to separate from other similar materials in the municipal waste stream, it is important that information be available on the projected levels of post-consumer plastic waste not only in terms of quantity and resin type, but also in terms of the sources of that waste. The sources of waste will help in the identification of those wastes that are defined in this report to be divertable plastic wastes.

2.3.1. Methodology

Post-consumer wastes are estimated and projected for nine product categories by resin type for the years 1990, 1995, and 2000. Product categories include automobiles, other transportation applications, packaging, building and construction, electrical and electronic goods, furniture, consumer and institutional goods, industrial machinery, and adhesives and other applications. The

TABLE 24. MANUFACTURING NUISANCE PLASTICS DISAGGREGATED BY RESIN TYPE (IN MILLIONS OF POUNDS) (LEIDNER (1981))

	1987 Composition of U.S. Resins (%)	1990	1995	2000
Thermosets				
Epoxy	0.8%	26	30	35
Polyester	2.5%	81	96	110
Urea and Melamine	2.9%	95	112	129
Phenolics	5.1%	170	201	232
Other Thermosets	4.9%	162	191	221
Total Thermosets	16.1%	534	630	726
Thermoplastics				
LDPE	17.2%	570	673	775
HDPE	14.3%	475	560	646
Polypropylene	11.9%	395	466	537
ABS and SAN	2.3%	77	91	105
Polystyrene	8.6%	284	335	386
Nylon	0.9%	30	36	41
PVC	14.3%	473	558	644
Thermoplastic Polyesters	2.5%	83	98	113
Other Thermoplastics	7.9%	261	309	356
Total Thermoplastics	80.0%	2647	3125	3602
Polyurethane Foam	3.9%	552	651	751
Total	100.0%	3310	3907	4504

TABLE 25. MANUFACTURING NUISANCE PLASTICS BY RESIN TYPE (IN MILLIONS OF POUNDS) (50% LEIDNER (1981))

	1987 Composition of U.S. Resins (%)	1990	1995	2000
Thermosets				
Epoxy	0.8%	13	15	17
Polyester	2.5%	41	48	55
Urea and Melamine	2.9%	47	56	64
Phenolics	5.1%	85	101	116
Other Thermosets	4.9%	81	96	110
Total Thermosets	16.1%	267	315	363
Thermoplastics				
LDPE	17.2%	285	336	388
HDPE	14.3%	237	280	323
Polypropylene	11.9%	197	233	268
ABS and SAN	2.3%	39	46	53
Polystyrene	8.6%	142	167	193
Nylon	0.9%	15	18	20
PVC	14.3%	237	279	322
Thermoplastic Polyesters	2.5%	41	49	56
Other Thermoplastics	7.9%	131	154	170
Total Thermoplastics	80.0%	132.4	1562	1801
Polyurethane Foam	3.9%	65	76	88
Total	100.0%	1655	1954	2252

TABLE 2.6. MANUFACTURING NUISANCE PLASTIC ESTIMATES FROM SPAAK (1988a)
(IN MILLIONS OF POUNDS PER YEAR)

Type	1988 Pounds Per Year
Wire insulation, PVC, HDPE and PE	1,500
Multilayer film, including coextruded film	500
Polyethylene coated paper	1,000
PVC coated fabrics	500
Polyethylene film	200
Rubber-backed poly-pro carpet aaste	4
Carpet waste, polyester, nylon and polypropylene	250
Diaper trim and rejects - plastic and paper	6
Acrylonitrile-PVC foam	2
Acrylics - mixture router shavings	2
Chopped polyethylene and polypropylene mixed	3
Plastics and rubber mixed	2
Polyethylene terephthalate	6
Flexible PVC regrind	12
Polyethylene regrind	1
Polypropylene regrind	3
Polystyrene regrind	1
NPE show conversation with various attendees, mixed plastic materials	90
Total	4,081

Total production 1988 - 54 billion pounds
Projected total production 1990 - 78 billion pounds
Therefore, Manufactured Nuisance Plastic Waste
  estimated in 1990 equals - 5.89 billion pounds.
BTU value at average 12 x 10^3 BTU's/pound = 7.068 x 10^{13} BTU's

specific methodology used to project wastes from each product category differed slightly depending on the availability of data. The general methodology was the same as used in Curlee (1986, Chapter 4 and Appendix B). Variations from that general methodology are described in the appendix of this report.

The general methodology involved two main steps. First, information was obtained on the average life spans of the products in the nine product categories. Second, information was obtained on the historical use of specific plastic resins in those product categories. Projections of future waste streams by product type and resin were subsequently made.

Table 2.7 gives information on average product life spans and is taken from Curlee (1986, page 80). Average product life spans range from less than one year for packaging to 25 years for building and construction materials. When the data permit, three year averages have been used to estimate the flow of plastics from any particular product category. For example, the 1995 projections of plastic waste from the furniture and fixtures category is given by averaging the quantities of plastic resins used in that category during the years 1984, 1985, and 1986.

Data on the historical use of plastic resins in specific product categories was obtained from various issues of the Society of the Plastics Industry's Facts and Figures of the U.S. Plastics Industry whenever possible. Other information on resin usage was obtained from the January issues of Modern Plastics, which give similar information but in a slightly different form. Polyurethane foams were considered separately from other resins. See the appendix to this report for additional information on data and methodology.

TABLE 2.7. ESTIMATED LIFE SPANS OF
SELECTED PRODUCTS (IN YEARS)

Product Category	Estimated Life
Transportation	11
Packaging	<1
Building and construction	25
Electrical and electronics	15
Furniture and fixtures	10
Consumer and Institutional	5
Industrial machinery	15
Adhesives and Other	4

Source: Curlee (1986, page 80)

2.3.2 Results

2.3.2.1. Projections for 1990

Table 2.8 gives projected post-consumer plastic waste by product category and resin type. Figure 2.7 presents 1990 summary information. It is projected that total 1990 post-consumer wastes will total about 35.8 billion pounds. Thermoplastics will contribute 87.2%, followed by thermosets and polyurethane foams at 7.9% and 4.9%, respectively.

Note that the largest single source of waste is the packaging sector, which is projected to contribute 46.2% of the total. Further note that the vast majority (99.6%) of plastic packaging materials are made from thermoplastics, with LDPE and HDPE accounting for 68.8% of the total.

Following the adhesives and other category is the consumer and institutional goods category, which is projected to account for 11.2% of the total. Thermoplastics also dominate this category, accounting for 93.3% of the total.

Each of the remaining categories account for less than 8% of the total. The furniture and electrical and electronics sector account for about 7% each. Automobiles are expected to contribute 4.9% and other transportation applications about 1.4%.

2.3.2.2 Projections for 1995

Post-consumer plastic wastes are projected to increase in 1995 to 43.7 billion pounds, a 22% increase over the 1990 projected level. Thermoplastics are projected to contribute about 88.8% of the total, with thermosets and polyurethane foams accounting for 6.9% and 4.3%, respectively. Table 2.9 presents detailed information on wastes streams by product type and resin. Figure 2.8 presents summary information.

The packaging sector is projected to remain the largest contributor to the waste stream at 45.1% of the total. The adhesives and other category is again followed by the consumer and

TABLE 2.8 POST-CONSUMER PLASTIC WASTE (1990)

	Automobile	%	Transportation/Other	%	Packaging	%	Building and Construction	%	Electrical and Electronic	%	Furniture	%	Consumer and Institutional	%	Industrial Machinery	%	Adhesives and Others	%	Total	%
THERMOSETS																				
Epoxy	2.3	0.2%	.....	.....	6	0.0%	46	6.5%	32	1.4%	.....	.....	75[a]	1.9%	.....	.....	219	3.3%	260	0.7%
Polyester	143.4	9.9%	111.3	27.3%	6	0.0%	.....	.....	263	11.5%	19	1.1%	146	3.6%	26	6.9%	173	2.6%	837	2.3%
Urea and Melamine	.....	.....	.....	.....	33	0.2%	91	12.8%	59	2.6%	66	3.8%	50	1.2%	58	15.5%	320	4.8%	650	1.8%
Phenolics	20.9	1.4%	15.5	3.8%	14	0.1%	7	1.0%	241	10.6%	275	15.8%	.....	.....	.....	.....	285	4.3%	1051	2.9%
Other Thermosets	.....	.....	.....	.....	.....	.....	.....	.....	21	0.9%	.....	.....	.....	.....	.....	.....	.....	.....	28	0.1%
Total Thermosets	166.6	11.5%	126.8	31.1%	60	0.4%	144	20.3%	616	27.1%	361	20.8%	271	6.7%	84	22.4%	996	15.0%	2826	7.9%
THERMOPLASTICS																				
LDPE	56.8	3.9%	.....	.....	6647	40.4%	72[b]	10.1%	868[a]	38.1%	19[a]	1.1%	587	14.6%	97	26.0%	1134	17.1%	8497	23.7%
HDPE	.....	.....	.....	.....	4666	28.4%	.....	.....	.....	.....	76	4.4%	636	15.8%	53	14.2%	792	11.9%	7078	19.8%
Polypropylene	308.2	21.3%	94.9	23.3%	1564	9.5%	.....	.....	.....	.....	38	2.2%	779	19.3%	.....	.....	1324	19.9%	4199	11.7%
ABS and SAN	327.1	22.7%	42.3	10.4%	30	0.2%	3	0.4%	.....	.....	474	27.3%	64	1.6%	.....	.....	526	7.9%	1031	2.9%
Polystyrene	1.9	0.1%	.....	.....	1533	9.3%	41	5.8%	16	0.7%	.....	.....	1367	33.9%	25	6.6%	944	14.2%	4362	12.2%
PBT/PET	38.3	2.7%	.....	.....	1063	6.5%	.....	.....	59	2.6%	.....	.....	.....[d]	.....	18	4.8%	78	1.2%	1017	3.1%
Nylon	133.3	9.2%	16.4	4.0%	55	0.3%	330	46.5%	.....	.....	.....	.....	.....	.....	97	26.0%	580	8.7%	367	1.0%
PVC	249.7	17.3%	81.1	19.9%	662	4.0%	119[b]	16.8%	670	29.4%	731	42.1%	289	7.2%	.....	.....	265	4.0%	3610	10.1%
Other Thermoplastics	162.1	11.2%	45.7	11.2%	154	0.9%	.....	.....	48	2.1%	38	2.2%	35	0.9%	.....	.....	.....	.....	965	2.7%
Total Thermoplastics	1277.5	88.5%	280.4	68.9%	16375	99.6%	565	79.6%	1662	72.9%	1376	79.2%	3757	93.3%	290	77.6%	5643	85.0%	31220	87.2%
Polyurethane foam	304.7	.....	85.9	.....	128	.....	1[d]	0.1%	79	.....	914	.....	.....	.....	.....	.....	251	.....	1764	4.9%
TOTAL	1748.8	100.0%	493.2	100.0%	16563	100.0%	710	100.0%	2358	100.0%	2651	100.0%	4028	100.0%	373	100.0%	6890	100.0%	35816	100.0%
PERCENT	4.9	.....	1.4	.....	46.2	.....	2.0	.....	6.6	.....	7.4	.....	11.2	.....	1.0	.....	19.2	.....	100.0	.....

[a] Data do not distinguish between LDPE and HDPE.
[b] Disaggregate information not available. Mostly insignificant quantity.
[c] Data do not distinguish between other thermosets and other thermoplastics.
[d] Data do not distinguish between polyurethane foam and other polyurethane uses.

Note: Low Density Polyethylene (LDPE); High Density Polyethylene (HDPE); Acrylonitrile-Butadiene-Styrene (ABS); Styrene-Acrylonitrile (SAN); Polyvinyl Chloride (PVC).

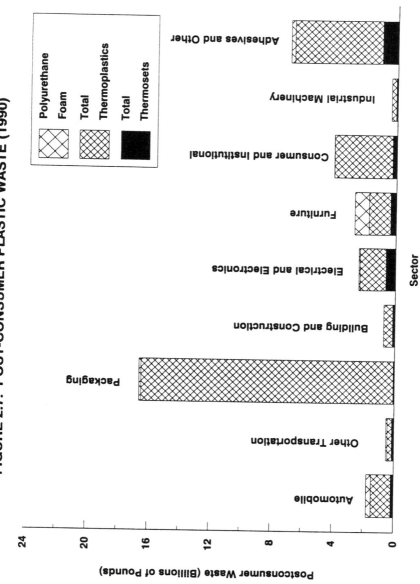

FIGURE 2.7. POST-CONSUMER PLASTIC WASTE (1990)

TABLE 29. POST-CONSUMER PLASTIC WASTE (1995)

	Automobile	%	Transportation/Other	%	Packaging	%	Building and Construction	%	Electrical and Electronic	%	Furniture	%	Consumer and Institutional	%	Industrial Machinery	%	Adhesives and Others	%	Total	%
THERMOSETS																				
Epoxy	8.1	0.5%	—ᵇ		8	0.0%			41	1.5%	—ᵇ		91	2.1%			210	3.0%	267	0.6%
Polyester	123.3	7.9%	106.6	24.3%	8	0.0%	258	6.7%	255	9.4%	14	0.6%	14	0.3%	30	6.9%	165	2.4%	1021	2.3%
Urea and Melamine			—ᵇ		40	0.2%			52	1.9%	54	2.5%	14	0.3%			294	4.3%	484	1.1%
Phenolics	24.3	1.6%	41.5	9.4%	17	0.1%	228	5.9%	236	8.7%	108	5.0%	36	0.8%	67	15.5%	265	3.8%	1023	2.3%
Other Thermosets							173	4.5%	34	1.2%									207	0.5%
Total Thermosets	155.8	10.0%	148.1	33.7%	72	0.4%	659	17.0%	619	22.7%	175	8.2%	141	3.2%	97	22.4%	935	13.5%	3001	6.9%
THERMOPLASTICS																				
LDPE	88.7	5.7%	20.7ᵃ	4.7%	7900	40.4%	351ᶜ	9.1%	604	22.2%	46	2.2%	566	12.9%	113	26.0%	967	14.0%	10543	24.1%
HDPE					5546	28.4%			187	6.9%	58	2.7%	831	19.0%			741	10.7%	7476	17.1%
Polypropylene	358.9	23.0%	63.9	14.5%	1859	9.5%	195	5.0%	19	0.7%	1401	65.5%	896	20.5%	62	14.2%	1165	16.8%	5825	13.3%
ABS and SAN	255.6	16.4%	73.6	16.7%	36	0.2%	137	3.5%	64	2.3%	5	0.2%	51	1.2%			1099	15.9%	1779	4.1%
Polystyrene	66.9	4.3%	8.5	1.9%	1822	9.3%			394	14.5%	59	2.8%	1547	35.4%			568	8.2%	4528	10.4%
PBT/PET			42.7	9.7%	1263	6.5%			34	1.2%			33	0.8%	29	6.6%			1372	3.1%
Nylon	173.9	11.2%	44.8	10.2%	65	0.3%			49	1.8%	6	0.3%			21	4.8%	51	0.7%	450	1.0%
PVC	233.4	15.0%	37.2	8.5%	787	4.0%	1900	49.1%	664	24.4%	388	18.1%	246	5.6%	113	26.0%	871	12.6%	5155	11.8%
Other Thermoplastics	224	14.4%			183	0.9%	470ᶜ	12.2%	90	3.3%			58	1.3%			524	7.6%	1699	3.9%
Total Thermoplastics	1402.7	90.0%	291.5	66.3%	19463	99.6%	3053	78.9%	2103	77.3%	1963	91.8%	4229	96.8%	337	77.6%	5986	86.5%	38828	88.8%
Polyurethane foam	268.1		75.6		159		156ᵈ	4.0%	76		888						253		1877	4.3%
TOTAL	1826.5	100.0%	515.2	100.0%	19694	100.0%	3868	100.0%	2798	100.0%	3027	100.0%	4370	100.0%	434	100.0%	7174	100.0%	43707	100.0%
PERCENT	4.2		1.2		45.1		8.9		6.4		6.9		10.0		1.0		16.4		100.0	

ᵃ Data do not distinguish between LDPE and HDPE.

ᵇ Disaggregate information not available. Mostly insignificant quantity.

ᶜ Data do not distinguish between other thermosets and other thermoplastics.

ᵈ Data do not distinguish between polyurethane foam and other polyurethane uses.

Note: Low Density Polyethylene (LDPE); High Density Polyethylene (HDPE); Acrylonitrile-Butadiene-Styrene (ABS); Styrene-Acrylonitrile (SAN); Polyvinyl Chloride (PVC).

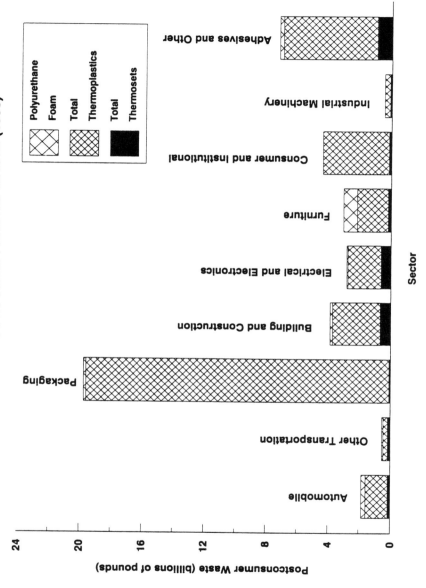

FIGURE 2.8. POST-CONSUMER PLASTIC WASTE (1995)

institutional goods category at 10.0%. By far the largest growth category is building and construction, which is projected to increase from about 2% of the total in 1990 to almost 9% of the total in 1995. Only moderate changes are projected in the other product categories.

2.3.2.3. Projections for 2000

Table 2.10 and Figure 2.9 give projections of post-consumer wastes in 2000. Total waste is projected to increase to 48.7 billion pounds, an increase of 11.4% above the 1995 projected level. Thermoplastics are projected to account for 88.8% of the total, with thermosets and polyurethane foams accounting for 6.7% and 4.5% respectively.

No significant shift in the mix of waste according to product type is projected between 1995 and 2000. Packaging remains the largest source of waste at 46.9% of the total.

2.3.2.4. Summary

Figures 2.10, 2.11, and 2.12 summarize the findings concerning plastic waste projections. Figure 2.10 shows the projections for both manufacturing nuisance plastics and post-consumer plastics for 1990, 1995, and 2000. Note that the nuisance waste projections reflect the "50% Leidner" case. The total quantity of waste is expected to grow from the 1990 level of 37.5 billion pounds to 45.7 billion pounds in 1995 to 51.0 billion pounds in 2000. Manufacturing nuisance plastics are projected to remain at a relatively constant percentage of total plastic waste -- about 4.4%.

Figures 2.11 and 2.12 illustrate how the composition of waste by product category is projected to change. Packaging is expected to remain the largest single source of waste. The category with the largest projected percentage increase is the building and construction sector, increasing from 2.0% in 1990 to 8.0% in 2000.

TABLE 2.10. POST-CONSUMER PLASTIC WASTE (2000)

	Automobile	%	Transportation/Other	%	Packaging	%	Building and Construction	%	Electrical and Electronic	%	Furniture	%	Consumer and Institutional	%	Industrial Machinery	%	Adhesives and Others	%	Total	%
THERMOSETS																				
Epoxy	20.4	1.2%	—	—	9	0.0%	11	0.3%	77	2.9%	—	—	102	2.1%	—	—	233	3.0%	350	0.7%
Polyester	183.6	11.1%	177.2	38.1%	9	0.0%	217	5.8%	65	2.4%	28	1.2%	16	0.3%	—	—	184	2.4%	966	2.0%
Urea and Melamine	—	—	—	—	46	0.2%	407	10.9%	37	1.4%	49	2.1%	41	0.8%	—	—	327	4.3%	882	1.8%
Phenolics	30.0	1.8%	6.8	1.5%	19	0.1%	402	10.8%	136	5.1%	118	5.1%	—	—	10	2.7%	295	3.8%	1058	2.2%
Other Thermosets	—	—	—	—	—	—	—	—	—	—	—	—	—	—	—	—	—	—	—	—
Total Thermosets	234.0	14.2%	183.9	39.5%	83	0.4%	1037	27.8%	316	11.8%	195	8.4%	159	3.2%	10	2.7%	1038	13.5%	3256	6.7%
THERMOPLASTICS																				
LDPE	106.2	6.4%	24.4	5.2%	9154	40.4%	26	0.7%	430	16.1%	53	2.3%	637	12.9%	—	—	1074	14.0%	11504	23.6%
HDPE	—	—	4.2	0.9%	6426	28.4%	398	10.7%	163	6.1%	—	—	936	19.0%	—	—	823	10.7%	8955	18.4%
Polypropylene	386.5	23.4%	118.5	25.4%	2154	9.5%	11	0.3%	291	10.9%	1625	69.9%	1009	20.5%	205	54.7%	1294	16.8%	6905	14.2%
ABS and SAN	208.8	12.6%	60.8	13.1%	42	0.2%	199	5.3%	257	9.6%	8	0.3%	57	1.2%	16	4.3%	1221	15.9%	2053	4.2%
Polystyrene	0.4	0.0%	—	—	2111	9.3%	168	4.5%	329	12.3%	52	2.2%	1742	35.4%	—	—	631	8.2%	5034	10.3%
PBT/PET	89.7	5.4%	13.5	2.9%	1463	6.5%	—	—	41	1.5%	—	—	—	—	—	—	—	—	1607	3.3%
Nylon	174.1	10.5%	4.2	0.9%	76	0.3%	—	—	71	2.7%	13	0.6%	38	0.8%	47	12.4%	57	0.7%	480	1.0%
PVC	205.4	12.4%	56.1	12.1%	912	4.0%	1662	44.5%	511	19.1%	378	16.3%	277	5.6%	44	11.6%	967	12.6%	5013	10.3%
Other Thermoplastics	245.7	14.9%	—	—	212	0.9%	234	6.3%	267	10.0%	—	—	66	1.3%	54	14.3%	582	7.6%	1660	3.4%
Total Thermoplastics	1416.7	85.8%	281.6	60.5%	22550	99.6%	2699	72.2%	2359	88.2%	2128	91.6%	4761	96.8%	365	97.3%	6649	86.5%	43211	88.8%
Polyurethane foam	295.9	—	83.5	—	191	—	140	—	111	—	1062	—	—	—	—	—	318	—	2202	4.5%
TOTAL	1946.6	100.0%	549.0	100.0%	22825	100.0%	3876	100.0%	2786	100.0%	3386	100.0%	4920	100.0%	375	100.0%	8005	100.0%	48669	100.0%
PERCENT	4.0	—	1.1	—	46.9	—	8.0	—	5.7	—	7.0	—	10.1	—	0.8	—	16.5	—	100.0	—

ᵃ Data do not distinguish between LDPE and HDPE.
ᵇ Disaggregate information not available. Mostly insignificant quantity.
ᶜ Data do not distinguish between thermosets and other thermoplastics.
ᵈ Data do not distinguish between polyurethane foam and other polyurethane uses.

Note: Low Density Polyethylene (LDPE); High Density Polyethylene (HDPE); Acrylonitrile-Butadiene-Styrene (ABS); Styrene-Acrylonitrile (SAN); Polyvinyl Chloride (PVC).

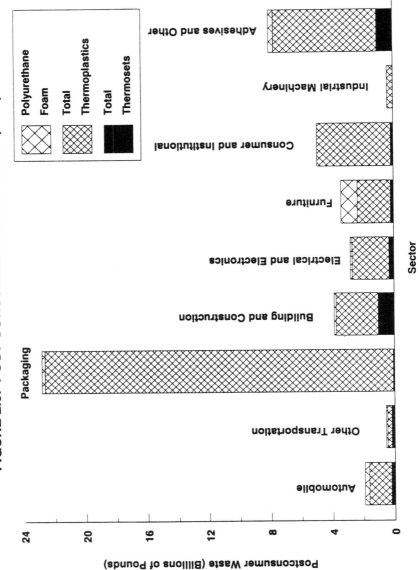

FIGURE 2.9.   POST-CONSUMER PLASTIC WASTE (2000)

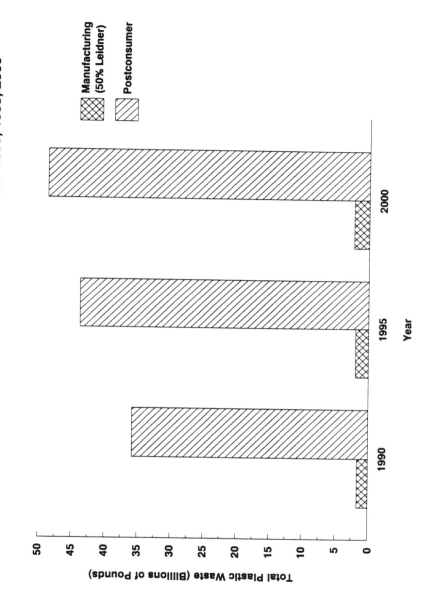

FIGURE 2.10. TOTAL PLASTIC WASTE PROJECTIONS: 1990, 1995, 2000

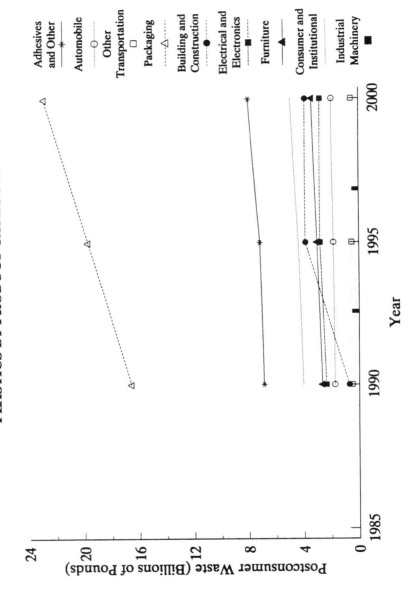

FIGURE 2.11. PROJECTED POST-CONSUMER PLASTICS BY PRODUCT CATEGORY

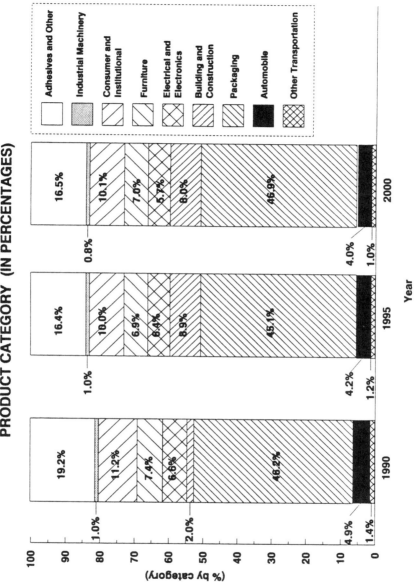

FIGURE 2.12. PROJECTED POST-CONSUMER PLASTICS BY PRODUCT CATEGORY (IN PERCENTAGES)

Given that packaging items have average life times of less than one year, the composition of the packaging sector by packaging type is projected to be the same as presented in Table 2.1 and Figure 2.3. The packaging sector in 1995 is projected to be made up of the following components: containers and lids, 52.4%; films, 33.5%; coatings, 8.0%; and closures, 6.1%. The composition of the packaging sector is important to the later discussion of the potential size and composition of DPW.

3. The Energy Content of Plastic Wastes

Given that plastics are made for the most part from petroleum and natural gas, plastics recycling has been argued for on the basis of energy conservation. Quaternary recycling could directly retrieve the heat energy of the waste resins. Tertiary recycling could potentially retrieve pre-polymers that embody significant amounts of energy. Secondary recycling, if used to manufacture products that would otherwise be made from virgin resins, could reduce the overall demand for virgin resins and the energy embodied in those resins. This section presents estimates of the energy required to manufacture plastics, the energy that can be retrieved from plastics when burned, and the aggregate energy contained in manufacturing nuisance plastics and post-consumer plastics for the years 1990, 1995, and 2000.

3.1. THE ENERGY CONTENT OF PLASTICS

Gaines and Shen (1980) present estimates of the energy contained in various plastic resins. Four energy estimates are given: net heat of feed combustion, net processing energy, total energy input, and product heat of combustion. Gaines and Shen define the terms as follows: "Heat of combustion of feed is the sum of the heats of combustion of all feedstocks entering into a process sequence, starting with oil and gas. Net process energy is the total fuel required to complete all steps of a manufacturing process minus the heat of combustion of any by-product fuels not burned within that process sequence. Total energy input is the sum of the heat of combustion of the feed and the net process energy. It is the total energy embodied in the final product. Product heat of combustion is defined as the sum of the heats of combustion of all process products and nonfuel by-products. It is the energy that would be recovered if the final products were burned." (page 2).

Table 3.1 presents energy content information for the various resin types discussed in Section 2 of this report. Note that energy content estimates were not available from Gaines and Shen for all the resin types discussed in Section 2. For those resins not covered in the Gaines and

417

Shen report, information was obtained from Thiessen (1989). The specific assumptions made concerning these additional resins are listed in Table 3.1. Note that, with the exceptions of PVC and nylon, the energy contents of most resins do not vary significant.

3.2 THE AGGREGATE ENERGY CONTENT OF MANUFACTURING AND POST-CONSUMER PLASTICS

Tables 3.2 and 3.3 present estimates of the energy contents of manufacturing nuisance plastics and post-consumer plastic wastes for the years 1990, 1995, and 2000. Total energy inputs and product heat of combustion are given for each waste resin category and each year and are defined as given in the previous subsection. Total energy inputs for all manufacturing nuisance plastics are projected to be 61.6×10^{12} BTUs in 1990 and increase to 83.8×10^{12} BTUs in 2000. Total energy inputs for all post-consumer plastics in 1990 is projected to be about $1,317 \times 10^{12}$ BTUs and increase to $1,779 \times 10^{12}$ BTUs in 2000. If incineration is used to retrieve the heat energy of the waste, manufacturing nuisance plastics are projected to contain 26.8×10^{12} BTUs in 1990 and increase to 36.5×10^{12} BTUs in 2000. Post-consumer wastes are projected to contain 617×10^{12} BTUs in 1990 if incinerated. In 2000 the incineration heat value of post-consumer plastics is projected to increase to 843×10^{12} BTUs. For comparison purposes, note that the U.S. consumed a total of 73.82 quads of energy (1 quad $= 1 \times 10^{15}$ BTUs). In 1985 total energy consumption was disaggregated as follows: transportation, 20.01 quads; residential and commercial, 26.80 quads; and industrial, 27.01. Further note that the U.S. currently burns only about 2% of all solid waste in incinerators that allow for heat recovery. Waste incineration with heat recovery currently accounts for only 0.04% of the United States' primary energy supply.

TABLE 3.1. ENERGY CONTENTS OF PLASTIC WASTES

	Net Heat of Feed Combustion (Btu/lb)	Net Processing Energy (Btu/lb)	Total Energy Input (Btu/lb)	Product Heat of Combustion (Btu/lb)
Thermosets				
Epoxy	24,700	24,000	48,700	11,400
Polyester	24,700	24,000	48,700	11,400
Urea and Melamine[a]	24,700	24,000	48,700	11,400
Phenolics[a]	24,700	24,000	48,700	11,400
Other Thermosets[a]	24,700	24,000	48,700	11,400
Thermoplastics				
LDPE	28,100	10,400	38,500	20,000
HDPE	27,300	9,200	36,500	20,050
Polypropylene	28,000	6,200	34,200	20,000
ABS and SAN[b]	24,570	8,280	32,850	18,045
Polystyrene	23,600	10,700	34,300	17,800
PBT/PET	24,700	24,000	48,700	11,400
Nylon	27,700	62,700	90,400	13,200
PVC	12,600	13,000	25,600	7,700
Other Thermoplastics[c]	27,300	9,200	36,500	20,050
Polyurethane Foam[d]	21,840	7,360	29,200	16,040

[a]Assumed to approximately the same as polyester.
[b]Assumed to be 90% of the estimated values for HDPE.
[c]Assumed to be the same as HDPE.
[d]Assumed to be 80% of the estimated values for HDPE.

Source of information on resin other than those footnoted: Gaines and Shen (1980).
Source of information on resins with footnotes: Thiessen (1989).

TABLE 3.2. ENERGY CONTENT (IN BILLIONS OF BTUS) OF MANUFACTURING
NUISANCE PLASTICS: 1990, 1995, 2000
(50% LEIDNER CASE)

	1990		1995		2000	
	(1)*	(2)**	(1)*	(2)**	(1)*	(2)**
Thermosets						
Epoxy	626	147	739	173	852	199
Polyester	1976	463	2332	546	2688	629
Urea and Melamine	2304	539	2719	637	3135	734
Phenolics	4147	971	4894	1146	5642	1321
Other Thermosets	3948	924	4660	1091	5372	1258
Total Thermosets	13001	3043	15345	3592	17689	4141
Thermoplastics						
LDPE	10968	5698	12946	6725	14923	7752
HDPE	8661	4758	10222	5615	11784	6473
Polypropylene	6747	3946	7963	4657	9180	5368
ABS and SAN	1268	697	1497	822	1726	948
Polystyrene	4866	2525	5743	2980	6621	3436
Nylon	1360	199	1606	234	1851	270
PVC	6056	1822	7148	2150	8240	2478
Thermoplastic Poly	2015	472	2378	557	2741	642
Other Thermoplastic	4771	2621	5631	3093	6492	3566
Total Thermoplastic	46713	22735	55135	26835	63557	30934
Polyurethane Foam	1884	1035	2224	1222	2564	1408
Total	61598	26814	72704	31648	83810	36483

*(1): Total Energy Input
**(2): Product Heat of Combustion

TABLE 3.3. ENERGY CONTENT (IN 10^{12} BTUS) OF POST-CONSUMER
PLASTIC WASTES: 1990, 1995, 2000

	1990		1995		2000	
	(1)*	(2)**	(1)*	(2)**	(1)*	(2)**
Thermosets						
Epoxy	13	3	13	3	17	4
Polyester	41	10	50	12	47	11
Urea and Melamine	32	7	24	6	43	10
Phenolics	51	12	50	12	52	12
Other Thermosets	1	0	10	2	0	0
Total Thermosets	138	32	146	34	159	37
Thermoplastics						
LDPE	327	170	406	211	443	230
HDPE	258	142	273	150	327	180
Polypropylene	144	84	199	116	236	138
ABS and SAN	34	19	58	32	67	37
Polystyrene	150	78	155	81	173	90
PBT/PET	54	13	67	16	78	18
Nylon	33	5	41	6	43	6
PVC	92	28	132	40	128	39
Other Thermoplastics	35	19	62	34	61	33
Total Thermoplastics	1128	557	1393	685	1557	771
Polyurethane Foam	52	28	55	30	64	35
Total	1317	617	1594	750	1779	843

*(1): Total Energy Input
**(2): Product Heat of Combustion

4. The Cost of Recycling Versus Disposal

Curlee (1986) discusses the concept of economic costs as compared to accounting costs. Economic costs include an opportunity cost component, whereas accounting costs do not. A distinction is also made between private costs and social costs. Social costs and benefits include monetary and nonmonetary costs and benefits that accrue to individuals not directly involved in economic transactions. The costs of environmental degradation that may result from disposal or recycling activities are included in social costs and may be the determining factor in selecting the optimal technology to process plastics from a social perspective. The external costs associated with oil import disruptions -- which may be reduced by recycling plastics -- are also considered under social costs. Accounting costs do not consider external costs and benefits of these types and therefore are not recommended as a means for judging the optimal social response to waste disposal questions.

Although accounting costs pose problems, the assessment of accounting costs can be beneficial when viewed from the proper perspective. Curlee (1986, Chapter 5) presents various estimates of the accounting costs of recycling as well as disposing of plastic wastes. This section briefly reviews the conclusions of Curlee (1986, Chapter 5) and presents new information on the estimated costs of disposal and the estimated costs and revenues associated with recycling technologies not available at the time of that book's publication.

4.1. SUMMARY OF PREVIOUS ASSESSMENT

The assessment presented in Curlee (1986) included secondary, tertiary, and quaternary recycling technologies. Also included were landfill and incineration without heat recovery. Specific recycling technologies considered included a polyester bottle recycling technology developed by Goodyear, the Mitsubishi Reverzer (which converts plastic wastes to lumber-like products), a U.S.S.

422

Chemicals pyrolysis process, and an incineration process developed by Industronics Incorporated to retrieve the heat energy from polyester bottles.

Four main conclusions emerged from the study. First, the data available at that time were considered to be poor, suggesting that the data could not support definitive conclusions about the competitiveness of recycling with disposal. Second, given the caveat of the first conclusion, recycling of plastics as relatively uncontaminated wastes appeared to be competitive with disposal in many parts of the country where disposal costs were high. Third, the recycling of municipal waste, in which plastics are a relatively small part, is generally more costly than disposal. Finally, the available data did not suggest that any one type of recycling -- i.e., secondary, tertiary, or quaternary -- was superior from an economic perspective.

4.2 RECENT ESTIMATES OF THE COST OF DISPOSAL

The costs of disposal have increased enormously in recent years. In 1987 the national average cost of landfilling had increased to $20.36 per ton and waste-to-energy incineration had increased to $33.64 per ton in 1987 dollars. In 1986 the average cost of landfilling was $13.43 and the cost of waste-to-energy incineration was $30.42 in 1986 dollars. What may be more shocking is the range of landfilling costs across regions in 1987 -- ranging from a low of $3.15 per ton to $75.00 per ton. Nominal costs of landfilling by region were: West $10.01 (1986) $10.75 (1987); South $10.95 (1986) $12.27 (1987); Midwest $10.86 (1986) $12.71 (1987); and Northeast $20.59 (1986) $39.23 (1987).[3]

4.3. RECENT ESTIMATES OF THE COST OF RECYCLING

Three additional revenue/cost estimates have been identified. These include the ET/1 extruder, the Recyclingplas process, and a materials recovery process developed by Bezner Systems

---

[3]This information was obtained from the March 1988 issue of Waste Age magazine, which conducts a yearly survey of disposal facilities around the country.

and operated in the U.S. by CR Inc. All estimates are given in 1988 dollars. The cost estimates are disaggregated into operating and capital costs. To avoid the problem of dealing with different rates of capital depreciation, capital costs are dealt with as though all capital is rented. In other words, it is assumed that capital investments must earn some real annual gross rate of return. While there is no general agreement on what this rate should be, the commonly used 15% rate is adopted as the base case. A low rate of 10% and a high rate of 20% are given as alternatives to assess the sensitivity of changing the capital cost component.

4.3.1. The ET/1 Extruder

The ET/1 Extruder process was developed in Belgium and has been used commercially in Europe for more than two years. The process has recently been adopted in the United States. Mid-Atlantic Plastic Systems, Incorporated, the U.S. agent for the ET/1 system, provided technical information about the system and cost and revenue information from which the cost/revenue estimates given in Table 4.1 were derived.

The process can accommodate a waste stream that contains various plastic resins and can tolerate as much as 50% non-thermoplastic contamination, although no more than 20% contamination is preferred. The ET/1 is based on a specially designed adiabatic extruder feeding a set of linear tubular molds housed in a water filled cooling bath. The resulting products of the process are claimed to be substitutes for similar products made from lumber or concrete.

Table 4.1 presents an assessment of the estimated costs and revenues associated with the ET/1. Note that it is assumed that input materials can be purchased at $0.06 per pound. Further

TABLE 4.1. COST/REVENUE ESTIMATES FOR THE ET/1 PROCESS
(IN 1988 DOLLARS)

Capacity	2,304,000 pounds/year			
Capital Cost	$513,865			
			Interest rate	
		10%	15%	20%
Interest on capital		$ 51,387	$ 77,080	$ 102,773
Production cost/year		346,586	346,586	346,586
General and administrative costs/year		113,386	113,386	113,386
Materials purchased @ $.06/pound		138,240	138,240	138,240
Total cost		$ 649,599	$ 675,292	$ 700,985
Total revenues				
@ $.30/pound		$ 691,200	$ 691,200	$ 691,200
@ $.50/pound		$1,152,000	$1,152,000	$1,152,000
Net revenue				
@ $.30/pound		$ 41,601	$ 15,908	$ -9,785
@ $.50/pound		$ 502,401	$ 476,708	$ 451,015
Net revenues per ton				
@ $.30/pound		$ 36.11	$ 13.81	$ -8.49
@ $.50/pound		$ 436.11	$ 413.81	$ 391.58

NOTE: Calculations are based on single ET/1 machine.

SOURCE: Data on cost and revenues from Maczko (1988).

note that two revenue assumptions are tested -- product sales at $0.30 per pound and at $0.50 per pound. At $0.50 the ET/1 is estimated to be quite profitable at all assumed rates of interest, with net revenues per ton ranging from about $392 to $436. Note that net costs are estimated in only one case -- a 20% interest rate in combination with a $0.30 price for the product sold. However, even in this worse case, the net cost is less than the $20.36 average cost of landfill in the United States.

4.3.2 The Recyclingplas Process

The Recyclingplas process was developed in West Germany and has been used commercially in Europe. The process was recently acquired by a new company in Atlanta, Georgia -- Innovative Plastic Products, Incorporated -- from which the base information presented in this subsection was obtained.

Like the ET/1, the Recyclingplas process can accommodate commingled plastics that contain at least 50% thermoplastics. The Recyclingplas process differs, however, from the ET/1 in that a significantly higher output rate is possible. The process also differs in that the products are extruded and compression molded. Multiple sizes of sheet and shapes are possible with the Recyclingplas process, whereas the ET/1 is more limited to plastic lumber.

Table 4.2 presents summary information about the estimated costs and revenues associated with the process. Note that input material costs were quoted at $0.10 per pound, as compared to $0.06 assumed for the ET/1. Also note that the assumed price at which the products can be sold is somewhat higher than in the case of the ET/1 -- $0.50 and $1.00 per pound. Given the assumptions made, the net revenues associated with the Recyclingplas technology are estimated to be higher than those associated with the ET/1. At $0.50 per pound, net revenue estimates range from $560 to $634 per ton. At $1.00 per pound, net revenue estimates range from $1,560 to

TABLE 4.2. COST/REVENUE ESTIMATES FOR THE RECYCLINGPLAS PROCESS
(IN 1988 DOLLARS)

			10%	15%	20%
Capacity	14,000,000 pounds/year				
Capital Cost	$4,000,000				
				Interest rate	
Interest on capital			$ 400,000	$ 600,000	$ 800,000
Mold cost*			60,000	120,000	180,000
Operating costs			700,000	700,000	700,000
Materials purchased					
@ $.10/pound			1,400,000	1,400,000	1,400,000
Total cost			$ 2,560,000	$ 2,820,000	$ 3,080,000
Revenues					
@ $.50/pound			$ 7,000,000	$ 7,000,000	$ 7,000,000
@ $1.00/pound			$14,000,000	$14,000,000	$14,000,000
Net revenues					
@ $.50/pound			$ 4,440,000	$ 4,180,000	$ 3,920,000
@ $1.00/pound			$11,440,000	$11,180,000	$10,920,000
Net revenues per ton					
@ $.50/pound			$ 634.28	$ 597.14	$ 560.00
@ $1.00/pound			$ 1,634,.28	$ 1,597.14	$ 1,560.00

*Mold life depends on the type of material being processed and can vary from
three-four years for rough materials to about ten years for smooth, less-
contaminated materials. In this calculation, mold cost is treated as a capital cost,
as is the case with the assessment of the ET/1 technology.

SOURCE: Data on costs and revenues from Kelly (1988).

$1,634 per ton. Note that the capacity of each Recyclingplas unit is 14 million pounds per year, as compared to 2.3 million pounds per year for the ET/1 units.

4.3.3. The Bezner Systems Processs

CR Inc. is a company that currently recycles commingled municipal wastes in the New England area. Recently the company purchased and is the North American representative for a new sorting technology by Bezner Systems, a European company. The design is currently employed in 16 materials recovery facilities (MRFs) worldwide. Given that this technology is designed to separate PET and HDPE bottles from other curbside-collected municipal wastes, the technology could lead to significant quantities of relatively clean plastic waste that could be used in processes such as the ET/1 or Recyclingplas. CRInc. currently employs the ET/1 to recycle the plastics retrieved from their recycling operations. Other materials separated by the sorting technology include newspaper, glass bottles, aluminum cans, and tin cans. Note that the incoming waste is from curbside collection programs in which these recyclable materials are collected as a common commingled waste stream at the household level. The main benefit of this technology is that the recyclable materials can be collected as a common waste stream, rather than collected individually, as is currently done in some curbside collection programs.

Cost information was obtained from Torrieri (1988). That information was used to arrive at the cost/revenue estimates for the sorting technology presented in Table 4.3. Note that it is assumed that the incoming commingled waste is obtained at zero cost. Further note that the segregated plastic waste is assumed to be sold for $0.06 per pound, which is consistent with the assumed input price for plastics entering the ET/1 process given in Table 4.1. Other materials are sold according to the prices given in the table.

TABLE 4.3. COST/REVENUE ESTIMATES FOR THE BEZNER SYSTEM SEPARATION PROCESS
(IN 1988 DOLLARS)

Capacity*   40,000 tons/years
Capital Cost $3,000,000

	Interest rate		
	10%	15%	20%
Interest on capital	$ 300,000	$ 450,000	$ 600,000
Operating costs	1,000,000	1,000,000	1,000,000
Total costs	$ 1,300,000	$ 1,450,000	$ 1,600,000
Revenues**			
Newspaper @ $.015/pound	$ 456,000	$ 456,000	$ 456,000
Glass bottles @ $.02/pound	456,000	456,000	456,000
Aluminum cans @ $.50/pound	760,000	760,000	760,000
Plastic bottles @ $.06/pound	364,800	364,800	364,800
Tin cans @ $.01/pound	152,000	152,000	152,000
Total revenues	$ 2,188,800	$ 2,188,800	$ 2,188,800
Net revenues	$ 888,800	$ 738,800	$ 588,800
Net revenues per ton	$ 22.22	$ 18.47	$ 14.72

*Approximately 5% of the incoming curbside waste must be disposed of. Therefore, the 40,000 tons/year capacity implies that 38,000 ton/year is recyclable.

**The incoming curbside waste is assumed to have the following composition by weight: newspaper, 40%; glass bottles, 30%; aluminum cans, 2%; plastic bottles, 8%; tin cans, 20%.

Source of revenue estimates: Plastic Recycling Foundation, Annual Report 1988, under "Highlights of Activities."

Source of other information: Torrieri (1988).

Given the available information, the sorting process is estimated to result in net revenues of between about $15 and $22 per ton, depending on the assumed interest rate.

5. An Update of Recent Institutional Changes That May Promote Recycling

Curlee (1986) concluded that plastics recycling is often hindered by institutional constraints that discourage the formation of required markets for waste materials and recycled products. Those same institutional constraints were also concluded to slow or prohibit the flow of information about available technologies, environmental impacts, economic viability, and regulatory issues. This section of the report summarizes recent developments in public-sector and private-sector institution building that may have significant positive implications for the future of plastics recycling.

Plastics recycling often requires the cooperation of plastic manufacturers, consumers, waste processors, and the public sector to overcome nontrivial barriers. And there are numerous arguments why each of these parties when acting individually has both incentives and disincentives to contribute to recycling efforts. If any one of the required parties faces a net disincentive to recycle, that party can, in effect, block a recycling effort that from a social perspective would result in positive net benefits. The formation of institutions through which cooperation and bargaining can occur is one means of "smoothing out" the incentives and disincentives among the various parties, such that the true social benefits of recycling can be realized.

The recent development of institutional structures that facilitate market formation and the flow of information is a major step towards encouraging plastics recycling. Several specific examples can be cited for both the public and private sectors.[4]

5.1 INDUSTRY SPONSORED GROUPS

Industry sponsored groups formed in recent years have greatly facilitated the flow of information about various aspects of recycling, provided financial support for the development of

---

[4]Note that recent developments in the area of government legislation and regulation are discussed in Section 6 of this report.

new technologies and the preparation of reports on topics such as environmental effects and market assessments, and helped solidify industry's position on some controversial issues. For example, The Plastics Recycling Foundation (PRF), which is centered at Rutgers University and is funded by both industry and state government, has been instrumental in developing and publicizing technology to recycle PET bottles. More recently, PRF has entered the commingled, secondary-recycling field with their purchase of an ET-1 machine and initiation of a pilot program for curbside collection of plastic containers in two New Jersey municipalities. The containers, which are made from various resins, are being manufactured into lumber-like products using the ET-1 machinery. Another industry sponsored group, the National Association for Plastic Container Recovery based in Charlotte, North Carolina, has set a goal of recycling 50% of all PET Bottles by 1992. The group, which is sponsored by several large resin producers, will indirectly promote the use of PET for beverage bottles by facilitating post-consumer uses.

Industry sponsored organizations, such as the Society for the Plastics Industry (SPI), are providing detailed information about the opportunities for plastics recycling and about the firms currently involved in recycling activities. The SPI has also proposed that industry adopt a voluntary digital and letter code that would identify the resin contained in a product. The code, which may facilitate separation in some cases, would affect bottles exceeding 16 ounces and other containers exceeding 8 ounces.

The Council for Solid Waste Solutions, another recently formed industry-sponsored group, will sponsor technical research and public education to increase plastics recycling. The group, sponsored by several major resin producers, recently announced it will spend $8.5 million to conduct research on plastics recycling and disposal and to promote government relations. Among the R&D topics to be addressed are minimizing collection costs, plastics separation, emissions from plastics when incinerated, and methods to characterize degradability.

Actions by individual companies are also promoting technology and market development. For example, General Electric Plastics and Luria Brothers have recently announced plans to collect and recycle engineering resins based on polycarbonate, thermoplastic polyester, and other polymers from scrapped automobiles. These parts will be collected before automobiles enter an automobile shredder -- the typical approach to separating the metallic and non-metallic components in automobiles. Another example is the Solid Waste Management Solutions Group formed at Mobil Chemical Company. The stated objective of that new group is to develop and implement methods for recycling and disposing of plastic wastes.

5.2 PUBLIC SECTOR ACTIONS

At the public-sector level, the EPA, through its Solid Waste Task Force, has recently set a goal of recycling at least 25% of all municipal waste by 1992. Although EPA has no power to enforce the goal directly, the stated objective acts as moral persuasion for industry and local and state governments to increase all recycling activities. The task force has also called for federal actions to promote markets for secondary goods, which may include additional federal procurement of secondary goods (note that some procurement currently occurs under the Resource Conservation and Recovery Act). Two additional actions are also being considered: the development of a national recycling council to explore international markets, and research to investigate how states might use tax incentives and loans for industries using or processing secondary materials.

An EPA official has stated that the plastics area is one of two areas in which the federal government should take the lead in promoting recycling. EPA has also directly encouraged the plastics industry to promote recycling by product design and possibly by color coding different types of plastics. Recently, the EPA announced plans for the establishment of a national clearinghouse for information relevant to all forms of recycling. Actions at the state and local levels are also setting recycling goals and creating or promoting institutions that facilitate plastics recycling. Specific examples are given in the following section.

6. Recent Legislative and Regulatory Actions

Local, state, and federal governments are passing legislation and implementing specific regulations that directly or indirectly affect plastics recycling. An assessment of these activities is important in assessing the viability of particular plastic recycling technologies and in evaluating the potential size of DPW. Curlee (1986) summarizes the activities prior to that book's publication. This section summarizes recent activities in this area.

Since 1986 legislative and regulatory activities have mushroomed at the local and state levels; and that movement is now reaching the federal level of government. Although a review of all the specific bills and regulations that affect plastics recycling is beyond the scope of this report, a review of some examples helps to identify the direction in which this movement is headed.

6.1. LOCAL AND STATE ACTIONS

Numerous actions have been taken at the local and state levels that, in general, promote recycling. For example, the National Solid Waste Management Association (NSWMA) reports that at lease six states currently require local jurisdictions to offer or provide recycling as an option to households. At least two of those states -- New Jersey and Rhode Island -- require some separation of waste materials at the source, meaning the household or business. Oregon was one of the first states to promote recycling. Its 1983 Opportunity to Recycle law requires that municipalities with populations over 4,000 must provide recycling drop-off centers and offer curbside collection of recyclables at least once per month. Household participation remains voluntary. Oregon also supports educational activities that appear to be successful. Although only 70 localities are covered by the law, more than 110 have established recycling programs. Many state laws set goals for recycling municipal solid waste -- usually about 25% of weight, but range between 15% to 50% -- and may have strong economic measures that take effect if the goals are not met.

434

A 1987 law passed in New Jersey labeled the "Mandatory Recycling Act" requires households to separate certain materials and gives municipalities until 1989 to achieve a recovery rate of 25%. The state government requires local governments to design their own programs and provides $8 million to those local governments in start-up aid. The program is funded by a tax on landfill use of $1.50 per ton and requires that a minimum of three materials be recycled -- with the specific materials to be selected by the local governments. While these actions have not in general been directed specifically at recycling plastics, they have indirectly encouraged plastics recycling by (1) fostering the formation of channels to collect recyclable materials and (2) providing moral suasion for consumers and plastic manufacturers to promote recycling.

Mandatory bottle deposits are probably the best known state measures that have directly promoted plastics recycling. At least 11 states currently have bottle deposits laws of some form. Typically, the deposits apply to beverage bottles of all types and are usually five cents per bottle.[5] These laws have been the key to the most publicized plastics-recycling success story -- i.e., PET beverage bottle recycling.

Several states offer incentives for firms involved in recycling activities. For example, Oregon offers income tax credits for the purchase of recycling equipment and facilities. New Jersey offers a 50% investment credit for recycling equipment. Indiana offers property tax exemptions for buildings, equipment, and land used for recycling operations. Wisconsin offers sales tax exemptions for equipment and facilities and some business property tax exemptions for some recycling equipment. North Carolina offers industrial and corporate tax credits and exemptions for recycling equipment and facilities. Other state actions include direct subsidies, grants, technical assistance, and low-interest loans.

---

[5]Roth (1985) reports that bottle reclamation rates average 90% or better in states with bottle deposit laws.

In addition some states give preference to recycled goods through government procurement and other programs. For example, Oregon allows their state departments to pay up to 5% more for recycled products that contain either 50% industrial waste or 25% post-consumer waste, as compared to products made from virgin materials. Vermont has set goals for purchasing recycled goods -- 25% by 1990 and 40% by 1993. The procurement program in New York provides a 10% price preference and requires a recycled content of at least 40% to qualify for the preference. California's law provides a 5% price preference and requires recycled content to be 50%, including 10% post-consumer waste. The National Solid Waste Management Association (NSWMA) reports that at least 18 states have some type of procurement laws for recycled products. Keller (1988) reports that 19 states and four local governments have laws in place that favor recycled products, covering more than 60% of the total U.S. population. Further, at least 13 additional states have considered legislation in 1988 to establish or expand their procurement programs.

A recently passed law in Florida requires localities to reduce landfilling by 30% by 1993, mostly by increased recycling. Taxes and fees on a variety of products will be used to encourage recycling projects. The heart of the new legislation is, however, the provision that requires that a one cent charge be assessed on every type of retail container sold (i.e., plastic, glass, plastic-coated paper, aluminum, and other metals) that does not reach a 50% recycling rate by October 1, 1992. The fee will rise to two cents if the target is not met by October 1, 1995. California recently passed a beverage bottle law that has similar provisions.

While the focus of most states has been on recycling in general, some states are now focusing specifically on plastics recycling. For example, the state of Massachusetts recently issued a report calling for a sustained, aggressive effort to make plastics recycling work in that state. [See Brewer (1988).] Although other states have mandated curbside collection of separated wastes, Massachusetts is the first state to call for the separate collection of plastics. Glass, cans, and newspapers will also be collected. The plan calls for 45% of all rigid plastic containers to be recycled by the year 2000. To get the plan started, the state will help fund construction by 1990 of at least two production-sized plants, one for recycling polyolefins and one for recycling mixed plastics.

At the regional level, several Northeastern states have joined together to promote recycling. The June 1988 issue of <u>Waste Age</u> reports that "The Coalition of Northeastern Governors (CONEG) is advocating a coordinated, comprehensive solid waste management policy incorporating source reduction and recycling, refuse-to-energy, and landfilling." (page 8). The coalition will work to promote secondary markets; provide information to residents about risks, choices, and benefits; investigate which materials can be recycled; set standards for waste facilities; and establish regulatory schemes for incinerator ash.

Another important set of state and local actions has indirectly affected the overall viability of plastics recycling -- i.e., actions to reduce or prohibit the use of plastics in particular applications or to mandate that some plastics be degradable. A common argument used against degradables is that mixing degradables with non-degradables may severely constrain the types of recycling technologies that can be used.

Most of these actions have been directed at banning or limiting the use of plastic packaging and/or proposing packaging or product taxes. Measures are being considered in several states that would require some or all packaging to be biodegradable. Although several measures have passed, many more bills are currently pending.

The recently passed bans on selected plastic packaging in Suffolk County, New York and Berkeley, California are prime examples of local initiatives. These laws can be criticized for being somewhat arbitrary and for being inconsistent across retail markets. In the opinion of most experts, the benefits of the laws will be insignificant in terms of reducing either the size or toxicity of the municipal waste stream. These particular laws are, however, most important for the general message they send. Local governments in some cases view plastics as a major problem in the municipal waste stream -- either because of the quantity of waste contributed by plastics or because of perceived environmental problems. And in some cases, plastics have become a scapegoat for the severe problems some localities are currently experiencing in disposing of their municipal wastes.

Several states are proposing a tax, ranging from one to five cents per package, on materials used to package consumer products. For example, some legislators in Massachusetts have recently called for a packaging disposal tax of three cents per layer on non-food products retailed in that state. Others in that state are calling for bans on the use of certain plastics in packaging and for restricting all packaging to contain only one resin. The recently passed Florida waste bill requires as of January 1, 1990 that any plastic shopping bags used by retailers in that state must degrade within 120 days.

Several additional states are considering legislation that would require all or most packaging materials to be degradable. Some examples: A proposed Vermont law would establish a five cent per package tax on goods sold at the wholesale level if the wholesale dealer does not certify that at least half of the packaging sold in the state by that firm is manufactured from recycled materials. A Missouri bill would prohibit any manufacturer, retailer, or wholesaler from selling products transported in containers using any petroleum-based, non-biodegradable materials. A proposed California law would require all one-time plastic containers and packaging to be either recyclable or biodegradable. A recently passed Maine law prohibits the use of non-degradable individual food and beverage containers by food services at state or local municipal facilities or functions. Sales and use tax incentives for degradables are provided in Iowa.

Several states, including Washington and Oregon, have proposed to require all disposable diapers sold in their states to be biodegradable. And at least 16 states now ban non-biodegradable plastic yokes on six-pack beverage containers.

6.2 FEDERAL ACTIONS

Incentives at the federal level have been much less specific than at the state and local levels. In fact, previous to the passage of the Resource Conservation and Recovery Act in 1976, the federal government had little to do with plastics recycling or municipal waste management in general. Prior to 1976 the federal role was defined by the 1965 Solid Waste Disposal Act, which authorized federal involvement in R&D in solid waste management, and by the 1970 Resource Recovery Act, which strengthened the federal government's R&D activities.

RCRA's subtitle D addresses municipal waste by, for example, mandating regulations on landfills and incinerators, establishing procedures for states to develop solid waste management plans, and calling for procurement guidelines for recycled materials.[6] Yet, under RCRA the states retain primary responsibility of municipal waste management.

Another statute that has implications for state and local MSW management is the Clean Air Act, which in its reauthorized form may impose stricter standards on incinerator emissions. The Public Utility Regulatory Policies Act (PURPA), which requires utilities to purchase electric power from independent suppliers, such as electricity-producing incineration facilities, indirectly promotes incineration. Finally, the Energy Act of 1978 provided an investment tax credit for recycling equipment from 1978 to 1983. That particular incentive has been discontinued.

---

[6]EPA recently established guidelines for federal procurement of recycled paper, re-refined oil, remanufactured tires, and building insulation made from recycled materials. Note that the standard for insulation may include some plastics. For example, McDonalds restaurants has recently initiated a program to recycle its polystyrene containers for non-food purposes -- one product being insulation. See Keller (1988) for additional information on recent federal procurement actions.

Action at the federal level to address the problem of municipal waste has quickened recently. For example, new legislation has been introduced in the U.S. Senate to reauthorize RCRA, which officially expired in September 1988. The Baucus Bill [Senator Max Baucus (D-MT)].calls for more federal involvement in MSW management and will probably be debated in the current session of Congress. Labeled "The Waste Minimization and Control Act," the bill sets ambitious goals, such as 25% recycling in four years, a 10% reduction in municipal solid waste within four years, waste minimization performance standards to be implemented within ten years, federal assistance to states to promote waste reduction and recycling opportunities, and federal procurement of recycled goods. Each state would be forced to develop a solid waste plan. The legislation would also establish a $7.00 per ton fee on new, unused materials to be utilized in packaging, including plastics.

Another example of proposed federal statutes is the Recyclable and Degradable Materials Act of 1988. If the provisions of this act should become law, they would mandate that within ten years all nondurable consumer goods made or sold in the U.S. be recyclable or composed of degradable materials. This legislation is an example of the recent movement against the use of conventional plastics in packaging.

Other federal actions have also been directed at degradable plastics.[7] For example, recent legislation introduced by Senator John Glenn (D-OH) would stimulate the market for biodegradables by forcing the federal government to give preference to buying degradable plastic products. In addition, Senator Sam Nunn (D-GA) supports degradables and amended the

---

[7]At the request of Senator John Glenn, the General Accounting Office (GAO) published in September 1988 a report on degradable plastics. The GAO report found that the federal government and the private sector are only making limited efforts to develop standards for degradable plastics, which in the opinion of the authors has seriously hurt R&D efforts. The report says that "virtually no testing of degradable plastics has been done...." Testing remains necessary to resolve two basic technical uncertainties about the performance of degradable plastics in the environment: the rate of degradation and the safety of the end products. The report also states that several bills, including one by Senator Glenn, have been introduced in the U.S. Congress to promote or mandate the use of degradable plastics.

Department of Defense (DOD) authorization bill for 1989 to require DOD to study the feasibility of using biodegradable plastics made from corn. Recently enacted federal legislation requires that within two years any plastic beverage yokes be degradable, unless the EPA determines that the by-products of degradation pose a greater threat to the environment than non-degradable yokes.

The recently passed United States-Japan Fishery Agreement Approval Act of 1987 places restrictions on the dumping of plastics at sea and calls for two government studies -- one by EPA to study methods to reduce plastics pollution in the environment, with emphasis on recycling, degradability, and the development of incentives, and another by the Department of Commerce to study the effects of plastic materials on the marine environment and provide recommendations to prohibit, tax, or regulate all sources of plastic materials that enter the marine environment.

Forthcoming regulations concerning incinerator ash management and air emissions either from EPA or as mandated in a revised Clean Air Act could have implications for the acceptability of plastics in the waste stream, and thereby influence the viability of plastics recycling. In particular, regulations on dioxin emissions from incinerators may be forthcoming, which will bring additional emphasis to the hotly debated relationship between PVC and dioxins. Potential regulations concerning heavy metal emissions will also bring additional attention to plastics.

6.3. IMPLICATIONS FOR PLASTICS RECYCLING

The numerous government actions that impact on plastics recycling have in most cases promoted, but in other cases discouraged, additional recycling. Some measures directly mandate that recycling occur, others indirectly make the option of recycling more attractive, and yet others promote alternative responses to the "plastics problem" -- i.e., measures such as degradability and source reduction.

It is increasingly clear that all levels of government view plastics as a components of the waste stream that requires some type of public-sector attention. Unfortunately, the rules and regulations currently being imposed are in some cases inconsistent; and the public sector has not as yet established what the overall goal should be with respect to plastic wastes. This lack of consensus is a reflection of the great technological, environmental, and institutional uncertainties currently faced by public-sector decision makers. And until more credible information is available about the option of recycling as compared to the options of disposal, degradability, and bans on the use of plastics, it can be expected that government actions will continue to vary in terms of purpose and scope. Although additional recycling will likely result from public-sector incentives, the uncertainties associated with future government programs will make the adoption of recycling by private firms a risky venture.

7. The Potential for Divertable Plastic Waste

This section discusses the potential size and composition of divertable plastic waste (DPW) -- i.e., plastic wastes that could be diverted from the general municipal waste stream and be a candidate for recycling in a secondary or tertiary sense. The first step in this process is to develop a set of realistic scenarios, which are combinations of technical, economic, institutional, and regulatory conditions. In the second step, these scenarios are used in combination with plastic waste projections presented in Section 2 to assess the potential for DPW, given future technological, economic, institutional, and regulatory developments.

7.1. SCENARIO DEVELOPMENT

For the purposes of this section, five general scenarios are developed. The technical, economic, institutional, and regulatory assumptions underlying these scenarios range from very optimistic to very pessimistic in terms of the relative constraints on the supply of and demand for plastic wastes of different types. Note that at this point in time at least some information is available to argue for the validity of each scenario. Scenario 1 presents the most restrictive assumptions; scenario 5 presents the least restrictive assumptions.

7.1.1. Scenario 1

In scenario 1 it is assumed that technological and/or economic conditions are such that plastics recycling is viable only for those thermoplastic wastes that are available in a clean, single-resin form. It is also assumed that institutional and regulatory conditions are such that no curbside collection of plastic wastes occurs. Neither is there a national bottle deposit bill. In this scenario, plastics recycling faces severe supply-side and demand-side constraints.

443

7.1.2. Scenario 2

In scenario 2 it is assumed that technological and/or economic conditions dictate that clean, single-resin thermoplastic wastes must be obtained in order for recycling to be viable. It is assumed, however, that viable separation technologies exist that can separate a clean stream of two or three commingled resins into individual resin streams. The institutions or regulations that provide for curbside collection of PET and HDPE packaging are assumed to exist. All PET and HDPE containers and lids are labeled according to the resins used for manufacture. Other household plastics are assumed to be collected as part of the general municipal waste stream. In this scenario both supply-side and demand-side constraints are eased slightly.

7.1.3. Scenario 3

In scenario 3 it is assumed that technical and economic conditions are such that commingled thermoplastics can be recycled without further separation. The commingled plastics can either be cleaned and separated into individual resins or the mixture can be recycled without further cleaning and separation into products such as plastic lumber. Non-thermoplastic contamination of as much as 50% can be accommodated. Institutional and regulatory conditions are such that no curbside collection of segregated plastics exists. In addition, there are no viable technologies available to separate plastics from other materials in the general waste stream. As compared to Scenario 2, this scenario represents less severe demand-side constraints.

7.1.4. Scenario 4

It is assumed that any clean commingled plastics are applicable for recycling in scenario 4. Commingled wastes consisting of at least 50% thermoplastics can either be recycled in their contaminated form or further cleaned and separated to produce higher-valued products. Institutions and regulations are assumed to provide curbside collection of a segregated stream of household and commercial plastics. It is assumed that households and commercial operations do not separate individual resins. Any required separation and/or cleaning is done after collection. In this scenario, supply-side restrictions are eased as compared to Scenario 3.

7.1.5. Scenario 5

In this least restrictive scenario, it is assumed that technical and economic conditions exist such that any dirty, commingled plastic wastes are candidates for secondary or tertiary recycling. It is assumed that thermosets can be recycled either as fillers or processed by tertiary means. Technologies are assumed to exist that can technically and economically separate the majority of plastics from other materials in the municipal waste stream into a form acceptable for secondary or tertiary recycling. Plastics need not be collected as a segregated waste stream. In this scenario, improvements in the collection of solid wastes are not required to promote additional plastics recycling.

7.2. ESTIMATED QUANTITIES OF DPW

In this subsection, estimated quantities of DPW are given for each of the scenarios described above. The estimates given are for the year 2000 and are derived from information given in Section 2. Both manufacturing and post-consumer wastes are considered.

7.2.1. Scenario 1

The restrictive assumptions of Scenario 1 imply that few plastics will meet the requirements of DPW, in addition to those plastics already being diverted and recycled. In this scenario, no additional manufacturing waste will be a candidate for recycling. It can be assumed that any clean manufacturing waste containing only one resin is already being recycled in some form. Recall that about 75% of all manufacturing wastes is currently recycled.

The possibilities for DPW from the post-consumer waste stream are also very limited. DPW will be limited to plastics collected as part of state bottle deposit programs and single-resin plastics that are delivered to a collection center by households. Without significant changes in institutions and regulations, the quantities of plastics delivered to collection centers by households are likely to be very small. The potential for DPW in this scenario is therefore no greater than the quantity of post-consumer plastics currently being recycled in this country -- about 1% of all post-consumer plastics.

7.2.2 Scenario 2

It is unlikely that easing the restrictions on plastics recycling as depicted in Scenario 2 will significantly increase the percentage of manufacturing waste that can be classified as DPW. Most clean plastic wastes containing two or three thermoplastics are currently being recycled. Neither will the easing of the restrictions contribute to the recycling of post-consumer wastes, with the exception of PET and HDPE packaging.

The marginal contribution of easing supply-side and demand-side restrictions as given in Scenario 2 will therefore be limited to PET and HDPE packaging -- in particular the container portion of the packaging sector. (Note that other packaging applications, such as films and coatings, will not be retrievable under the assumptions of this scenario.) A recent study by Robert A. Bennett (1989) estimated that about 130 million pounds of PET were recycled in 1987. About 58 million pounds of HDPE were recycled. Modern Plastics estimates that in 1987 a total of 3,331 million pounds of HDPE was used in the manufacture of containers. A total of 900 million pounds of PET was consumed in the production of containers. Therefore, the marginal contribution of collecting and recycling all post-consumer HDPE and PET containers would be 3,273 million pounds of HDPE and 770 million pounds of PET -- or a total of 4,043 million pounds given 1987 data. Modern Plastics estimates that total resin usage for containers in 1987 was 6,433 million pounds. A total of about 63% of all containers could therefore be recycled, given the conditions of Scenario 2. The total usage of plastics for all packaging was estimated at 12,649 million pounds in 1987.

It is estimated in Section 2 of this report that a total of 11,187 million pounds of plastic waste will come from plastic containers in the year 2000. Total plastic waste from all packaging is projected to total 21,090 million pounds in 2000. If we assume that the marginal contribution of recycling HDPE and PET containers in 2000 is in percentage terms the same as for 1987, the marginal contribution of easing recycling restrictions as given in Scenario 2 is equal to 7,031 million pounds of additional plastic waste available for recycling. The projected total quantity of plastics entering the post-consumer waste stream in 2000 is 48,669 million pounds. Therefore, a marginal improvement in the quantity of plastics available for recycling of 7,031 million pounds would be equivalent to 14% of all post-consumer plastics.

7.2.3. Scenario 3

Scenario 3 presents a situation in which commingled wastes are acceptable for recycling if at least 50% of the waste stream is composed of thermoplastics. There is, however, no curbside collection program to provide post-consumer waste as a segregated waste stream. Neither is there a viable separation system that can separate plastics from other materials in the municipal waste stream. Under these conditions, the availability of waste from the post-consumer waste stream will not change from the current situation.

The demand-side improvements represented in this scenario do, however, make it more likely that some or all manufacturing nuisance plastics will become divertable plastic wastes. The exact percentage of current manufacturing nuisance plastics that would be affected cannot be estimated because detailed information about the contamination levels of different parts of that waste stream is not currently available.

Significant additions to DPW are obtained if it is assumed that all of the thermoplastic portion of manufacturing nuisance plastics becomes recyclable as a result of technical and/or economic improvements. The thermoplastic portion of the manufacturing nuisance waste stream is estimated to total 1,801 million pounds in 2000 if the assumptions adopted in Table 2.5 are adopted. The estimates for 2000 increase to 3,602 million pounds if more pessimistic assumptions -- those represented in Table 2.4 -- are adopted.

7.2.4. Scenario 4

In Scenario 4 the demand-side constraints are the same as in Scenario 3; however, the potential for supplying the required waste stream is increased tremendously by the addition of curbside collection of all household and commercial items that are made predominantly from plastics. The potential for DPW in this scenario is therefore all the manufacturing nuisance plastics discussed in the above assessment of Scenario 3, as well as any household and commercial wastes that could be collected as part of a curbside collection program.

It is anticipated that a curbside collection program will target the collection of packaging and consumer and institutional goods. It is assumed that virtually all packaging items, with the exception of coatings, could be collected by a curbside recycling program. In addition, the vast majority of consumer and institutional goods could be collected if a 50% contamination level is acceptable.

By the year 2000 it is projected that the use of plastics for closures, films, and containers will total 19,455 million pounds. Consumer and institutional goods will total 4,920 million pounds. Combined, the two product sectors could contribute 24,375 million pounds to secondary and tertiary recycling streams. Given that post-consumer plastics are projected to total 48,669 million pounds in 2000, recycling all consumer and institutional goods and all packaging, with the exception of coatings, could reduce the quantities of post-consumer plastics going to landfills and/or incinerators by 50% of its weight. (Note that while thermosets may not be recyclable in this scenario, it is likely that thermosets would be collected with thermoplastics in a curbside collection program. Thermosets would be considered a contaminant in this case. In the case of packaging, thermosets are projected to represent only 0.4% of the total waste stream.   In the case of consumer and institutional goods, thermosets are projected to represent 3.2%.)

7.2.5. Scenario 5

Scenario 5 allows for the technical and economic feasibility of recycling dirty commingled plastics. Collection is not particularly important in this scenario, because it is assumed that either separation technologies exist that can economically separate plastics from other waste materials or processes exist that can recycle the commingled waste stream in either a secondary or (more likely) tertiary sense.

In this scenario, all manufacturing nuisance plastics can be classified as DPW. These less restrictive assumptions also open several large post-consumer product sectors to DPW consideration. In particular, plastics from automobiles and other transportation applications may qualify as DPW. Automobiles are usually recycled for their metallic content by shredding the vehicles into fist-size pieces and separating the metallic from the non-metallic pieces. The non-metallic residue contains most of the plastics, which can account for 10% to 20% of the residue's total weight.   Plastic wastes from the automobile sector are projected to total 1,947 million pounds in 2000.   Wastes from other transportation applications are projected to contribute another 549 million pounds. The two sectors combined are projected to account for 5.1 percent of all post-consumer plastics in 2000.

The less restrictive assumptions may also increase the potential for recycling plastics from the building and construction sector, electrical and electronic goods, furniture, and industrial machinery. These product categories are projected to contribute the following percentages to the post-consumer plastics wastes stream in 2000:   building and construction, 8.0%; electrical and electronic goods, 5.7%; furniture, 7.0 %; and industrial machinery, 0.8%.

The potential for recycling plastics from these sectors under the conditions of Scenario 5 are not clear because of insufficient information on the usual methods used to recycle or dispose of wastes from these product categories. Industrial machinery and electrical and electronic goods are sometimes recycled for their metallic contents. The remaining residue may contain significant percentages of plastics. The use of plastics in the building and construction sector is growing at a rate exceeded only by the packaging sector. New methods for collecting plastics from the building and construction sector may make those plastic wastes candidates for DPW classification.

The only product classifications that would not appear to be candidates for DPW classification in Scenario 5 are adhesives and coatings. In 1987, adhesives and coatings accounted for only 1,797 million pounds or 3.6% of all plastics consumed in that year.

7.3. SUMMARY

Table 7.1 summarizes the demand-side and supply-side constraints assumed in the five scenarios. In addition, the table summarizes the potential marginal impacts that altering the supply-side and demand-side limitations to plastics recycling may have on the availability of DPW.

The most likely scenario for the coming decade depends for the most part on legislation and regulation at the state and federal levels. Scenario 1, which represents no easing of demand-side or supply-side restrictions, is not probable, given the current level of interest in plastics recycling. Neither is Scenario 5, which represents the development of an economically viable technology that can separate and recycle plastics, given that plastics are collected as part of a single solid waste stream. In the authors' subjective judgement, Scenario 4 is most probable during the coming decade. Commingled plastics containing at least 50% thermoplastics can be recycled currently into products that compete with lumber and concrete. Other more sophisticated technologies are under development that could produce higher valued products. The main uncertainty on the supply-side with respect to Scenario 4 is the economic viability of these existing and proposed processes. Significant uncertainties exist about the economic viability of plastic

TABLE 7.1. SCENARIO SUMMARY AND MARGINAL IMPACTS ON DPW

SCENARIO	DEMAND SIDE CONSTRAINTS	SUPPLY SIDE CONSTRAINTS	MARGINAL CONTRIBUTION TO DPW
1	Technological and/or economic conditions limit recycling to clean, single-resin waste. No viable separation technologies.	No curbside collection of plastics.	Zero. Quantities of DPW from manufacturing and post-consumer sectors remain at current levels.
2	Technological and/or economic conditions limit recycling to clear, single-resin waste. Viable separation technologies exist to separate two or three commingled resins from a clean waste stream.	Curbside collection of PET and HDPE packaging containers.	No marginal impacts on manufacturing wastes. Assuming all PET and HDPE conditioners are recycled: PET, 770 million pounds (year 2000) HDPE, 3,273 million pounds (year 2000). Equivalent to 14% of all post-consumer plastics.
3	Technological and/or economic conditions limit recycling to waste streams containing a minimum of 50% thermoplastics. No viable technology exists to remove plastics from the general waste stream.	No curbside collection of plastics.	No marginal impacts on post-consumer wastes. Most manufacturing nuisance plastics added to DPW. Depending on assumptions, marginal DPW additions range from 1,971 million pounds to 3,943 million pounds in 2000.
4	Same as Scenario 3.	Curbside collection of all items made predominantly from plastics.	Marginal impacts on manufacturing waste are same as in Scenario 3. Margin impacts on post-consumer waste equal to 24,375 million pounds in 2000, or the equivalent of 50% of all post-consumer plastic wastes.
5	Dirty, commingled plastic wastes are acceptable for recycling. Thermosets can be used as fillers or processed in a tertiary sense. Viable technologies exist to separate plastics from other materials in the general waste stream.	No curbside collection required.	Marginal impacts on manufacturing waste are same as in Scenario 3. Most post-consumer plastics become candidates for DPW, with exceptions of adhesives and coatings--which account for only 3.6% of all post-consumer plastics.

lumber and proposed tertiary processes both in terms of costs of operations and future product prices. [See Curlee (1989) for more details].

The availability of plastic waste from curbside collection programs is becoming less uncertain as the cost of disposal and incineration escalate and the institutions are put into place to facilitate such collection. The average cost of landfill in the U.S. has increased in nominal terms from $13.43 per ton in 1986 to $26.92 per ton in 1988. Certain parts of the country, especially the Northeast, have experienced drastic increases in the cost of landfill in recent years. In 1988 the average cost of landfill in the Northeast region was $45.48 per ton, as compared to $20.59 per ton in 1986. (Disposal cost estimates are from various issues of Waste Age magazine, which conducts a yearly survey of disposal costs.) A recent article in the February issue of Waste Age estimated that a curbside recycling program in an average U.S. municipality is economically feasible if the cost of disposal is greater than $20 per ton. Plastics were included in the analysis.

It is likely that the combination of economic, political, and institutional incentives will combine to increase significantly the curbside collection of commingled plastics within the next decade. If Scenario 4 is representative of the future, virtually all thermoplastic nuisance plastics and about 50% of all post-consumer plastics can be added to the classification of divertable plastic wastes.

8. Plastics Recycling from the Current Industry's Perspective

The Society of the Plastics Industry (SPI) publishes an annual listing of all the firms in the United States currently involved in one or more segments of the plastics recycling industry [Society of the Plastics Industry (1988)]. That listing was obtained, and individuals at selected firms were contacted by phone to informally assess the current industry's perspective of the technical, economic, institutional, and regulatory incentives and barriers currently facing their industry. Note that this process did not involve any questionnaire nor did it involve a formal interview format. The findings of the semi-structured interview format which was used cannot therefore be employed in any formal analytical way to assess the status or opinions of the current industry. However, given the rapidly changing nature and composition of the plastics recycling industry and the rapidly changing technical, economic, institutional, and regulatory environment in which that industry must operate, the authors are of the opinion that information gathered through informal conversations with industry representatives can be both informative and revealing.

According to the SPI data, 22 firms are currently involved in the manufacture of end products from plastic wastes. An additional 43 firms are involved in one or more of the other phases of recycling, i.e., equipment manufacture, brokering, processing, and third-party pickup. Figures 8.1, 8.2, and 8.3 show the geographical dispersion of the firms in the various phases of recycling. When the 22 firms involved in end-product manufacture are disaggregated by type of resin processed, 13 firms are primarily involved with PET, 15 with HDPE, and 10 with other resin types. Only firms involved in end-product manufacture were selected for our informal, semi-structured phone conversations. One additional firm involved in end-product manufacture was identified from a recent publication, bringing the total number of firms in that phase of recycling to 23. Of the 23 firms, representatives of 20 of the firms were contacted by phone.

453

FIGURE 8.1.   GEOGRAPHICAL DISTRIBUTION OF FIRMS IN BROKERING AND THIRD-PARTY PICK-UP

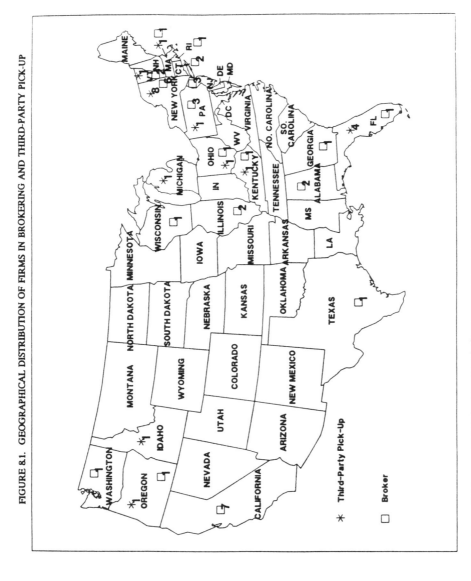

★   Third-Party Pick-Up

☐   Broker

Source of data: PLASTIC BOTTLE RECYCLING DIRECTORY AND REFERENCE GUIDE, 1988, The Society of the Plastics Industry, Inc., Washington, D.C.

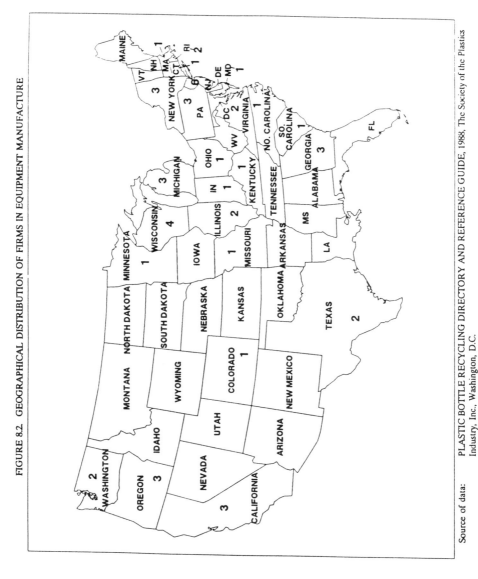

FIGURE 8.2 GEOGRAPHICAL DISTRIBUTION OF FIRMS IN EQUIPMENT MANUFACTURE

Source of data: PLASTIC BOTTLE RECYCLING DIRECTORY AND REFERENCE GUIDE, 1988, The Society of the Plastics Industry, Inc., Washington, D.C.

FIGURE 8.3.   GEOGRAPHICAL DISTRIBUTION OF FIRMS IN PROCESSING AND END-PRODUCT MANUFACTURE

○   Processor

△   End-Product Manufacturer

Source of data: PLASTIC BOTTLE RECYCLING DIRECTORY AND REFERENCE GUIDE, 1988, The Society of the Plastics Industry, Inc., Washington, D.C.

Table 8.1 presents summary information about the firms contacted. Two firms employed tertiary processes in the recycling of PET bottles, while the remaining 18 firms used some type of secondary process. The source of the firms' waste material varied widely, from 100% manufacturing waste to 100% post-consumer waste. Eight of the firms contacted are involved in the manufacture of products that can substitute for products made from wood. In terms of capacity, a wide variation was found, with about half of the respondents in the 1 million to 10 million pounds per year range. Although no statistically defensible conclusions can be drawn from the informal phone conversations, some informal observations are worthy of discussion. In particular, the suggested relationships between perceptions about the industry's strengths and weaknesses and the type of recycling process used and type of waste processed are interesting.

First, firms primarily involved in the recycling of PET beverage bottles do not generally cite lack of demand for their products as a problem of concern. The exception is with firms involved in the tertiary recycling of bottles, in which case the products being produced are not very competitive with the same products produced from virgin materials. PET recycling firms do often cite problems with obtaining sufficient quantities of waste for processing. In some cases, the firms advocate stronger incentives for consumers to provide a larger stream of segregated PET bottles.

Second, firms involved in the recycling of commingled post-consumer plastics -- usually into products that substitute for wood -- typically cite the reverse problems. In other words, these firms often cite problems with marketing their products, but seldom cite problems with obtaining sufficient quantities of plastic materials. In some cases, the firms suggest that supply of waste is no problem.

Third, firms primarily involved in recycling manufacturing waste cite problems with marketing their products most often. The availability of waste materials is not usually mentioned as a severe problem.

TABLE 8.1.   SUMMARY INFORMATION ABOUT FIRMS MANUFACTURING
END-PRODUCTS FROM RECYCLED PLASTICS

	Number of Firms
Type of Recycling Operation:	
Secondary	18
Tertiary	2
Source of Firm's Plastic Waste:	
100% Post-consumer	5
100% Manufacturing	3
>50% Post-consumer	6
50%-50% Split	2
>50% Manufacturing	4
Type of Product Produced:	
Wood substitutes	8
Pipes	3
PET derived products	5
Other	2
Capacity:	
1 Million pound/year or less	5
Between 1 million and 10 million pounds/year	9
Greater than 10 million pounds/year	4
Not disclosed	2

Several other general observations are worth noting. It may be surprising that at least 8 firms are involved in the production of lumber substitutes. What may be more surprising, however, is that several of these firms use proprietary processes. Although the technologies currently available from Europe to recycle relatively contaminated commingled plastics are receiving a lot of attention in the press, there are other proprietary technologies available in the U.S. that perform much the same function.

Several firms recycling commingled wastes into lumber-like products cited the advantages that larger scale operations would offer. It is interesting to note that recent entries in this area plan relatively large scale operations.

Biodegradable plastics were not favored by those firms currently processing commingled plastics. Individuals mentioned that biodegradables mixed with conventional plastics could severely degrade the physical properties of their products. One individual commented that "Biodegradables are the worst."

An interesting observation was made by one individual about a potential environmental benefit of recycling commingled plastics. In processing PET, HDPE, and PVC containers, some cleaning is usually required. The residual contents of the containers are to some extent removed and collected and therefore diverted from the general municipal waste stream. The individual suggested that the collected wastes could subsequently be processed in a sound environmental way. As a side note, the EPA has estimated that about 0.5% of municipal waste entering landfills could be classified as toxic waste. The degree to which the residual contents of plastic containers contribute to this percentage is unknown, at least to the knowledge of the authors.

In some instances, individuals provided information on the price paid for incoming waste materials and the price charged for recycled products. Costs per pound of incoming wastes were as follows: HDPE, $.06 to $.25 for baled and slightly higher (2 to 5 cents) for regrind; PET bottles, $0.04 to $0.12; and commingled, $0.00 to $0.20. The selling price of the products produced obviously depends on the particular product. Prices per pound for products ranged as follows: HDPE, $0.40 to $1.20; PET, $0.35 to $0.50; and commingled, $0.30 to $2.00.

9. Conclusions

This report addresses numerous issues related to plastics recycling in the industrial sector: manufacturing and post-consumer plastic waste projections, the estimated energy content of plastic wastes, the costs of available recycling processes, institutional changes that promote additional recycling, legislative and regulatory trends, the potential quantities of plastics that could be diverted from the municipal waste stream and recycled in the industrial sector, and the perspectives of current firms in the plastics recycling business. This work updates and extends the findings presented in Curlee (1986).

The production and use of plastics are expected to increase at a rapid rate during the coming decade, increasing from an estimated 56 billion pounds in 1987 to almost 72 billion pounds in 2000. Post-consumer wastes are projected to increase from 35.8 billion pounds in 1990 to 48.7 billion pounds in 2000. Plastic packaging is expected to account for about 46% of all post-consumer plastics during the coming decade and remain the by-far largest single contributor to the waste stream. The product category with the largest projected percentage increase is, however, the building and construction sector, which is projected to increase from 2.0% of the total in 1990 to 8.0% in 2000.

Information on the percentage of manufacturing throughput that becomes a manufacturing nuisance plastic was obtained from several industry experts. While there is significant variance between the high and low estimates provided by the experts, estimates of the throughput that becomes a nuisance plastic are generally lower than those used in Leidner (1981) and Curlee (1986). Manufacturing nuisance plastics are projected to account for about 4% of total plastic wastes. These manufacturing wastes, which are currently landfilled or incinerated, may represent a unique opportunity for recycling because they can often be collected independently of other waste materials. Expensive and technically difficult separation processes can thus be avoided.

461

The energy content of plastic wastes depends on whether one means embodied energy or retrievable energy. Embodied energy refers to the energy required to manufacture plastics. Retrievable energy refers to the energy that could be obtained from burning the plastics. Estimates presented in Section 3 of this report indicate that the total energy inputs required to manufacture the resins that are projected to appear in the solid waste stream in 1990 will exceed $1,378 \times 10^{12}$ BTUs. If all manufacturing nuisance plastics and post-consumer plastics are incinerated, the product heat of combustion from those plastics is projected to be about 644×10^{12} BTUs in 1990. The estimate for total energy inputs is equivalent to about 1.8% of what total U.S. energy consumption was in 1985. The estimate for product heat of combustion is equivalent to about 0.8% of the 1985 estimate for total energy consumption. While small as a percent of total U.S. energy consumption, the energy quantities available from plastics are nonetheless large in absolute terms.

Recent estimates of the costs of recycling plastic wastes reconfirm the conclusions presented in Curlee (1986). As was the case in 1986, the cost and revenue data are not sufficiently detailed or validated to draw definitive conclusions. Given the caveats, both the ET/1 and the Recyclingplas secondary processes appear to be economically viable if the appropriate materials are available. It is interesting to note, however, that informal discussions with representatives of the current recycling industry indicate that secondary recyclers are experiencing some difficulty in marketing their products. Although the costs of secondary technologies appear favorable, no good information is available to suggest the potential size of the market for the lumber-like materials being produced. Neither is there information about how the prices of lumber substitutes may decrease as the production of those products increases.

The lack of information about potential secondary markets is one of the most severe handicaps faced by government and private-sector decision makers. If, on the one hand, lumber-like materials have a large potential market, regulators may select to divert plastics away from the municipal waste stream in a form that will be appropriate to these technologies. In that case, alternative technologies, such as tertiary processes may be de-emphasized. On the other hand, if the potential market for these lumber-like materials is small, the development of alternative recycling technologies becomes more important. It may also be the case that regulations that influence the type of waste available for recycling outside of the municipal waste stream should be designed to cater to the specific technologies that have the greatest technical and economic potential.

Significant changes have occurred recently in terms of the institutional structures and regulations that impact on plastics recycling. In the majority of cases, these changes promote additional recycling. For example, mandatory moves to curbside collection of plastics and subsidies for the development and use of new recycling technologies have made plastics recycling more economically attractive. In some cases, however, the institutional and regulatory trends have questionable implications for plastics recycling and are reflective of the continuing uncertainty about plastics in general. For example, recent bans on plastic products and required adoption of biodegradable plastics may hinder some future recycling operations.

Section 7 presents several scenarios in which specific technical, economic, institutional, and regulatory conditions are assumed. Given the most probable scenario, the quantity of plastics available for recycling will grow tremendously. If curbside collection of segregated household plastics becomes the norm and if economically viable technologies are available to recycle commingled plastics containing 50% contaminants, most manufacturing nuisance plastics will become recyclable. In addition, about 50% of the post-consumer plastic waste stream -- i.e., about 24 million pounds in 2000 -- will be a candidate for some kind of recycling.

This assessment suggests that there are significant opportunities for the industrial sector to recycle plastic wastes during the coming decade. Most technical, economic, institutional, and regulatory trends point toward more possibilities and greater incentives for additional recycling activities. There are, however, potential barriers. Technologies that can accommodate dirty commingled plastics require further development, especially tertiary technologies that could produce high-valued pre-polymers. Separation technologies also require further development if some of the larger waste streams are to be recycled in a secondary or tertiary sense. Market constraints may limit significantly the potential for secondary products, especially those products that compete with wood or concrete. Finally, regulatory programs, which will impact on the collection of plastics and the economic viability of recycling operations, are currently being developed at the federal, state, and local levels. Unfortunately, the specifics of these regulations are highly uncertain at this time. The appropriate regulatory structure will depend in part on the outcomes of additional work to further develop technological and economic options for plastics recycling.

Appendix—Plastic Waste Projections: Methodology and Assumptions

A.1. INTRODUCTION

This appendix discusses the methodology and assumptions that underlie the plastic waste projections presented in Section 3 of this report. The methodology is similar to that used in Curlee (1986). Please see Appendix B of that book for a further description of the methodology.

A.2. HISTORICAL AND PROJECTED U.S. RESIN PRODUCTION AND USE

All projections were developed using time series analyses. In the interest of studying the historical relationships between resin production and use and the level of production in certain product categories, the regressions included appropriate industrial production index as an independent variable. Included in this appendix are projections of future values of these production indexes, which are derived using simple time series analyses. Note that the inclusion of these production indices in the regression equations does not alter the projections of resin production and use, since future values of the indices are based solely on time. Ordinary least squares was used for all regressions.

Data on the production and use of plastics were obtained from either the SPI's Facts and Figures of the U.S. Plastics Industry or Modern Plastics. Production indices were obtained from the Survey of Current Business.

Table A.1 defines the variables used in this appendix. Table A.2 gives summary information on the regression results for the projection of the production indices used in subsequent projections of resin usage. Table A.3 gives summary information on regression results to project the use of resins in various product categories. Table A.4 presents regression results for the use of polyurethanes in selected product categories.

465

Table A.5 presents projected values for the relevant production indices. Detailed numerical results from the projection of the use of resins in product categories are given in Table A.6. Projections of the use of polyurethanes are given in Table A.7.

Table A.8 presents summary information from regressions to project the use of resins in selected packaging categories. Table 3.1 in the text gives the actual projections for the different packaging sectors.

A.3. POST-CONSUMER PLASTIC WASTE PROJECTIONS

Data on the historical use of plastic resins in specific product categories was obtained from various issues of SPI's Facts and Figures of the U.S. Plastics Industry, wherever possible. Other source information was obtained from the January issues of Modern Plastics. The average product life spans were assumed to be the same as given in Curlee (1986, Table 4.3).

A.3.1. Transportation

The three-year average method was used to estimate the use of plastics in the transportation sector. Estimates of the use of plastics in 1990 were made using data from 1978, 1979, and 1980; for 1995 using data from 1983, 1984, and 1985; and for 2000 using projections for 1988, 1989, and 1990. The total use of plastics in the sector was disaggregated between automobiles and other transportation categories in the ratio of 78% to 22%, respectively. This ratio corresponds to the use of plastics in these two categories of the transportation sector in 1987 (Source: Facts and Figures of the U.S. Plastics Industry, 1988 Edition).

Projections for 1990:

The use of plastics in the automobile sector was disaggregated according to the usage of particular resins in the production of automobiles in 1982 (Source: Automotive Plastics Report - 1987, Market Search Inc., Toledo, Ohio 43615, 1987). Data from 1982 was used as proxy data

TABLE A.1. DEFINITION OF VARIABLES

Variable Name	Definition of Variable
RESPRO	Total resin production
RESPAK*	Total resins used in packaging
RESC&I*	Total resins used in consumer and institutional goods
RESTRA*	Total resins used in the transportation sector
RESE&E*	Total resins used in the electrical and electronics sector
RESBUI*	Total resins used in the building and construction sector
RESFUR*	Total resins used in furniture and fixtures
RESIND*	Total resins used in industrial equipment
RESADH*	Total resins used in adhesives
RESEXP*	Total resins exported
RESOTH*	Total resins used in goods not included in other product categories
PFTRAN	Polyurethane used in the transportation sector
PFPACK	Polyurethane used in packaging
PFBUIL	Polyurethane used in building and construction
PFELEC	Polyurethane used in electrical and electronic goods
PFFUR	Polyurethane used in furniture and fixtures
PFOTH	Polyurethane used in products not included in other product categories
PITOT	Production index for all goods
PIMAN	Production index for all manufacturing goods
PINON	Production index for all nondurable goods
PIDUR	Production index for all durable goods
PIC&I	Production index for consumer goods
PIINDU	Production index for industrial equipment
PIFUR	Production index for furniture and fixtures
PITRAN	Production index for transportation equipment
PICONS	Production index for construction supplies
PIELEC	Production index for electrical machinery
CLOSUR	Total resins used in closures for packaging
COATNG	Total resins used in coatings for packaging
CONTNR	Total resins used in containers and lid for packaging
FILM	Total resins used in films for packaging

Note: *Excludes the use of polyurethane

TABLE A.2 SUMMARY OF REGRESSION RESULTS: PRODUCTION INDEXES COEFFICIENT

DEPENDENT VARIABLE	R-SQUARED	TIME INTERVAL	CONSTANT	T-STAT	TIME	T-STAT
PITOT	0.85	1973-1987	-5345.89	-8.58	2.75	8.75
PIMAN	0.85	.	-6006.32	-8.34	3.09	8.50
PINON	0.92	.	-6406.77	-11.97	3.29	12.17
PIDUR	0.78	.	-5713.41	-6.59	2.94	6.71
PIC&I	0.89	.	-5307.60	-10.28	2.73	10.48
PIINDU	0.84	.	-7832.91	-8.06	4.01	8.18
PIFUR	0.82	.	-8239.09	-7.49	4.22	7.59
PITRAN	0.55	.	-4642.47	-3.87	2.40	3.96
PICONS	0.52	.	-4244.45	-3.64	2.20	3.73
PIELEC	0.93		-13932.97	-12.76	7.10	12.88

TABLE A.3. SUMMARY OF REGRESSION RESULTS: RESIN USE IN SELECTED PRODUCT CATEGORIES

DEPENDENT VARIABLE	R-SQUARED	TIME INTERVAL	CONSTANT (T-STAT)	T-STAT (T-STAT)	PRODUCTION INDEX	COEFFICIENT (T-STAT)
RESPRO	0.98	1974-1987	-629821.09 (-1.22)	306.04 (1.16)	PITOT	581.72 (6.99)
RESPAK	0.98	"	-750516.08 (-4.98)	379.83 (4.92)	PIMAN	77.67 (3.61)
RESC&I	0.84	"	35010.84 (.34)	-18.35 (-0.34)	PIC&I	46.97 (2.69)
RESTRA	0.71	"	53556.84 (1.68)	-27.16 (-1.67)	PITRAN	19.68 (4.33)
RESE&E	0.66	"	175624.62 (3.16)	-89.25 (-3.13)	PIDUR	34.10 (4.32)
RESBUI	0.97	"	-762416.09 (-9.02)	385.72 (8.93)	PICONS	56.32 (4.42)
RESFUR	0.83	"	54231.64 (1.13)	-27.46 (-1.12)	PIFUR	17.25 (3.59)
RESIND	0.64	"	67366.51 (4.47)	-34.23 (-4.44)	PIINDU	6.78 (4.04)
RESADH	0.59	"	387316.47 (3.85)	-196.91 (-3.82)	PIMAN	44.25 (3.08)
RESOTH	0.94	"	-162853.36 (-1.80)	81.28 (1.75)	PITOT	48.10 (3.30)
RESEXP	0.90	"	-490391.03 (-3.88)	249.40 (3.85)	PIMAN	-2.89 (-0.16)

TABLE A4. SUMMARY OF REGRESSION RESULTS: THE USE OF POLYURETHANE IN SELECTED PRODUCT CATEGORIES

DEPENDENT VARIABLE	R-SQUARED	TIME INTERVAL	CONSTANT	T-STAT	TIME	T-STAT	PRODUCTION INDEX	COEFFICIENT	T-STAT
PFTRAN	0.79	1973-1985	33798.41	5.40	-17.21	-5.39	PITRAN	6.19	5.53
PFPACK	0.92	"	-10173.34	-4.38	5.15	4.33	PIMAN	0.37	1.00
PFBUIL	0.98	"	-45357.05	-16.57	22.97	16.44	PICONS	1.84	3.42
PFELEC	0.46	"	-1998.82	-0.70	1.03	1.47	PIDUR	0.48	1.07
PFFUR	0.32	"	-25788.90	-0.74	13.34	0.75	PIFUR	1.92	0.45
PFOTH	0.51	"	-10968.13	-0.66	5.46	0.64	PITOT	2.71	0.93

TABLE A.5. HISTORICAL AND PROJECTED VALUES FOR SELECTED PRODUCTION INDEXES

	PITOT	PIMAN	PINON	PIDUR	PIC&I	PIINDU	PIFUR	PITRAN	PICONS	PIELEC
1973	94.4	94	90.8	96.3	91.2	92.4	100.5	99.1	102.3	90.7
1974	93	92.6	90.2	94.3	88.4	96.5	93.5	90.1	95.8	89.8
1975	84.8	83.4	84.5	82.6	84.9	86.1	80	81	82.3	77.2
1976	92.6	91.9	93.1	91.1	93.3	89.3	89.4	92.2	92	86.8
1977	100	100	100	100	100	100	100	100	100	100
1978	106.5	107.1	105.5	108.2	104.3	112.2	109.1	106.3	106.9	112.9
1979	110.7	111.5	108.2	113.9	103.9	124.7	111.7	108.3	108.7	125.7
1980	108.6	108.2	107	109.1	102.7	125.1	108.9	96.9	100.6	130.3
1981	111	110.5	109.7	111.1	104.1	127.6	109.9	95.1	98.6	134.1
1982	103.1	102.2	105.5	99.9	101.4	113.6	104.5	87.6	88.3	128.4
1983	109.2	110.2	113.7	107.7	109.3	115.4	118.2	99.2	100.6	143.8
1984	121.4	123.4	122.3	124.2	118	134.2	134.3	112.2	114	170.5
1985	123.7	126.4	124.6	127.6	119.8	140.2	138	122.8	119.2	168.4
1986	125.1	129.1	130.1	128.4	124	139.4	143.8	127.5	126.4	165.7
1987	129.8	134.7	136.8	133.1	127.8	144.5	152.8	129.2	131.5	172.3
1988*	121.1	136.6	133.7	131.3	119.6	139.0	150.27	128.73	129.15	181.83
1989	123.9	139.7	137.0	134.3	122.4	143.0	154.49	131.13	131.35	188.93
1990	126.6	142.8	140.3	137.2	125.1	147.0	158.71	133.53	133.55	196.03
1991	129.4	145.9	143.6	140.1	127.8	151.0	162.93	135.93	135.75	203.13
1992	132.1	149.0	146.9	143.1	130.6	155.0	167.15	138.33	137.95	210.23
1993	134.9	152.1	150.2	146.0	133.3	159.0	171.37	140.73	140.15	217.33
1994	137.6	155.1	153.5	149.0	136.0	163.0	175.59	143.13	142.35	224.43
1995	140.4	158.2	156.8	151.9	138.7	167.0	179.81	145.53	144.55	231.53
1996	143.1	161.3	160.1	154.8	141.5	171.0	184.03	147.93	146.75	238.63
1997	145.9	164.4	163.4	157.8	144.2	175.1	188.25	150.33	148.95	245.73
1998	148.6	167.5	166.6	160.7	146.9	179.1	192.47	152.73	151.15	252.83
1999	151.4	170.6	169.9	163.7	149.7	183.1	196.69	155.13	153.35	259.93
2000	154.1	173.7	173.2	166.6	152.4	187.1	200.91	157.53	155.55	267.03

Note: *Starting year for projections

TABLE A.6. HISTORICAL AND PROJECTED PRODUCTION AND USE OF PLASTICS (IN MILLIONS OF POUNDS)

	RESPRO	RESPAK	RESC&I	RESTRA	RESE&E	RESBUI	RESFUR	RESIND	RESOTH	RESADH	RESEXP
1974	29274	6720	3168	1725	2524	4327	1791	488	2215	2150	1585
1975	22828	5579	2875	1248	1787	3736	1360	313	1771	1942	1351
1976	29196	7342	2801	1808	2524	4555	1617	319	2160	2330	2168
1977	33948	7899	3242	1911	2756	6008	1391	472	2728	2566	2142
1978	37605	9044	3592	2015	2952	6965	1686	380	2939	2902	2588
1979	41577	10334	3753	1934	3043	7573	1894	517	3443	2794	3432
1980	37347	10003	3553	1605	2453	6424	1646	391	3054	2387	3670
1981	39867	10465	3670	1573	2670	7259	1670	393	3259	2572	3425
1982	36607	10497	3269	1392	2275	7154	1556	241	3232	1584	3909
1983	42777	11813	3816	1896	2514	8552	2007	318	3636	1800	4150
1984	46336	12398	3986	2109	2757	9691	2117	400	4080	1993	3796
1985	47946	12774	3975	1989	2659	10038	2107	364	5122	2142	4170
1986	50849	13267	4123	1988	2609	10085	2191	361	4555	1638	4206
1987	55751	15234	5063	2029	2779	11285	2235	332	4665	1797	4593
1988*	49039	15196	4151	2096	2673	11669	2233	259	4557	1904	5021
1989	50944	15816	4260	2116	2684	12179	2279	252	4770	1844	5262
1990	52850	16435	4370	2136	2695	12688	2324	245	4984	1784	5502
1991	54756	17055	4480	2156	2706	13198	2369	238	5197	1723	5743
1992	56662	17675	4590	2176	2717	13707	2415	231	5411	1663	5983
1993	58567	18295	4700	2197	2728	14217	2460	224	5624	1603	6224
1994	60473	18915	4810	2217	2739	14727	2505	217	5838	1543	6464
1995	62379	19534	4920	2237	2750	15236	2551	210	6052	1483	6705
1996	64285	20154	5030	2257	2761	15746	2596	203	6265	1423	6945
1997	66190	20774	5139	2277	2772	16256	2641	196	6479	1362	7186
1998	68096	21394	5249	2297	2783	16765	2687	189	6692	1302	7426
1999	70002	22014	5359	2317	2794	17275	2732	182	6906	1242	7667
2000	71908	22634	5469	2337	2805	17784	2777	175	7119	1182	7907

Note: *Starting year for projections

TABLE A.7. HISTORICAL AND PROJECTED USE OF POLYURETHANE IN SELECTED PRODUCT CATEGORIES
(IN MILLION OF POUNDS)

	PFTRAN	PFPACK	PFBUIL	PFELEC	PFFUR	PFOTH
1973	410	25	140	75	660	107
1974	367	21.5	140	77.5	602.5	99
1975	324	18	140	80	545	91
1976	390.5	33.5	188	80.5	748	87
1977	457	49	236	81	951	83
1978	466	57	283	79	980	96
1979	418	62	322	84	973	71
1980	288	57	315	69	872	111
1981	268	77	328	76	898	97
1982	248	77	328	80	859	82
1983	301	85	379	87	872	185
1984	330	90	400	100	895	235
1985	400	80	430	125	790	305
1986*	409	102	494	108	980	214
1987	402	110	526	112	1011	233
1988	382	115	545	112	1020	215
1989	379	122	572	114	1041	227
1990	377	128	599	117	1062	240
1991	375	134	626	119	1084	253
1992	372	141	653	122	1105	266
1993	370	147	680	124	1127	279
1994	368	153	707	126	1148	292
1995	365	159	734	129	1170	305
1996	363	166	761	131	1191	318
1997	361	172	788	134	1213	331
1998	358	178	815	136	1234	344
1999	356	185	842	139	1255	357
2000	354	191	869	141	1277	370

Note: *Starting year for projections

TABLE A.8 SUMMARY OF REGRESSION RESULTS: PLASTICS USE IN PACKAGING

Dependent Variable	R-Squared	Time Interval	Constant	T-Stat	Time	T-Stat
CLOSUR	0.94	1981-1987	-105159.86	-8.50	53.25	8.54
COATNG	0.86	1981-1987	-82585.14	-5.45	42.11	5.51
CONTNR	0.96	1981-1987	-754232.71	-10.43	382.71	10.50
FILM	0.94	1981-1987	-389652.43	-8.90	198.29	8.98
TOTAL	0.96	1981-1987	-1331630.10	-10.54	676.36	10.62